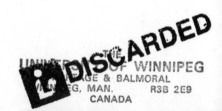

Springer Series in Microbiology

Editor: Mortimer P. Starr

Armand Maggenti

General Nematology

With 135 Figures

Springer-Verlag New York Heidelberg Berlin

Armand Maggenti
Department of Nematology
University of California
College of Agricultural and Environmental Sciences
Davis, California 95616
U.S.A.

Series Editor:
Mortimer P. Starr
Department of Bacteriology
University of California
Davis, California 95616
U.S.A.

Front cover illustration. Left: Heth, a parasite of myriapods and diplopods. The genus was first described by Cobb in 1898. Right: A composite, three-dimensional drawing of a plant parasitic nematode.

Library of Congress Cataloging in Publication Data

Maggenti, Armand R.
  General nematology.

  (Springer series in microbiology)
  Bibliography: p.
  Includes index.
  1. Nematoda.   I. Title.   II. Series.
QL391.N4M213   595.1'82   81-8863
                                    AACR2

9 8 7 6 5 4 3 2 1

ISBN 0-387-90588-X   Springer-Verlag   New York   Heidelberg   Berlin
ISBN 3-540-90588-X   Springer-Verlag   Berlin   Heidelberg   New York

To Mary Ann

Preface

This text is an overall view of nematology because I believe the science should be treated as a unified discipline. The differences in the biological habits of nematodes do not justify the separation of plant nematologists and animal nematologists, since the separation is not a reflection of any differences inherent to nematodes. Therefore, the book is arranged with a format that in the beginning chapters illustrates the similarities and sequence of development of morphological characters among nematodes regardless of their biological habits. The later chapters illustrate the integration of the evolutionary development of the parasitic habit from related free-living forms.

Nematology is probably the last major discipline to establish its independence from the parent science of zoology. This natural evolvement of nematology has occurred because of the overwhelming accumulation of sophisticated information and research that reflects the unique relationships of nematodes to other forms of plant and animal life as well as their relationships in other facets of the environment. Nematodes are invertebrate animals that, like insects, are unusual in their great numbers and varieties, their small size (generally microscopic), their high degree of internal organization, and their virtually ubiquitous distribution. They occupy almost every ecological niche, often causing disease of humans, other animals, and plants. These activities often result in debility, death, or in the impairment and loss of food supply with consequent loss to producers and consumers.

Hopefully this book will intrigue teachers, students, nematologists, plant pathologists, parasitologists, and zoologists. Each will approach the book from their own level of needs; some will read it superficially, some will delve into its speculation, and all, I hope, will learn to appreciate the science itself.

In presenting my understanding of nematology, I hope to comply with the admonition of Thomas Huxley: "My business is to teach my aspirations to conform themselves to fact, not to try and make facts harmonize with my aspirations. . . . Sit down before fact as a little child, be prepared to give up every preconceived notion, follow humbly wherever and to whatever abysses nature leads, or you shall learn nothing." May the reader approach my effort with the same attitude.

I wish to acknowledge but not blame for this effort: D. J. Raski, who first introduced me to nematology; M. W. Allen for his teaching; and more immediately my wife Mary Ann to whom the book is dedicated, who put aside her own work to bring this book to fruition; W. H. Hart, Nahum Marban-Mendoza, and Ella Mae Noffsinger for their reviews and comments; Gaylen Paxman, nematologist, librarian, critic, and friend; my graduate students Fawzia Adbel-Rahman and Steve Martinez, who were interested and involved; all my students in Introductory Nematology, on whom the concepts presented here were first tried; R. Giblin; I. Cid del Prado; and especially my son Peter, who in critical moments eased the pressure by preparing pencil sketches for inking. My obligation is unbounded to B. G. Chitwood. He is a constant stimulus and his memory an abiding inspiration.

In 1967 B. G. Chitwood offered me the opportunity to revise the classic *An Introduction to Nematology* (B. G. and M. B. Chitwood). I felt then and I still believe that classic should stand. To modify it is to redo the Mona Lisa in crayons. I urge all students of nematology to read, read, and reread Chitwood. To have known the man was a pleasure, to have had him as a mentor exasperating, to have had him as a friend magnificent.

This is a short preface because the purpose of the book is to read it. Enjoy now these magnificent animals as I have.

Davis, California   ARMAND MAGGENTI
July 1981

Contents

Chapter 1

History of the Science

I. Introduction

> "The trails of the world be countless, and most
>   of the trails be tried;
> You tread on the heels of the many, till you
>   come where the ways divide;
>
> By the bones of your brothers ye know it, but
>   oh, to follow you're fain.
> By your bones they will follow behind you, till
>   the ways of the world are made plain."
>
> Robert Service

Too often history is considered an irrelevant study, but this is an injustice to past and present contributors of knowledge and experience. Paleontology is devoted to elucidating the history of animal and plant development on earth; so also is the history of a science an evolution of events leading to current concepts. Like a proper taxonomic classification, history allows us to understand not just the evolution of a phylum, but the internal individual phylogenies temporally and spacially. As Robert Service points out, the trails to truth are marked by the pioneers of the past, and those of the future by our own works.

Nematodes are the most numerous multicellular animals on earth. No other group of animals, save Arthropoda, have had such an impact on humans either directly or through agriculture. Nematodes are categorized as being free-living in a marine, freshwater, or soil environment, and as parasites of plants and animals. Often those working with plant parasites also study the free-living forms. Historically, the animal parasitologists were separated from nematologists as helminthologists, but this is now changing, thereby bringing the science of nematology into unity. Subjects such as taxonomy, morphology, and physiology require that all information be integrated and not fractionated. Cobb realized this and was the first to suggest that we call ourselves nematologists. With this proposal he acknowledged nematology as probably the last zoological science to deserve separate distinction.

The history of nematology, in this text, is not limited to this chapter. Readers will find many events and stories related in chapters throughout the book, wherever the telling is more relevant.

II. Ancient Times to the Eighteenth Century

The earliest reference to nematodes relates to animal parasites, specifically human parasites. Most often recorded were the large parasites visible to the naked eye. It is interesting, but not surprising, that written references were made to nematodes some 2000 years ago in records recovered from the great prevailing civilizations of the Mediterranean, Middle East, and Orient.

A. China

The oldest reference to parasitic nematodes is found in Huang Ti Nei Ching or *The Yellow Emperor's Classic of Internal Medicine* from China ca. 2700 B.C. This account is quite sophisticated in as much as it designates foods to be avoided as well as symptomatology and treatment. The symptoms of the giant intestinal worm (*Ascaris*) are surprisingly accurate. Hoeppli and Ch'iang in 1940 translated the following: "The symptoms of the disease are cardiac and intestinal pain, malaise, moving masses in the abdomen with intermittent pain, sense of heat in the abdomen with thirst and salivation."

Chinese medicine developed along empirical lines with little change until recently. Their philosophies greatly influenced their medical approach and prevented advancement until the twentieth century. However, the remarkable observations, as well as the herbal and acupuncture remedies that were developed, were at a more sophisticated level than in the Western world for thousands of years. Chang Chi (or Chang Chung-ching) noted ca. 217 A.D. that "During ordinary abdominal pains, the pulse becomes feeble and thready. If, on the contrary, it is full and bounding, it indicates the sure presence of Huei Ch'ung [*Ascaris*] in the abdomen."

Perhaps even more startling is the report of Ch'en Yen ca. 1174 A.D. in his work *The Three Causes and One Effect of Disease*: "Some people become parasitized by worms through eating fruits and vegetables or animal's viscera, which contain their progenies." This is a remarkable acknowledgment and represents a concept not accepted until the nineteenth century in the Western world. Bear in mind that the belief in spontaneous generation prevailed until Louis Pasteur's experiments in 1864. It is only fair, however, to point out that others such as Redi, von Leeuwenhoek, Spallanzani, Cagniard-Laton, and Schwann had much earlier proposed that the theory of spontaneous generation be discarded but were unable to satisfactorily convince the scientific world.

B. Mediterranean and Middle East Civilizations

The oldest record of nematodes among the ancient civilizations of the Mediterranean and Middle East occurs in the Ebers' Papyrus dated 1553–1550 B.C. This legacy of an Egyptain physician, discovered by Ebers in 1872, indicates that *Ascaris* (the giant intestinal worm) and *Dracunculus medinensis* (the guinea worm) were known at the time. However, it should be pointed out that it is impossible to know with certainty whether the intestinal worm referred to is an *Ascaris*, hookworm, tapeworm, or some other helminth. In addition to symptoms of "bowel worm" an anthelmintic (a drug used against helminths) made from the bark of the pomegranate tree (*Punica granatum*) was prescribed for explusion of the worm.

The next references to nematodes are found in the Bible, and some interpret the passages of Moses relating to Hebrew Laws of sanitation and hygiene as emanating from his early learnings from Egyptian physicians about parasites. Moses' probable knowledge of the guinea worm is found in Numbers 21:6–9. This reference to the fiery serpent and the likeness Moses made by winding the serpent on a staff are believed to have served as an example for the people to extract the worm from their tissues by winding it around a stick. This method of extraction is still practiced in many parts of North Africa and the Middle East.

Moses not only categorized animals as "clean" or "unclean" on the basis of visible parasites, but warned the people to beware of infected water. Moses could not have been aware of *Cyclops* as the intermediate host of *Dracunculus* or of the schistosome cercaria (flukes) in water; however, he certainly would have seen the cloudlike release of nema larvae from dracunculoid tumors when infected people stepped into water.

However one interprets biblical accounts concerning parasites, there can be little doubt of the reference in Plutarch attributed to Agatharchides of Cnidus (181–146 B.C.). In this account he clearly describes the guinea worm: "the people taken ill on the Red Sea suffered . . . worms, upon them, which gnawed away arms and legs, (tumorous ulcers) and when touched, retracted themselves up in the muscles and there gave rise to the most unsupportable pains. . . ." It is from this record that the generic name *Dracunculus* is derived: Plutarch used the Greek *dracontia micra* meaning "little snake." These observations also lend credence to the fiery serpent being *Dracunculus* because the Bible states that the Israelites passed through the region of the Red Sea on their way from Hor to Oboth. It has been estimated that this migration could have taken up to 12 months, which corresponds to the developmental period of 10–12 months for the guinea worm.

There certainly was no great impetus for the physicians of the day to pursue greater knowledge of parasites, including nematodes. No parallel development of symptomatology or therapeutics, such as occurred in the Orient, was proceeding in the Western world. This period is characterized by writings and observations without investigation. This stagnation persisted into the Christian Era and did not change until the nineteenth century.

Some notable events during these times are worthy of mention. Hippocrates, ca. 430 B.C., was aware of nemas as parasites and was likely the first to record knowledge of the pinworm *Enterobius vermicularis*. In his *Aphorisms* he mentions the presence of worms in the vagina of women (a common occurrence with pinworm) and of similar worms in horses. The latter represents the first veterinary observation.

Aristotle also knew of nemas as parasites, especially *Ascaris* and the pinworm. Unfortunately, he stated: "These intestinal worms do not in any case propagate their kind." For the world of science this engendered the theory of "spontaneous generation." No one before or since has held such a disastrous influence over science. I do not believe that Aristotle desired to stifle scientific development, but he did, and science was plunged into a dark age for almost 2000 years. During these dark ages little advance was made and most observations were isolated and of little import. Celsus (53 B.C.–7 A.D.) distinguished roundworms from flatworms; Columella, ca. 100 A.D., mentioned an *Ascarid* from a calf, probably *Neoascaris vitulorum*; Galen (130–200 A.D.) was the first to record nematodes of fish; Vegetius (ca. 400 A.D.) was the first to mention the horse ascarid *Parascaris equorum*. This period of history terminates with Albertus Magnus (1200–1280 A.D.) who provided the first record of nematodes from birds, namely, falcons. The science of nematology was advancing slowly with nearly 100 to 200 years between finds.

The sixteenth century marks a rewakening of science, but discoveries were still 50 to 100 years apart—encouraging, but only barely an improvement. As Chitwood noted: "the period from this time (16th century) until the latter part of the 18th century may be regarded as the medieval period of our subject." Caesalpinus (1519–1630) discovered the giant kidney worm *Dioctophyma renale* from a dog kidney. This parasite (1 m × 1.5 cm) remained the largest known nema until *Placentonema gigantisma* (8 m) was discovered by Gubanov in 1951 from the placenta of a sperm whale. Vinegia (1547) was the first to discover a filarid (Spiruria) in birds, again a falcon, as well as two intestinal nematodes.

III. History and Development in Nineteenth and Twentieth Century Europe

The major contribution to the advancement of nematology was the microscope. Tyson (1683) broke from the traditional recording period and was the first to study nemic anatomy and describe a nematode egg. Borellus (1656) discovered the first free-living nematode. From this period the science began to flourish both among zoologists, whose interest was concentrated on the free-living forms, and parasitologists who now could observe the lesser nemas. It was in this era that the first plant parasitic nema was discovered by Needham (1743) in wheat. This nema (*Anguina tritici*) continues to inflict economic losses in many regions of the world.

This early and exciting phase of discovering a wondrous variety of nemas shifted in the nineteenth century to anatomical studies by Bojanus (1817–1821) and Cloquet (1824), and life history and transmission studies by Owen (1835) and Leidy (1846). Owen discovered trichinosis and Leidy showed the role of rats and pigs in the transmission of the disease. These discoveries generated further research into nemic biology and in the waning years of the nineteenth century such startling discoveries as alteration of generations between free-living and parasitic phases were elucidated by Leuckhart (1865) and Metchnikoff (1865). These discoveries led, in turn, to the discovery that invertebrates often act as intermediate hosts for nemic parasites of higher vertebrates and humans. As a result, the mystery of the transmission of the fiery serpent (*Dracunculus*), which had eluded science for thousands of years was solved by Fedtschenko (1871) when he discovered that the small aquatic crustacean *Cyclops* was the intermediate host and that the disease was transmitted by drinking water contaminated with these animals.

The science of nematology finally took its rightful place among the other zoological sciences of Europe. In addition to biological studies, the art of taxonomy continued to flourish and was upgraded by scientists such as Dujardin (1845), Deising (1851), and Schneider (1866). Outstanding among these was Bastian who in 1865 described 100 new species of free-living nematodes. This work was followed by Bütschli and de Man.

Bütschli was not only a distinguished nematologist, but also a histologist whose development of paraffin embedding for thin tissue sections opened an entirely new avenue of research for all biological sciences.

Bütschli (1875) is also credited with the first observation of polar body formation during subdivision of the nucleus of the ovum. Nematodes then proved to be useful tools for the study of embryology and genetics. Van Beneden in 1883 discovered the mechanism of Mendelian heredity. Other contributors to the line of research were Boveri, zur Strassen, Martini, Müller, and Pai.

A great discovery for Nematology and all parasitology was made in 1878 by Manson who discovered that mosquitoes were the intermediate hosts and vectors of elephantiasis (*Wuchereria bancrofti*). This discovery was a significant catalyst to the discovery of mosquitoes as vectors for such diseases as malaria and yellow fever.

Plant nematology was not idle during this time either. Probably most significant to its growth and recognition was the introduction of sugar beets into Europe. It was not long until the industry was experiencing severe economic losses. The condition was not immediately attributed to nematodes, but to "beet tired soil." In 1859 Schacht discovered that the decline in beet production was always associated with a cyst-forming nematode now recognized as *Heterodera*. However, the nema was not accepted as the causal agent for some time and was first described and named by Schmidt in 1871. The official name became *Heterodera schachtii* after the original discoverer. During this period, the golden nematode of potato was also first seen, but not recognized as a separate species.

Investigations directed toward controlling these nemas dominated plant nematological research in Europe from 1870 to 1910. The control developed for sugar beet nematode is still used, i.e., crop rotation. The first trials to control nemas with soil fumigation were attempted by Kühn (1871) using carbon disulfide. Kühn also studied the feasibility of trap crops, which are plants that attract the nema but in which it cannot develop, or plants that are removed after invasion but before egg laying begins.

Root-knot nematode was first recognized in 1855 when it was discovered by Berkeley on cucumbers in an English glasshouse. This discovery soon led to the recognition of other plant parasitic nemas.

An important contribution, often overlooked, was made by Oerley in 1880, who in a compilation paper describing 202 nematodes in 27 genera, organized nemas into a classification and gathered related genera into families, most of which stand today.

In the late nineteenth and early twentieth century nematology experienced rapid growth and attracted some of the great biologists of Europe. One cannot read and study the works by Goldschmidt in the early 1900s without absolute amazement. His work on the nervous system remain magnificent monuments to his genius. Only now, with the electron microscope and computer analysis of sections, is similar work being produced. What is amazing is that very little is shown to differ from Goldschmidt's findings with the light microscope and paraffin sections.

In the 1930s two outstanding Russian scientists, Paramanov and Filipjev, were influential in establishing philosophies that brought nematology to maturity as a zoological science. Paramanov was a theoretician devoted to the evolutionary concept and the first to offer hypotheses concerning nemic relationships, evolution, and phylogenies. Many of his proposals still form the basic foundation of our understanding of nematode evolution. Filipjev was a taxonomist and offered a sound classification that was not widely accepted outside of Russia, but now forms the basis of current classifications of Nemata.

This is a brief history of the development of nematology in Europe and all contributors cannot be discussed. Among the many who should be remembered are Schuurmans Stekhoven, Jr., de Coninck, Steiner, Fuchs, and Goffart. In addition T. Goodey deserves special mention because his books still serve the science and he was instrumental to the development of nematology in Great Britain.

IV. History and Development in America

The first active study of free-living nemas in America was conducted by Joseph Leidy in 1851, but it was not until 1889 that the science really obtained some national recognition. The source of this impetus came from J. C. Neal and G. F. Atkinson who independently published on root knot in America. Neal's work covered the then active agricultural areas of the United States and this publication is well worth reading. In it Neal notes that, though not recognized by the scientific community, the presence of root knot in Florida extended back as far as the early Spanish explorers.

The most important person to the development of nematology in America was N. A. Cobb. His scientific contributions are notable, but even more important to the science was his personality. Cobb publicized nematology and obtained independent recognition for the science in the United States Department of Agriculture. His first paper on nematodes in America was published in 1913. A few years later he became associated with W. E. Chambers, whose illustrations distinguish Cobb's papers from all others. The quality of Chambers' art has never been equaled; however, they had a profound influence on the quality of illustrations by nematologists throughout the world. Because of Chambers the best illustrations among nematologists are produced by those workers interested in free-living and plant nematodes. Seldom do those who work with animal parasites achieve the quality so common in these other branches of the science. As a result, the classification and the species identification of animal parasites is chaotic.

Cobb surrounded himself with a nucleus of people who were the real founders and architects of the science in the United States, people such as A. L. Taylor, B. G. Chitwood, G. Thorne, J. R. Christie, and G. Steiner. Chitwood is likely the most outstanding nematologist of all time. Few even comprehended the broad scope of the science as he did. His book *An Introduction to Nematology, Section 1* is a classic; the information is as relevant today as when he first published the book in 1937. It combined personal research with the most complete compilation of nematode knowledge ever put together by an individual. A problem that evolved from this book was the overshadowing of Filipjev's classification. However, that seems to have been reversed in more recent years. Chitwood's contribution was to the whole science and not to any particular branch. The only comparable work is *Traité de Zoologie, Tome 4, Fasicule 2–3* edited by Pierre P. Grassé, published in 1965.

Through socratic teaching, these scientists, inspired by Cobb, trained the nematologists of the 1930s, 1940s, and 1950s. In the spring of 1948 M. W. Allen, trained by G. Thorne, taught the world's first formal university course in nematology at the University of California at Berkeley. Among the students in the first class were H. Jensen, of Oregon State University, and W. H. Hart. Hart in 1951 was the first nematologist appointed to a state position in the California Department of Food and Agriculture. In 1959 Hart moved to the University of California Cooperative Extension Service as the first specialist in nematology. Allen was also instrumental in 1954 to the forming of the first Department of Nematology at a university. The first chairman of the statewide department was D. J. Raski at the University of California at Davis with branch departments at Berkeley and Riverside. Since that time classes and departments devoted to the science have formed throughout North America.

A period of great expansion occurred after World War II. The impetus came from the discovery by W. Carter in the early 1940s of a safe, economical, and highly effective soil fumigant. This allowed nematologists to demonstrate with practical field control the great economic losses that were occurring in agricultural crops. The fumigant was a dichloropropene–dichloropropane mixture, which in a refined state is still extensively used throughout the world.

Chapter 2
Nematodes and Their Allies

The relationship of nematodes to other organisms remains unclear even after 100 years of zoological arguments. Nematodes have been assigned to no less than four phyla. Perhaps the most generally accepted has been that of Aschelminthes, Grobben 1909. This group, adhered to by Hyman, includes 6 classes: Rotifera, Gastrotricha, Kinorhyncha (Echinodera), Priapulida, Nematoda, and Nematomorpha. The Priapulida, because their musculature is longitudinal and circular and because of variations in their body cavity, have been excluded from the phylum Aschelminthes. In other schemes, nematodes are placed in the phylum Nemathelminthes, which generally includes just the Nematoda and Nematomorpha, thus leaving Aschelminthes to hold Rotifera, Gastrotricha, and Kinorhyncha. In this text I will hold to the concept that nematodes belong in a phylum of their own, Nemata, as first proposed by Cobb in 1919, and reinstated by Chitwood in 1958; and that each of the so-called related groups, that is, rotifers, gastrotrichs, kinorhynchs, and nematomorphs, are to be placed in their own separate phyla.

The primary reason for disagreement arises from the concept of the pseudocoelom (body cavity). The pseuodcoelom is a nonmorphological zoological term. The fact remains that probably no other structure has been submerged in more vagueness, pseudodefinitions, and misinterpretations than the coelom, whether it is pseudo (partially lined by mesoderm), or true (completely lined by mesoderm). It is distressing that in the past 100 years we have learned nothing more about the embryological development of the so-called body cavity of pseudocoelomates. Some believe that it is a remnant of the original blastocoel, others define it as a gymnocoel, and still others put it in the classification of a mesenchymocoel or schizocoel.

The **blastocoel** is the primary cavity formed during the embryological development of animals. It is believed by some that in the pseudocoel groups the blastocoel persists into the adult animal. A **mesenchymocoel** is a body cavity within a mesenchymal mass of tissue. A **schizocoel** is a body cavity formed by spaces within a compact tissue of the embryo. A **gymnocoel** is a body cavity that has no special lining cells other than tissues bordering cavities such as epidermis or gastrodermis. There is little evidence to indicate that the pseudocoelom evolved in exactly the same manner embryologically in these diverse groups of animals with some superficial resemblance. The known embryology of the various groups indicates that each group has a rather distinct type of body cavity. Nematomorphs are filled with a mesenchyme-like tissue. Nematodes have a well-developed body cavity filled with fluid and with some evidence of mesodermal lining, if one considers the muscle sheath as mesoderm, and the epidermal layer around the gonads and the basal lamella of the intestine as being of mesodermal origin. In gastrotrichs each class differs in the type and manner of body cavity formation as well as in the number of cavities included within the body. It may be that all these embryological phenomena occur within and among these group. If this were so, this would further support the independence of these groups.

Most of the characteristics used to define the groups such as Aschelminthes are rather superficial, such as the **protonephridial** excretory system (an excretory system composed of tubules ending in flame cells), which is present in all the groups save one class of gastrotrichs and the entire Nemata. There is no evidence that the ventral excretory cell seen in Nemata has any relationship to a protonephridial excretory system.

The bodies of members of all these groups are covered by a noncellular elastic cuticle and this is given as a point of relationship, but the formation of the epidermis underlying this cuticle differs among all the groups. In most nematodes it is composed of discrete cells whose cell bodies lie in chords laterally, ventrally, and a portion of the body dorsally. In other pseudocoelomates the epidermis is syncytial (multinucleate), and occasionally discrete uninucleate cells are present, but seldom do they lie in chords in the same fashion seen in nematodes. The musculature of the various "Aschelminthes" also differs. There is no complete muscle sheath in rotifers, gastrotrichs, or kinorinchs. In these animals muscles are limited to scattered bands.

Most of the characteristics considered to show a relationship are not a function of evolutionary sequence but rather of evolutionary demands because small animals have problems in compensating surface area to volume. Small animals also are limited in the number of cells they can contain. Therefore, variety and modification based on these two restrictive elements allow for few or bizarre variations. The larger the animal the greater become the demands on surface-dependent functions. The coelom is necessary not only for complex locomotion, but also for an increase in size and cell numbers. Animals compensate for an increase in size in three general ways: They can differentially increase the surface body area by changing shape without complication of structure, that is, they can attenuate and flatten or they can develop various types of appendages. A second method is to incorporate inactive organic matter within the body. An example of this type of compensation is seen among Coelenterates, in which extensive complication of structure is avoided by having inert non-respiring matter in a central area that occupies the main volume much the same as in trees. As a consequence the bulk of the animal or plant can increase in size as the cube of the length while the active part served by surface-dependent functions does not increase faster than the surface itself. The third solution, the formation of a coelom, is characteristic of most metazoans either as a pseudocoelom or as a true coelom. The development of the body cavity enables the animal to differentiate its organs and tissues for specialized independent functions. For instance, in flatworms, where there is no body cavity, there is a ramification of the intestinal tract, excretory system, and reproductive system. This is the only manner in which metabolites can be effectively transported to all the vital body parts. Thus, the animal is limited in position and development of musculature, and therefore, limited in types of locomotion.

Complex locomotor ability, as is seen in pseudocoelomates and coelo-mates, can only come about when the musculature is isolated from ramifying organs of the body. Because of the branched excretory, digestive, and reproductive system of flatworms they are limited to creeping. In nematodes there is no enclosed circulatory system, and therefore, metabolite movement among tissues requires an essentially straight one-cell thick alimentary canal housed inside a cylindrical animal. This is exactly what occurs in nematodes and the other pseudocoelomates. Movement (circulation) of metabolites is facilitated by sinusoidal locomotion.

An important characteristic of many small animals is the **eutelic condition**, that is, a constant cell number and arrangement from hatched larva to adult. This condition may exist in rotifers, gastrotrichs, and kinorinchs, but it certainly does not exist among the majority of nematodes, which do increase in cell number with size. Their small size does limit the number of cells their body can accommodate and in turn restricts cell division, but the number of cells and nuclei does increase by division after hatching, and there is an increase in size and complexity through increasing cell numbers as nematodes approach maturity. Thus, there is a lack of cell constancy (eutely) in number and arrangement.

Each phylum of the pseudocoelomates is definable and has discrete unequivocal parameters. It is within these clear boundaries, not clouded by unproved homologies, that knowledge must be gained in order to propose meaningful relationships. It is unfortunate that at one time the genus *Rhabditis* was chosen to represent the ancestral nematode. It is really because of this choice that relationships with other pseudocoelomates are proposed. *Rhabditis* is a derived terrestrial nematode in the class Secernentea and should not be considered as the ancestral representative.

It is the superficial resemblance of the esophagus of this nematode with that of gastrotrichs and rotifers that is the basis of the arguments used in evolutionary relationships. The Rhabditida also have an H-formed tubular excretory system which also has superficial resemblances to the protonephridial excretory system of other pseudocoelomates. Those nemas considered by contemporary nematologists as being representatives of ancestral nematodes are among the marine Enoplia, which have no special valves or bulbs in the esophagus. There is no similarity between their esophagi and the mastax of rotifers, nor is there any tubular excretory system.

Gould observed, "however much we celebrate diversity and revel in the peculiarities of animals we must also acknowledge a striking lawfulness in the basic design of organisms. This regularity is most strongly evident in the correlation of size and shape." Until we have more research into the very basic facets of the embryology and development of these animals, the relationships often asserted among the pseudocoelomates should be considered reflections of physical and chemical mandates, rather than of any evolutionary sequence. Most of the similarities seem to be a consequence of the organisms coping with size restrictions. This text supports the belief that we best serve scientific knowledge by following the concepts of separate phyla for these groups. Current technology does not allow us to distinguish between similarities resulting from inheritance and those resulting from trying to solve the mandated problems of small size. It serves no purpose to place these diverse groups in the single phylum Aschelminthes or Nemathelminthes knowing full well that they are without meaningful definition. From the available information we can safely state that reliance cannot be placed on similarities of the pseudocoel, the protonephridial system, comparative morphology of the stomodeum (stoma and esophagus), cell constancy, or embryology. This leaves us without zoological basis for putting this loose assemblage of animals into a single phylum. Therefore, based on our present understandings in zoology and comparative morphology, nematodes are herein treated as a separate phylum.

I. Phylum Rotifera
(L. rota=wheel; ferre=to bear)

Rotifers or "wheel animacules" can be found throughout the world in marine, freshwater, and terrestrial habitats. Among the 1500 recognized species there are benthic, pelagic, soil, epizoic, and parasitic forms. Rotifers are colorless except for their eyespots (Fig. 2.1A) and the food within the digestive tract. They are among the smallest metazoans, many not exceeding the ciliate protozoans in length. The largest species may attain a length of 3 mm; however, the vast majority are less than 1 mm.

A. Shape

The body of rotifers can be divided into three regions: the short head, which is narrow or broad, rounded or truncate; the elongate, cylindrical, or saccate trunk; and the foot or tail, which generally ends in two "toes" (Fig. 2.1A). The latter contain cement glands that allow the animal to attach to various substrates.

B. External Covering

The body is covered by a noncellular elastic cuticle that is the product of an underlying syncytial epithelium (Fig. 2.1B). The cuticle may be marked by transverse striae (Fig. 2.1A), thus forming annules or rings that allow telescoping of the head and tail, or it may be thickened and sculptured on the trunk forming an encasement called the **lorica** (Fig. 2.1B). Other cuticular modifications and ornamentations include the corona, skipping blades, or spines (Fig. 2.2A), dorsal and lateral antennae (Figs. 2.1B, 2.2A), sensory receptors, as well as the lining of the stomodeum and rectum.

C. Corona

The ciliated corona is characteristic of the phylum. In its primitive form the corona is believed to have consisted of a ventral ciliated field around the mouth that was used primarily for locomotion. The **corona** is divided into two parts: the **buccal field** which surrounds the mouth, and a **circumapical band** encircling the margin of the head (Fig. 2.1B). The central anterior unciliated area is called the **apical field** (Fig. 2.1B). Modification of the corona characterizes the various groups of rotifers; it may be reduced to a broad transverse or triangular area leading into the mouth, or reduced to a small ciliated zone around the mouth. In the Collothecacea (Monogonata) the corona is extended into elongated lobes edged with setae or tufts of setae (Fig. 2.2B). The head thus forms a trapping funnel, with the mouth at the center of the bottom. Another modification is the formation of **auricles**, paired lateral coronal projections, that aid in swimming and can be retracted during creeping. The most advanced modification is the formation of conspicuous oral or lobed discs that are employed in food gathering and locomotion. In sessile Monogonata (*Limnias*) the circlets are employed

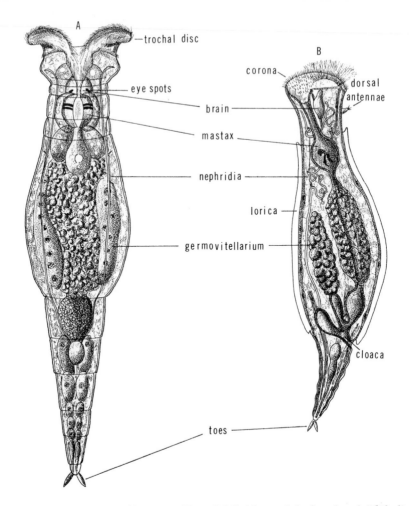

Figure 2.1. Phylum Rotifera. A. Class Bdelloidea; adult female of *Philodina roseola* (modified from Hyman). B. Generalized female rotifer (modified from Delage and Herouard).

in producing food currents. In the Bdelloidea where they are called trochal discs (Fig. 2.1A), they are raised on pedestals and are used in both swimming and the production of food currents.

D. Alimentary Canal

The ventroanterior mouth may open into the ciliated buccal tube (Fig. 2.1B) or directly into the ciliated **mastax** (=pharynx) (Fig. 2.1A). The mastax is characteristic of rotifers and often, without substantiation, is

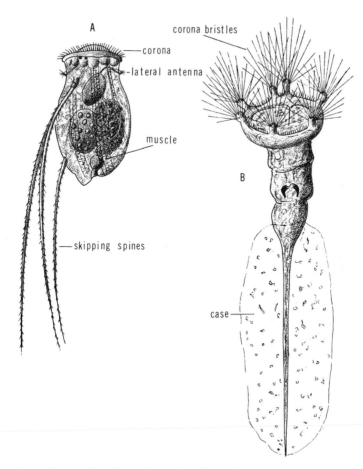

Figure 2.2. Phylum Rotifera. A. Order Flosculariacea; adult female, *Filinia longiseta*, common pelagic rotifer with skipping spines (based on Weber in Hyman). B. Order Collothecacea; a sessile rotifer (modified from Hyman).

homologized with the esophagus of nematodes. The rotifer mastax is a muscular rounded, trilobed, or elongate organ with internal cuticular pieces called trophi (Figs. 2.1B, 2.2B). The dorsal wall of the rotifer pharynx is either partially or completely ciliated, and lacks radial muscles. In suspension feeders the mastax apparatus is adapted for grinding; in carnivorous species, the buccal tube is absent, and the mastax is constructed like a pair of forceps that can be projected from the mouth to seize prey. Rotifers that feed on unicellular algae have short broad teeth in the mastax, and those that feed on multicellular plants have pointed teeth with which they pierce the cells. The muscular pharynx then acts as a pump to suck in the cell contents.

The pharynx is followed by a cylindrical esophagus, either ciliated or lined by cuticle (Fig. 2.1B). The mesenteron is composed of two parts: a stomach with attendant digestive glands and a short intestine. Posteriorly the intestine, oviduct, and excretory system join to form a common cuticularly lined duct called a **cloaca** (Figs. 2.1B, 2.2A). The cloacal opening (= anus) opens to the exterior on the dorsal surface near the posterior of the trunk.

E. Excretory System

The excretory system consists of two slender tubes (**nephridia**) distally branched, with each branch terminating in a **flame cell** (Fig. 2.1B). The presence of flame cells characterizes the excretory system of rotifers as protonephridial. Proximally the nephridia connect into an enlarged vesicle (bladder) that discharges into the cloaca (Fig. 2.1A, B). The main irregularly coiled nephridia extend lengthwise in the animal.

F. Somatic Musculature

Rotifers lack a well-defined subepidermal muscle layer. Somatic musculature is limited to a few scattered longitudinal and circular muscle bands (Fig. 2.2A). The longitudinal muscles function primarily as retractors for the head and foot. The circular muscles, three to seven in number, are widely spaced and best developed as sphincters on the head region, behind the corona and at the junction of the trunk and foot. In addition there are scattered cutaneovisceral muscles.

G. Nervous System

The central nervous system consists of a dorsal bilobed neural mass called the brain or cerebral ganglia (Fig. 2.1A, B) located above the pharynx. Several nerves run to and from the brain. A number of sensory nerves extend to the brain from the sense organs of the head region. Two main ventral nerve cords extend the length of the body from the brain mass. Also in a ventral position are the pharyngeal nerves that innervate the mastax region. In addition several motor nerves that innervate the somatic musculature emanate from the brain mass.

H. Pseudocoel

The body cavity or pseudocoel is fluid filled and contains a network of syncytial amoeboid cells (Fig. 2.1A), believed to be phagocytic and excretory in function.

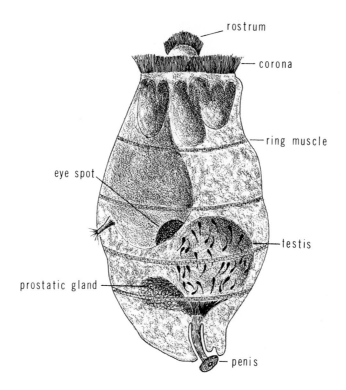

rostrum

corona

ring muscle

eye spot

testis

prostatic gland

penis

Figure 2.3. Phylum Rotifera; order Ploima; male (modified from Hyman).

I. Reproduction

Rotifers are dioecious, and all reproduction is sexual. A single ovary (germarium) combined with a yolk producing vitellarium to produce a **germovitellarium** is most common (Figs. 2.1A, B, 2.2A), although females with two gonads are known. Yolk is supplied to developing eggs by the vitellarium through cytoplasmic connections. Individual females produce approximately one dozen eggs. Eggs pass to the cloaca by way of the oviduct (Fig. 2.1B). Males, which are generally smaller than females, usually have a degenerate intestinal tract (Fig. 2.3). The single saclike testis opens to the exterior through a gonopore that is homologous to the anus of females and therefore technically a cloacal opening. In copulation the penal organ of the male penetrates the body wall of the female directly. Thus, sperm are passed into the body cavity of the female by hypodermic impregnation. Males are unknown among Bdelloidea where the diploid females are produced parthenogenetically.

Two types of eggs are produced by rotifers: amictic and mictic. **Amictic** eggs do not undergo a meiotic division and are therefore diploid. These thin-shelled eggs produce females. The **mictic** eggs, which have undergone meiosis, are haploid, and unfertilized they produce haploid males. When the males, always haploid, fertilize females producing mictic (haploid) eggs, a thick-shelled diploid dormant or "winter" egg is produced. These eggs can withstand long periods of desiccation and other adverse environmental conditions. Winter or dormant eggs when exposed to favorable conditions always hatch into females. The longest record for survival of dormant eggs is 59 years, but survival is generally measured in months.

In the initial steps of rotifer embryology egg cleavage is spiral and determinate. After hatching no larval development takes place; adult features and sexual maturity are attained in a few days without cuticular molts. Sessile rotifers hatch as free-swimming larvae that upon attachment soon assume adult characteristics. The sessile forms often have an elongated foot provided with a gelatinous case (Fig. 2.2B).

J. Biology

Rotifers occur worldwide and adapt to a wide range of ecological conditions. Most commonly, they live in freshwater ponds and lakes where they are found swimming or attached to the bottom, creeping among vegetation or feeding on small organisms. Only about 50 species are known to live in a marine habitat. Terrestrial forms are most often found in mosses and lichens, and are active when free water is available. The **epizoic** forms are generally associated with the gill structures of small crustaceans or externally on freshwater annelids. Endoparasitic species occur in snail eggs, within colonies of *Volvox*, or within the intestine and body cavity of annelids and slugs. *Proales wernecki* is parasitic on freshwater algae *Vaucheria* where it produces gall-like swellings within the filaments.

K. Classification

Phylum Rotifera
  Class Seisonacea. Corona reduced; lateral antennae and toes absent. Gonads paired, males well developed. Epozoic on marine crustacea.

  Class Bdelloidea. Corona modified into two trochal discs. Lateral antennae absent. Caudal toes present or absent (0–4). Female gonads paired; males unknown. Common in quiet freshwaters, locomotion by swimming or creeping.

  Class Monogonata. Corona variable; lateral antennae present. Toes present or absent (0–2). Female with single gonad. Males usually present but degenerate with one testis. Represents largest class of rotifers.

    Order Ploima. Corona normal, adapted to swimming. Foot, when present, with two toes.

Order Flosculariacea. Corona circular or lobed. Foot present, toes absent. Males degenerate.

Order Collothecacea. Anterior modified to trapping funnel. Foot without toes. Sessile.

II. Phylum Gastrotricha (Gr. gaster=belly; trichos=hair)

Gastrotricha are microscopic metazoans found in freshwater and marine environments. The known species vary in length from 0.05 to 4 mm, the majority of forms being less than 0.6 mm long. The phylum currently consists of 400 species, all free-living.

A. Shape

Gastrotrichs are generally dorsoventrally flattened with a bristly or spiny trunk. The slightly lobed anterior end, sometimes set off by a constriction, merges into the main body trunk, which ends generally in a forked posterior. In the order Macrodasyida the anterior usually is not sufficiently delimited to be designated a head (Fig. 2.4A). The body sides are most commonly parallel and the shape varies from short and broad to long and slender. The posterior extremity is often forked (Fig. 2.4A), but may be rounded, truncate, or attenuated in some forms to a long slender tail.

B. External Covering

The body covering is a noncellular elastic cuticle, underlined by a syncytial epidermis. The epidermis is often thickened laterally in the macrodasyoids. The cuticle bears a number of warty eminences and spines on cuticular scales characteristic of the phylum (Fig. 2.4A). The spined scale is characteristic of the chaetonotids. Spination may occur in bunches on the body; marginal spines are usually longer than those located on other parts of the body. Located along the length of the body are specialized structures called **adhesion tubes**, which are supplied with muscles and adhesion gland cells (Fig. 2.4B). Adhesion tubes may number as high as 250 or they may be limited to one or two on the forked tail, or situated terminally on elongated forks.

The ventral surface cuticle bears the locomotor ciliation. Gastrotrichs glide about on these ventral cilia. Trunk ciliation occurs in a variety of patterns. The cilia may be in one broad band, longitudinal bands, patches

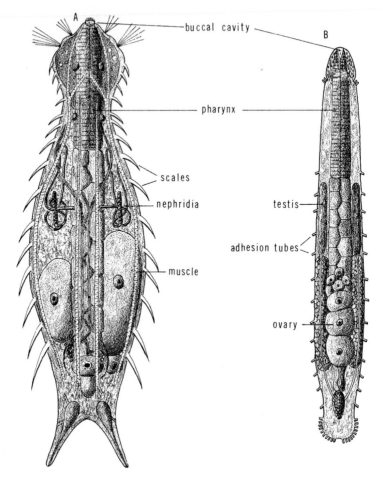

Figure 2.4. Phylum Gastrotricha. A. Order Macrodasyida; adult female, *Cephalodasys* (modified from Remane). B. Chaetonotoid gastrotrich; ventral view (modified from Zelinka and Pennak).

arranged in longitudinal rows, or transverse bands. In some forms trunk ciliation is completely absent or limited to the ventral surface of the head.

C. Alimentary Canal

The anterior mouth is terminal or slightly ventral. A short buccal capsule or stoma, to some extent protrusible, is located between the mouth and pharynx (Fig. 2.4A, B). The cuticularly lined stoma may be armed with longitudinal ridges or projecting teeth.

The gastrotrich pharynx (Fig. 2.4) has no resemblance to the rotifer mastax (=pharynx). In form and structure it is more similar to the nematode esophagus. The lumen of the esophagus (=pharynx) is cuticularly lined and triradiate in cross section. In Chaetonotida the triradiate arms follow the pattern of nematodes, that is, one midventral and two dorsolateral; however, in Macrodasyida the arrangement is reversed, one arm is middorsal and two are ventrolateral. The wall of the esophagus is composed of columnar epithelium, radial muscle fibers, and some gland cells. The form is that of an elongated tube with one to four swellings. In the Macrodasyida the "pharynx" opens to the exterior through a pair of pores through which excess water from ingested food is excreted. Posteriorly the esophagus connects with the midgut.

The mesenteron (=midgut) is not clearly differentiated into stomach and intestine (Fig. 2.4). The one-layered cuboidal or columnar epithelium of the intestine is covered externally by a fine layer of circular muscle fibers.

The cuticularly lined rectum is usually provided with a sphincter muscle. The anus is ventral and opens near the posterior of the trunk.

D. Excretory System

A defined nephridial excretory system is known only in the order Chaetonotida (Fig. 2.4B). Here it is a simple paired system with a single flame cell located distally; proximally the coiled nephridial tubes open through separate pores on the ventrolateral surface near the middle of the trunk. The system of excretion in Macrodasyida is not understood.

E. Somatic Musculature

Gastrotrichs, like rotifers, lack a subepidermal muscle sheath. Two to six bands of delicate longitudinal muscles occur along most or part of the dorsal walls (Fig. 2.4A). These bands insert at the level of the mouth region and pharynx. In addition there are two sets of muscles that operate the adhesion tubes and movable bristles: one set of musculature is circular and the other longitudinal. The longitudinal sets are ventrolateral and extend the length of the body and operate the anterior and posterior tubes. The circular muscles operate the lateral tubes and bristles.

F. Nervous System

The central nervous system consists of a dorsal pharyngeal nerve commissure (Fig. 2.4B) connecting paired lateral ganglionic masses. From the "brain mass" paired longitudinal ganglionated nerves extend throughout the body. Sensory organs are scattered over the body and include cilia on the head lobe, tactile hairs, bristles, and eyespots.

G. Pseudocoel

The body cavity and its compartments are very dissimilar between Chaetonotida and Macrodasyida. In Macrodasyida there are two lateral cavities containing the germ cells and a central cavity surrounding the intestine (Fig. 2.4A). At maturity eggs and sperm are found in the central cavity. The lateral cavities are bounded by longitudinal muscle bands, whereas the central cavity is lined by thin membranes and circular muscle fibers. The Chaetonotida are either without a lined cavity or have one that has no special lining except epidermis and gastrodermis. When present the cavity contains eggs and is called not a pseudocoel, but rather a **gymnocoel** (Fig. 2.4B).

H. Reproduction

On the basis of the diversity of reproductive habits and a propensity for a marine existence, the hermaphroditic Macrodasyida are considered the most primitive order of Gastrotricha. In the Chaetonotida the male system has degenerated; as such, only females occur and reproduction is by parthenogenesis. The Macrodasyida, on the other hand, are hermaphroditic, with some being markedly **protandric**, that is, the male system maturing before the female organs. In at least one genus of Macrodasyida one finds males, females, and hermaphrodites.

The female reproductive system contains one or two ovaries, consisting of a generative cell mass without a definitive capsule (Fig. 2.4B). The gonads are generally situated in the posterior part of the body. It appears that oocyte number is fixed during embryonic development and adult females produce only one to 25 eggs. A nutritive tissue, which may be termed yolk, is associated with the female gonad and is present as a single or paired mass. The connecting structures of ovary, uterus, and oviduct are not clear. As eggs mature they become free in a poorly defined space called the uterus, or in the gymnocoel alongside the gut. The female gonopore of Macrodasyida opens anterior to or in conjunction with the anus as a cloaca. In Chaetonotida there is no oviduct, instead eggs pass through the X-organ that opens ventrally.

The male reproductive system in hermaphroditic forms has one or two testes that open generally through a single pore (Fig. 2.4B). Sometimes the paired sperm ducts open through separate pores. The male system may open through the same pore as the female system. Sperm are either transferred directly to the female by a penal modification of the sperm duct or spermatophores are attached externally on the posterior region of the female trunk.

Among Chaetonotida two types of eggs, a thin-shelled egg and a thicker-shelled dormant egg, are both produced parthenogenetically. Egg development is poorly understood. Where studied, it is reportedly bilateral and determinate. No molting occurs after hatching; however, integumental structures such as adhesion tubes may increase in number with maturation.

I. Biology

Gastrotrichs are found in bottom debris of lakes, ponds, and slow moving streams. They glide or creep about, in a manner similar to leeches, on their ventral cilia. They eat dead or living organic matter, including bacteria and diatoms. The marine forms are found in shallow, calm areas where they feed on algae and bottom debris.

J. Classification

Phylum Gastrotricha

Class Macrodasyida. Protonephridia absent; adhesion tubes, anteriorly, laterally, and posteriorly. Hermaphroditic. Marine and brackish water.

Class Chaetonotida. Bottle-shaped. Adhesion tubes usually restricted to tail. Two protonephridia, reproduction hermaphroditic or parthenogenetic. Freshwater, some marine.

III. Phylum Kinorhyncha
(Gr. kineo=movement; rhynchos=snout)

Kinorhynchs are the smallest group among the pseudocoelomate phyla. The phylum is exclusively marine and only about 30 species are known. They are rarely seen by collectors or students of marine metazoa because their habitat is the mucky bottom areas of the shallow **littoral** (shore) zone.

A. Shape

The kinorhynch body is divisible into three parts: head, neck, and trunk. The head, armed with spines, is retractile, and when the animal is inactive it is withdrawn into the second or third **zonite** (body segment) (Fig. 2.5A). The second or third zonite is designated as the neck and in two of the

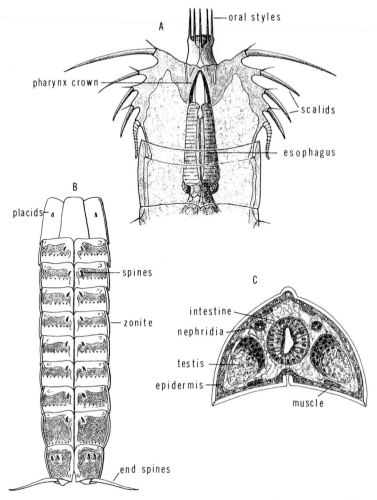

Figure 2.5. Phylum Kinorhyncha. A. Class Homalorhagea; *Pycnophyes*, anterior end extended (modified from Hyman). B. Adult female; *Pycnophyes frequens*; anterior end retracted. C. Cross section through the intestinal region of *Pycnophyes* (modified from Zelinka).

three classes is retractable. The segmented trunk is composed of 13 to 14 zonites. The last zonite usually bears conspicuous elongated **terminal spines** (Fig. 2.5B).

B. External Covering

The body is covered by a noncellular elastic cuticle. The cuticle is underlain by a syncytial epidermis (Fig. 2.5C) that extends into the body spines and projects into the body cavity. The epidermis appears as thickenings at the base of the **scalids** (head spines). Dorsally, laterodorsally, and laterally the epidermis forms longitudinal chords that are enlarged in each zonite.

Kinorhynchs are devoid of external body cilia. The cuticle is variously ornamented by spines, plates, and segmentation (Fig. 2.5B). The mouth is central, terminal, and situated on a protrusible cone (Fig. 2.5A). It is surrounded by oral styles. The area of the retractile head, posterior to the mouth cone, is distinguished by circlets of posteriorly directed spines called scalids. In those forms with a retractable neck the third zonite is covered with large plates (**placids**) (Fig. 2.5B); in Cyclorhagea the neck (second zonite) is nonretractable and bears the "closing apparatus" or placids. The remaining zonites in all groups constitute the trunk. Each trunk zonite overlaps the succeeding segment; intersegmental membranes of thin cuticle permits flexibility (Fig. 2.5B). Typically, each zonite is composed of three plates: one arched dorsal or **tergal plate**, and two ventral or **sternal plates** (Fig. 2.5C). Zonites generally bear lateral spines, bristles, or on the sternal plates of the third or fourth zonite, a pair of adhesion tubes. The tergal plate of the terminal zonite usually bears a conspicuous pair of movable lateral spines.

C. Alimentary Canal

The mouth cavity is followed by a thick cuticular ring, the **pharynx crown** (Fig. 2.5A), which projects into the cavity from the pharynx (=esophagus). The cuticularly lined pharynx has a syncytial epithelium and an outer layer or radial muscles arranged in longitudinal series (Fig. 2.5A). The lumen is triradiate, rounded, or flattened. The kinorhynch pharynx is more similar to the esophagus of nematodes than to the pharynx of rotifers. It would be less misleading, therefore, if it were referred to as an esophagus, especially since the so-called "esophagus" of kinorhynchs, posterior to the "pharynx," is no more than the junction between the pharynx and intestine and, accordingly, comparable to the esophago-intestinal valve of nematodes. The remainder of the alimentary canal is a simple straight tube leading to the cuticularly lined end-gut or **rectum.** The intestine is single layered, cuboidal, or columnar nonciliated epithelium without external glands (Fig. 2.5C). A loose network of circular and longitudinal muscles covers the intestine. The rectum opens through an anal pore located between the tergal and sternal plates of the last zonite.

D. Excretory System

The excretory system is protonephridial and is confined between the tenth and eleventh zonites (Fig. 2.5C). The paired distal enlarged flame bulbs are in the region of the tenth zonite, and lead by short tubes to separate nephridiopores on the tergal plate of the eleventh zonite.

E. Somatic Musculature

Like rotifers and gastrotrichs, the kinorhynchs lack a definite subepidermal muscle sheath (Fig. 2.5C). The somatic musculature is segmentally arranged according to the body division into zonites. Ring muscles occur in the first two zonites. These are modified in the trunk region to dorsoventral bands running from tergal to sternal plates. In the lateral trunk region musculature is diagonally arranged and extends from the anterior edge of one zonite forward to the preceding zonite. In addition there are paired longitudinal bands dorsolaterally and ventrolaterally. The various muscle units operate the head retraction, mouth protrusion, body elongation and contraction, as well as movement of the terminal spines.

F. Nervous System

The base of the mouth cone and/or the anterior portion of the "pharynx" is surrounded by a nerve commissure and attendant ganglia called the brain. Nerves run anteriorly to the head region and posteriorly there is a ventral ganglionated nerve cord running in the midventral groove between sternal plates. A ganglionic mass occurs in the middle of each trunk zonite. The most common sense organs are sensory bristles and, in at least one genus, anterior eyespots.

G. Pseudocoel

The pseudocoel is spacious and fluid filled. Amoebocytes are numerous in the fluid.

H. Reproduction

All known species of kinorhynchs are dioecious. In some species males are readily distinguished from females by the presence of adhesion tubes. No great external morphological differences distinguish males and females. The paired female gonads are saclike and each distally has an apical cell followed by germ cells, nutritive cells, and an epithelial wall. From each gonad an oviduct extends to its separate gonopore on the thirteenth zonite. The seminal receptacle is a diverticulum located on the dorsal aspect of the oviduct. The male gonad is also saclike, covered by epithelium, and houses spermatogonia in various stages of spermatogenesis (Fig. 2.5C). The male has two testes, which open by separate pores on the thirteenth zonite; the short sperm duct is armed with two or three penal spicules.

The embryology is unknown, but is presumed to be determinate. The youngest females found contain sperm; therefore, it is presumed that fertilization of eggs is required. Larval stages are terminated by a complete molting of all cuticle and cuticular structures, and have little or no resemblance to adults. They are without adult zonites and are also lacking head, scalids, placids, pharynx, and anus. Zonites are gained as larvae approach adulthood.

I. Biology

All known species of Kinorhyncha are marine. They are not free swimmers; instead they burrow in a wormlike fashion through the bottom mud and debris. Locomotion is accomplished by head scalids in cooperation with extension and retraction of the body zonites. Feeding is accomplished by thrusting the mouth cone and spread oral styles into the food mass, which is then rapidly ingested by muscular action of the pharynx.

Kinorhynchs feed on algae, diatoms, or in the slime of crabs and mollusks. Maintenance in the laboratory is best at temperatures below 25°C. They have, under this condition, been reared on microalgae for as long as 60 days. It is believed that the three species found along the north-central California coast were introduced through the introduction of the European oyster.

J. Classification

Phylum Kinorhyncha

Class Homalorhagea. Head and second zonite (neck) retractable. Third zonite with three ventral plates, retractable and closable against arched tergal plate. Oblique musculature absent.

Class Conchorhagea. Head and neck retractable. Third zonite with two bilateral flaps; oblique musculature absent.

Class Cyclorhagea. First zonite (head) retractable; second zonite closing apparatus, composed of 14–16 plates. Adults, 13–14 zonites.

IV. Phylum Nematomorpha
(Gr. nema=thread; morpha=form)

The known Nematomorpha are free-living as adults, but parasitic in arthropods and annelids as juveniles. About 230 species are recognized in Nematomorpha, which are commonly known as "hairworms," or "gordian worms." The phylum is divided into two groups, the Gordioidea which as adults live in freshwater, and a second class Nectonematoidea comprised

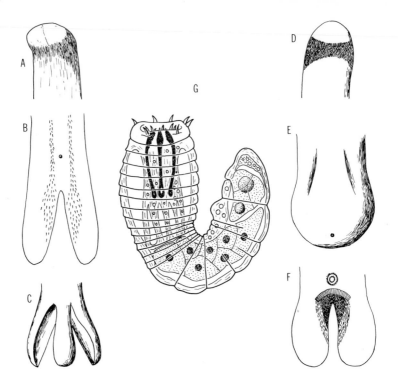

Figure 2.6. Phylum Nematomorpha. A. Anterior end of *Paragordius*. B. Posterior end of male *Paragordius*. C. Posterior end of female *Paragordius*. D. Anterior end of *Gordius*. E. Posterior end of female *Gordius*. F. Posterior end of male *Gordius*. G. Larva of *Parachordodes* (redrawn from Hyman). (A,B redrawn from May; D,F redrawn from Heinze; E redrawn from Hyman; C redrawn from Montgomery).

of a single marine genus. Nematomorphs often reach lengths exceeding 36 cm; however, they are generally very slender, being not much more than 1 mm in width. The adult worms are often found in ponds, quiet streams, rain puddles, or even horse troughs.

A. Shape

Adults are opaque and yellow-gray to brown or black. They are slender and greatly elongated. The anterior extremity is often marked by a white area followed by a darkened band, the **calotte** (Fig. 2.6A, D). In both sexes the cloacal opening is posterior and located either terminally or ventrally (Fig. 2.6B, C, F). However, males and females are readily distinguished. The female tail is usually bluntly rounded, whereas the male tail is deeply lobed (Fig. 2.6B, F). As in nematodes, males are generally shorter than the females, and the posterior part of the body is often coiled ventrally.

B. External Covering

The body wall, as in the other pseudocoelomates, consists of the non-cellular cuticle secreted by the underlying epidermis. The cuticle is divisible into two layers: an outer homogeneous layer and an inner fibrous layer. Externally the cuticle has a rough texture due to the presence of **areoles**, which project above the surface as papillae. The areolar patterns are diagnostic, and topped by one or more bristles, or perforated by pores. The fibrous layer is laminated of some 45 strata. The epidermis is a one-layered epithelium, varying from low cuboidal to columnar-shaped cells. In gordians, only the ventral chord protrudes into the body cavity. Sometimes the areoles are altered into adhesive warts on the cloacal region of the male.

C. Alimentary Canal

The mouth is situated terminally or ventrally in the calotte region. In both juveniles and adults the digestive tract is more or less degenerate, making the ingestion of food improbable. The anterior part of even the degenerate digestive tract presumably represents the esophagus, and is composed of a solid cord of cells often sufficiently formed to end in a bulb. The intestine or midgut, such as it is, is a simple epithelial tube lying free in the pseudocoelom. Toward the posterior end, the intestine connects with the genital ducts and thereafter forms a **cloaca** lined with cuticle. In the Nectonematoidea group, the intestine is incomplete and does not reach the cloaca, which serves sexual functions.

D. Somatic Musculature

The body wall musculature resembles that of nematodes, being longitudinal and forming a subcuticular sheath. The fibers in at least Nectonematoidea are similar to the coelomyarian type of muscle cell found in nematodes. The **coelomyarian** muscle fiber has the contractile elements not only on the area closest to the external cuticle, but extending part way up the sides of the cell. In Gordioidea the contractile fibers completely encircle the proto-

plasmic part of the cell, and therefore resemble the circomyarian musculature of nematodes.

E. Nervous System

The nervous system is similar to that found in kinorhynchs and consists of
a circumenteric cerebral mass laying in the head callote, with the major
body nerve extending as a midventral cord from the ring. The ventral cord
is closely associated with the epidermis, but moves inward in gordioids.
From the cerebral mass arises a pair of dorsal nerves that end in the so-
called eye. The nature of the callote and the function of the eye are poorly
understood, although some support for a photic nature is the transparency
of the callote and the presence of the dark pigment ring just behind it. The
callote is heavily innervated by the two dorsal nerves coming from the
nerve ring. Presumably the various bristles, spines, and warts of the cuticle
have a sensory function, and those at the posterior end of males may play
a sensory role in copulation.

F. Pseudocoel

The pseudocoelom differs in the two classes of Nematomorpha. In the
Nectonematoidea the pseudocoel is complete, extending the body length
between the body wall and the digestive tube. Anteriorly the pseudocoel
is separated from the rest of the body by a septum. In the class Gordioidea
the pseudocoel is filled with mesenchyme cells of a loose parenchymal
nature. The degenerate digestive tube is surrounded by a space throughout
its course, and the gonads lie in spaces bounded by mesenchyme.

G. Reproduction

In freshwater gordians both sexes have paired gonads. The nature of the
gonadal covering is questionable. Some report that the gonads are bounded
by epithelium, while others claim that they are confined only by mesenchymal tissue. The paired male gonads enter the cloaca separately by a
short sperm duct. The posterior portion of the testis cylinder is sometimes
considered to be a seminal vesicle. Though copulatory spicules are lacking,
the cloaca often bears spines or bristles. As the female matures the tubular
gonads begin to take on lateral diverticula; it is within these diverticula that
the eggs ripen. According to some authors the eggs do not pass back into
the central tube until just before entering the cloaca. Posteriorly the uterus
functions as an oviduct that opens separately into a common cloacal
chamber lined by cuticle.

In marine Nectonematoidea the reproductive system is poorly under-stood. In males with a single testis it opens posteriorly through the cloaca. There appears to be no definite ovary in the females; the oocytes are connected to the epidermis and later they become free and fill the pseudo-coel. The eggs are emitted to the exterior by a short genital tube, the remains of a cloaca, at the posterior end of the worm.

Gordians emerge from their insect host before reaching maturity, generally in the vicinity of water, in spring and early summer. Copulation begins immediately and is followed by egg laying. The motile sperm, which are placed externally on the female near the cloacal opening, make their way into the female reproductive system where they migrate into a seminal receptacle. Fertilized eggs are laid in long strands formed by a secretion from glands associated with the female reproductive system. Eggs undergo equal holoblastic cleavage.

Larval gordians are distinguished by the protrusible stomal spines or proboscis stylets (Fig. 2.6G). The larvae are capable of only a short free existence. They penetrate almost any small aquatic animal, but apparently can only undergo development in insects, centipedes, or millipedes, which are attacked at the time they come to water to drink. Larval forms are also found in many aquatic insects and annelids where they cannot develop. Molting has been noted in juvenile worms after they leave the host and just prior to complete adult structure formation. Among *Nectonema* there appear to be a series of molts, although these have not been counted.

H. Biology

Gordian worms are found throughout the world in the tropical and temperate zones and in every sort of aquatic habitat including high moun-tain streams. Since the adult worms do not feed, they do not require any special environment as long as it is sufficiently moist and has adequate oxygen. This is seldom a problem since the parasitized host seeks water when the worms are ready to emerge. Males are much more active than females, often swimming about actively or crawling on the bottom with serpentine undulations. Since gordian worms in water often tangle them-selves in very intricate knots, their name is most apt. Nectonematids parasitize hermit crabs and true crabs, though the full life cycle is not completely understood.

I. Classification

Class Gordioidea. Paired gonads; pseudocoel filled with mesenchyme parenchyma. Somatic musculature interrupted only ventrally. Fresh-water, parasites in juvenile stages of terrestrial and aquatic arthropods.

Class Nectonematoidea. Gonads unpaired; pseudocoel not filled or reduced. Double row of swimming (natatory) bristles. Somatic musculature interrupted dorsally and ventrally. Marine, parasites of crustacea.

V. Phylum Nemata
(Gr. nema=thread)

Nematodes occupy almost every conceivable habitat on earth. Among the 15,000 recognized species there are free-living forms in marine, freshwater, and terrestrial environments, as well as parasites of both plants and animals. The habitats occupied by nematodes are more varied than any other group of animals save arthropods. Nematodes are extremely varied in size, ranging from less than 0.3 mm to over 8 meters. Nematodes are colorless except for those few species having eyespots or food in the intestinal tract, or some animal parasites, which may be blood red.

A. Shape

Nematodes generally take on a cylindrical wormlike form, being somewhat rounded anteriorly and tapering posteriorly (Fig. 2.7). The body cannot easily be divided into head, neck, trunk, or tail, although the region posterior to the anus is often referred to as the **tail.** The greatest diversity in body shape occurs among the parasitic forms, where they are often pear-shaped, lemon-shaped, or globular. Sexual dimorphism does occur among some parasitic forms. Normally the males can be distinguished from the females by the secondary sexual characteristics and by the extreme ventral twisting of the male tail, due to the copulatory musculature. Males are also distinguished by the possession of a cloaca. Females have separate genital and digestive tract openings.

B. External Covering

Nematodes are covered by a noncellular elastic cuticle (Fig. 2.8). The cuticle is a complicated histological and chemical structure composed of several layers, but basically this can be divided into four: the **epicuticle,** **exocuticle, mesocuticle,** and **endocuticle.** Underlying the cuticle is the **hypodermis** (epidermis in the other so-called pseudocoelomates). The hypodermis in nematodes is divisible into four chords (main cell bodies): two major chords laterally, a dorsal chord, which is nucleated only in the region of the esophagus, and a ventral chord, which is composed of one

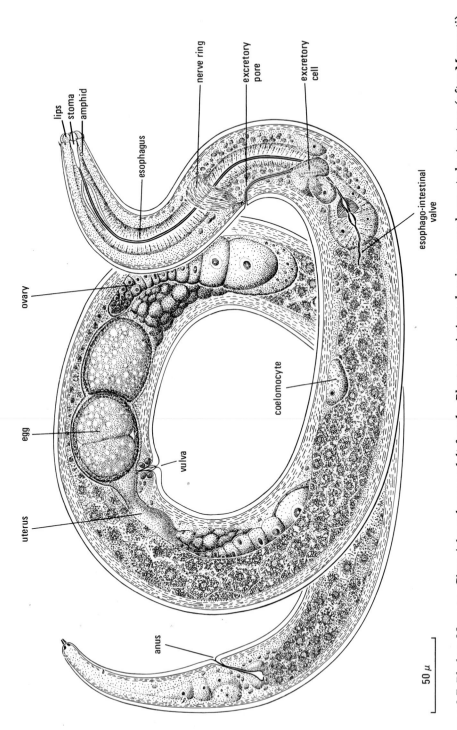

Figure 2.7. Phylum Nemata: Class Adenophorea; adult female, *Plectus parietinus* showing general nematode structure (after Maggenti).

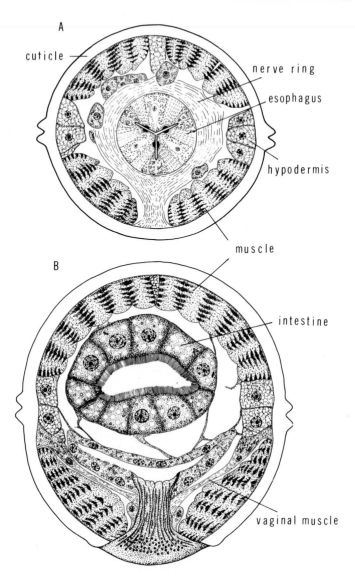

Figure 2.8. Phylum Nemata: *Plectus parietinus*. A. Cross section through esophageal region. B. Cross section through intestine at the level of the vulva. (A,B redrawn from Maggenti).

or two cells in cross section (Fig. 2.8). The chords run longitudinally the length of the body. The cytoplasmic portion of the hypodermis extends as a thin layer between the cuticle and the muscle sheath. The nuclei for the hypodermal cells are found in the chordal regions (cell body). The cuticle

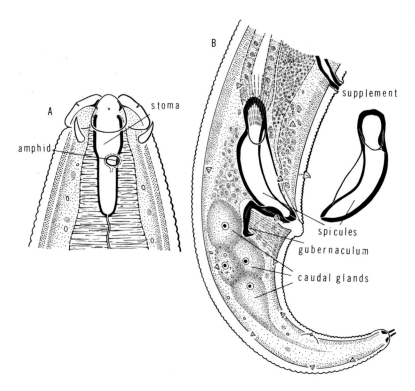

Figure 2.9. Phylum Nemata: *Plectus parietinus*. A. Female anterior end. B. Male tail. (A,B after Maggenti).

may be smooth or marked by transverse striae (Fig. 2.9A), forming a pseudoannulation, or it may be thickened and sculptured in various patterns. Along the body, especially in marine nematodes, a variety of somatic setae are located; these are presumed to have a tactile sensory function (Fig. 2.9B). At the so-called head end or lip region nematodes almost always have 16 sensory organs (sensilla); these may be papilliform, setiform, or a combination of papillae and setae (Fig. 2.9A). Just posterior to these sensory structures are two other sensory structures called **amphids**, which are presumed to be chemoreceptors. In some forms the tail ends in a **spinneret** which is connected to three caudal glands (Fig. 2.9B, C). Alae or wing areas are often found laterally on the head, along the main body, or on the tail of males. On males the alae of the tail are referred to as **caudal alae** or **bursae.** Adhesion tubes occur on some marine nematodes, especially those in the subclass Chromadoria.

The stomodeum (stoma and esophagus), rectum, vagina, and body pore tubes that lead to the excretory cell or hypodermal glands are cuticularly lined.

C. Alimentary Canal

The mouth is situated terminally and may or may not be surrounded by three to six lips (Figs. 2.7, 2.9A). The mouth is followed by the stoma, which is usually composed of two parts: the anterior **cheilostome** and posterior **esophastome.** The two together are generally referred to as the **buccal cavity**, which takes on a variety of forms according to feeding habits. Predaceous or carnivorous nematodes generally have a buccal cavity armed with teeth, which are movable or nonmovable. Bacterial feeders generally have a long, narrow, cylindrical stoma without teeth. The stoma of fungal feeders and parasites of plants is generally armed with a movable hollow spear.

The stomatal region is followed by the esophagus. In cross section the lumen of the esophagus is triradiate (Fig. 2.8A). The esophagus is composed of radial muscle tissue, epidermal tissue, and esophageal "salivary" glands. The radial arms of the esophagus are directed so that one is ventral and two are subdorsal. The esophagus is followed by an esophageal intestinal valve marking the junction of the esophagus and intestine (mesenteron). The intestine is composed of a single layer of cells with microvilli on the surface (Fig. 2.8B). When visible the area is referred to as the bacillary layer or brush border. In the order Dorylaimida the intestine is divisible into the intestine (stomach) and prerectum, which is distinguished by the microvilli or a bacillary layer. The junction between the intestine and rectum is often marked by a sphincter muscle and rectal glands (Fig. 2.7). The cuticularly lined rectum opens through a ventral anal pore, which in males joins with the reproductive system to form the cloaca.

D. Somatic Musculature

The somatic musculature of nematodes forms a complete subcuticular sheath interrupted only by the hypodermal chords (Fig. 2.8). Musculature is generally longitudinal. When contractile fibrils are found at the base of the cell closest to the cuticle, the muscles are called **platymyarian.** If the contractile fibrils extend up the side of the individual muscle cell, approaching the body cavity, they are called **coelomyarian** muscles. The specialized muscles of the body are generally **circomyarian**, that is, contractile fibrils surround the muscle cell. Specialized muscles are scattered throughout the body. They operate the spear in plant parasitic nematodes; they may be cutaneoesophageal, cutaneointestinal, or operate the male spicules or sphincters of the female vulva (Figs. 2.7, 2.9B).

E. Nervous System

The central nervous system is composed of a **circumesophageal commissure** with large ganglia attached laterally, dorsally, and ventrally (Fig. 2.7). Ventrally the circumesophageal commissure leads to the major nerve cord of the body, the ventral nerve cord, which is ganglionated at various points along the body. Anterior to the circumesophageal nerve ring, nerves run to the sensory structures of the cephalic region, including the amphids (Figs. 2.7, 2.9A). Throughout the length of the body, commissures occur that connect the ventral nerve cord with the lateral nerves or with the dorsal nerves, or dorsal nerves with the ventral nerve. A subcuticular peripheral nerve net has been described for marine nematodes.

F. Pseudocoel

The body cavity of nematodes is fluid filled and extends the length of the body between the muscle sheath and internal organs (Fig. 2.8). The pseudocoel contains a variety of cells, membranes, and fibrous tissue. In some forms a nucleated network covers the esophagus, intestine, and gonads, and may delimit the muscle bands and hypodermal chords.

G. Excretory System

No single type of excretory system characterizes Nemata; indeed, in some it is completely absent. In marine nematodes it generally is composed of a single ventral cell. In the terrestrial nematodes it is often connected with one or more tubules extending the length of the body, directed either posteriorly or both posteriorly and anteriorly. In either event, when present the excretory system opens through a single ventral pore located in the anterior region of the body. In those forms where the excretory system is not seen, the system of excretion is unknown. It is presumed that, at least in some nemas, the hypodermal glands or perhaps in Dorylaimida even the prerectum may function in excretion. This, however, has not been verified.

H. Reproductive System

A female gonad may be single or paired (Fig. 2.7). However, as many as 32 ovaries have been reported in some parasitic nemas. The female gonads are tubular, with an apical cell and a thin nucleated epithelium covering the entire gonad. The ovary is followed by the uterus, which connects to the vagina, which opens on the ventral surface of the body at the vulva.

The male reproductive system is composed of one or two testes, which empty posteriorly with the rectum forming a cloaca (Fig. 2.9B).

In the embryological development of nematodes the early cleavages follow a modified spiral plan, but the development is highly determinate. After full mature embryonation and development, the first stage larva may molt once within the egg, though in many free-living forms the larva emerges from the egg in the first stage and undergoes four molts prior to becoming an adult. At each molt the external body cuticle and the cuticle lining the esophagus and rectum are shed. Cell multiplication after hatching is restricted except within the reproductive tract, midgut, epidermis, and somatic musculature. Few or no changes occur in the esophagus or nervous system.

I. Biology

Nematodes are among the most common metazoan inhabitants of marine, freshwater, and soil environments. In numbers they exceed all other metazoa. Nematodes have been found from the deepest ocean floor to the highest mountains. They are known to occur from the Arctic to the Antarctic. Some have actually been revived from ice. In soil they are found as far as roots can penetrate. Marine nematodes are abundant in all the oceans from intertidal zones to the greatest depths. As parasites of plants and animals they are unequaled by any other metazoan, including insects.

J. Classification*

    Class: Adenophorea
      Subclass: Enoplia
        Order: Enoplida
          Suborder: Enoplina
            Superfamily: Enoploidea
            Superfamily: Oxystominoidea
          Suborder: Oncholaimina
            Superfamily: Oncholaimoidea
          Suborder: Tripylina
            Superfamily: Tripyloidea
            Superfamily: Ironoidea

---

* See Chapter 10 for an expanded classification with characters.

Order: Isolaimida
   Superfamily: Isolaimoidea
Order: Mononchida
  Suborder: Mononchina
   Superfamily: Mononchoidea
  Suborder: Bathyodontina
   Superfamily: Bathyodontoidea
   Superfamily: Mononchuloidea
Order: Dorylaimida
  Suborder: Dorylaimina
   Superfamily: Dorylaimoidea
   Superfamily: Actinolaimoidea
   Superfamily: Belondiroidea
   Superfamily: Encholaimoidea
  Suborder: Diphtherophorina
   Superfamily: Diphtherophoroidea
   Superfamily: Trichodoroidea
  Suborder: Nygolaimina
   Superfamily: Nygolaimoidea
Order: Trichocephalida
  Suborder: Trichocephalatina
   Superfamily: Trichuroidea
   Superfamily: Trichinelloidea
   Superfamily: Cystoopsoidea
Order: Mermithida
  Suborder: Mermithina
   Superfamily: Mermithoidea
   Superfamily: Tetradonematoidea
Order: Muspiceida
Subclass: Chromadoria
Order: Araeolaimida
  Suborder: Araeolaimina
   Superfamily: Araeolaimoidea
   Superfamily: Axonolaimoidea
   Superfamily: Plectoidea
   Superfamily: Camacolaimoidea
  Suborder: Tripyloidina
   Superfamily: Tripyloidoidea
Order: Chromadorida
  Suborder: Chromadorina
   Superfamily: Chromadoroidea
  Suborder: Cyatholaimina
   Superfamily: Cyatholaimoidea
   Superfamily: Choanolaimoidea
   Superfamily: Comesomatoidea

Order: Desmoscolecida
  Suborder: Desmoscolecina
    Superfamily: Desmoscolecoidea
    Superfamily: Greeffielloidea
Order: Desmodorida
  Suborder: Desmodorina
    Superfamily: Desmodoroidea
    Superfamily: Ceramonematoidea
    Superfamily: Monoposthioidea
  Suborder: Draconematina
    Superfamily: Draconematoidea
    Superfamily: Epsilonematoidea
Order: Monhysterida
  Suborder: Monhysterina
    Superfamily: Monhysteroidea
    Superfamily: Linhomoeoidea
    Superfamily: Siphonolaimoidea
Class: Secernentea
  Subclass: Rhabditia
    Order: Rhabditida
      Suborder: Rhabditina
        Superfamily: Rhabditoidea
        Superfamily: Alloionematoidea
        Superfamily: Bunonematoidea
      Suborder: Cephalobina
        Superfamily: Cephaloboidea
        Superfamily: Panagrolaimoidea
        Superfamily: Robertioidea
        Superfamily: Chambersielloidea
        Superfamily: Elaphonematoidea
    Order: Strongylida
      Suborder: Strongylina
        Superfamily: Strongyloidea
        Superfamily: Diaphanocephaloidea
        Superfamily: Ancylostomatoidea
        Superfamily: Trichostrongyloidea
        Superfamily: Metastrongyloidea
    Order: Ascaridida
      Suborder: Ascaridina
        Superfamily: Ascaridoidea
        Superfamily: Cosmocercoidea
        Superfamily: Oxyuroidea
        Superfamily: Heterakoidea
        Superfamily: Subuluroidea
        Superfamily: Seuratoidea

Suborder: Dioctophymatina
    Superfamily: Dioctophymatoidea
Subclass: Spiruria
  Order: Spirurida
    Suborder: Spirurina
      Superfamily: Spiruroidea
      Superfamily: Physalopteroidea
      Superfamily: Filarioidea
      Superfamily: Drilonematoidea
  Order: Camallanida
    Suborder: Camallanina
      Superfamily: Camallanoidea
      Superfamily: Dracunculoidea
Subclass: Diplogasteria
  Order: Diplogasterida
    Suborder: Diplogasterina
      Superfamily: Diplogasteroidea
      Superfamily: Cylindrocorporoidea
  Order: Tylenchida
    Suborder: Tylenchina
      Superfamily: Tylenchoidea
      Superfamily: Criconematoidea
    Suborder: Sphaerulariina
      Superfamily: Sphaerularoidea
  Order: Aphelenchida
    Suborder: Aphelenchina
      Superfamily: Aphelenchoidea
      Superfamily: Aphelenchoidoidea

Chapter 3
Nematode Integument

The **exoskeleton** serves as the interface between the nematode and its external environment. The combination of a noncellular elastic cuticle and an underlying hypodermis makes the exoskeleton one of the most complex histological and chemical structures of the nematode body. This complex organ acts as a barrier against detrimental conditions externally and functions to maintain, in part, the delicate chemical balance internally. The exoskeleton preserves the structural integrity of the body while protecting the internal organs. Through specialized organs, evolved from and in the cuticle, it receives sensory information from the environment to which the nematode responds. In Nemata it is also one of the agents of movement.

The cuticle is the product of the body wall cells (hypodermis) derived from the ectoderm. One cannot understand the tegumentary structure by treating the product (cuticle) as a separate entity from the generating source (hypodermis). The underlying cells retain the embryological potential of ectoderm, which is manifested in the specialized organs of the integument that function in excretion, sensation, and secretion. Furthermore, the glands of the body (hypodermal, esophageal, rectal, etc.), as well as the cuticle of the various body invaginations (stomodeum, rectum, vagina, gland ducts, spicules), are best understood when recognized as derivatives of the ectoderm, "hypodermis."

As systematists and morphologists we seldom appreciate that setae, bristles, tubes, teeth, surface sculpturing, and excretory system on which much of nematode taxonomy depends are reflections of mechanisms inherent in the hypodermis (ectoderm).

I. External Cuticle

The basis for our understanding of nema cuticle stems from the early studies of *Ascaris*, in which nine layers are described and placed into three strata: **cortical, median** (matrix), and **basal** (oblique fibers). Both electron microscopy and a survey of the literature show the error of attempting to use *Ascaris* as a foundation for cuticular nomenclature of Nemata (that implies chemical, functional, and homologous relationships of cuticular strata). It is misleading to assume that these ascarid strata are universally present among nematodes. Therefore, this text conforms to a topographical nomenclature, as employed for arthropods, which expresses relative distribution of cuticular laminations. *Deontostoma* (Enoplia) is used as the model cuticle because current knowledge shows it to have the most nearly complete complement of strata (Fig. 3.1). Furthermore, orderly changes among the classes and orders of nematodes can be derived from this model. Nema cuticle in this system is divisible into four fundamental strata: epicuticle, exocuticle, mesocuticle, and endocuticle. Anteriorly and posteriorly some nematodes have what appears to be an additional deposition called "infracuticle."

A. Epicuticle

The **epicuticle** seems to be a consistent feature of all nematode integuments (Fig. 3.1). Typically it is a trilaminate structure consisting of two electron dense layers separated by a less electron dense layer. The chemical nature of this stratum shows characteristics of being similar to **keratin** because of its sulfur content and similar to collagen because of x-ray diffraction patterns. Quinones and polyphenols have also been detected in the epicuticle, which correlates well with this layer being tanned in the formation of *Heterodera* cysts (plant parasite: cyst nematodes). The outer electron dense layer itself is recorded as being triple layered in *Meloidogyne javanica* (plant parasite: root-knot nematode) and *Trichinella spiralis* (animal parasite: causal organism of trichinosis). It has been proposed that this additional outermost layer is lipid.

B. Exocuticle

This stratum lies just below the epicuticle and is remarkably consistent in form throughout Nemata. In most adenophorean nemas it is completely transversed by **radial striae**, which in longitudinal section are ribbonlike. In more derived nemas, especially parasitic species, the exocuticle can be divided into the **external exocuticle**, which has no visible structure, and

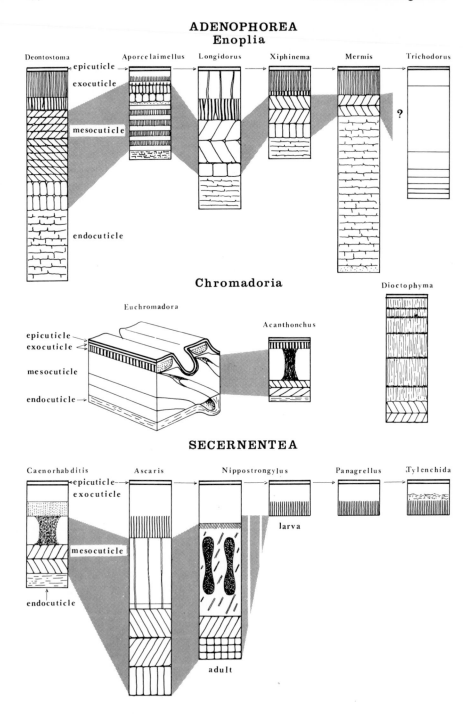

Figure 3.1. Schematic comparison of nematode cuticles and strata relationships. Mesocuticle comparisons are shown by shaded connections (after Maggenti).

the **internal exocuticle**, which is characterized by radial striae. In those few forms where radial striae are not seen, *Trichodorus* and adults of *Caenorhabditis*, it probably is due to faulty microscopic technique resulting in poor resolution rather than actual absence. Among many Diplogasteria and especially in the orders Tylenchida and Aphelenchida the cuticle consists only of this layer and the epicuticle (Fig. 3.1).

C. Mesocuticle

Among nematode cuticle strata, **mesocuticle** shows the greatest variation. Most often it takes the form of oblique fiber layers, variable in number and angular relationship to each other. Available studies indicate these layers are most numerous in marine free-living nematodes. In the classes Rhabditia and Chromadoria the mesocuticle can be modified into diagonal plates or combined fiber layers and structural struts in a homogenous matrix (Fig. 3.1). The latter is described as being fluid filled in *Nippostrongylus*. In parasitic forms such as *Ascaris* and *Nippostrongylus* it is this external modified mesocuticle that increases in thickness with age.

D. Endocuticle

This is probably the least understood stratum of nematode cuticle. Confusion exists since the endocuticle stratum is absent in *Ascaris* (Fig. 3.1). However, endocuticle is common throughout Adenophorea and in some members of the secernentean order Rhabditida. **Endocuticle** can most often be recognized by its fibroid nature. Though fibrillar in nature its structure never shows the organized oblique fiber latticework characteristic of mesocuticle. Often the pattern is disorganized and appears like overlapping feathers.

E. Cuticular Strata Significance

The classes and subclasses of Nemata can be distinguished on the basis of their cuticles (Fig. 3.1). Enoplia have all layers represented but differences are seen in the relative thickness of the mesocuticle and endocuticle. Among the marine Enoplida the number of oblique fiber layers contributing to the mesocuticle is much greater than in soil Enoplida. Whereas the endocuticle is greatly increased among the parasitic Enoplia, this is especially evident among the insect parasitic mermithids. The subclass Chromadoria is recognized by the specialized modifications within the mesocuticle, which may be platelike, or a combination of oblique fibers and "fluid" substratum encompassing structural struts (Fig. 3.1). Among Secernentea the phylogenetic trend appears to be toward a reduction in strata present in the cuticle. Within Rhabditia a complete array is represented from a complete complement of strata to only epicuticle and exocuticle. Spiruria are characterized by having only three strata: epicuticle, exocuticle, and mesocuticle. Diplogasteria currently may be distinguished by having their cuticles limited to epi- and exocuticle.

There are differences in cuticular strata along the length of the body. The foregoing discussion is limited to main body cuticle and does not necessarily reflect the cuticle condition of the anterior or posterior body. Insufficient observations have been made on anterior and posterior cuticle. However, an additional stratum termed **infracuticle** has been reported among Dorylaimida and Tylenchida.

An anomaly occurs among dioctophymatids (animal parasites, including the giant kidney worm), where the cuticle is unlike adenophoreans to which they are often assigned. In dioctophymatids there is no evidence of an endocuticle; in fact, the cuticle is reminiscent of most Secernentea (Fig. 3.1). As we shall see in future chapters, the uncertainty that systematists feel about the taxonomic placement of this group is substantiated by further morphological characters.

II. Internal Body Cuticle

The nematode's internal body cuticle, which lines the esophagus, body pores, vulva, anus, and cloaca, appears to be constituted of material similar to external epicuticle and exocuticle (Figs. 2.9, 3.2). According to differential staining the spicular cuticle in males appears chemically different among species, although it is likely to be exocuticle in nature.

Esophageal cuticle has been observed more than other internal integument. In *Deontostoma* both internal epicuticle and exocuticle are apparent (Fig. 3.3A). In fact, there appears to be a great similarity, even to the presence of "radial" striae along the radial arms of the esophageal lumen lining. However, there are physiological differences between external and internal cuticle. Differential isoelectric points can be demonstrated by differential staining. Differences are also manifest in those nematodes that are vectors of viruses. Virus particle location is partially explained by the availability of charged receptor sites on esophageal cuticle and an absence of such sites in cheilostomal or external cuticle.

The question in the esophagus that comes to mind is whether the "platelets" of *Xiphinema* are longitudinal manifestations of the radial striae. In Enoplia the radial striae of the external body exocuticle are ribbonlike or platelike in longitudinal section, having an appearance similar to the "sliding plates" in adenophorean esophagi (Fig. 3.3B).

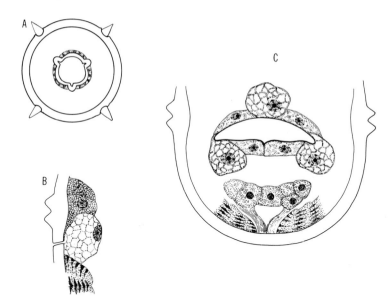

Figure 3.2. *Plectus parietinus*, cross sections showing internal structures lined by cuticle of external origin. A. Section through cheilostome. B. Section through lateral hypodermal gland and pore. C. Section through rectal region showing rectal glands and anterior anal ganglion (redrawn from Maggenti).

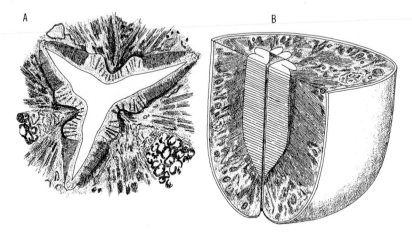

Figure 3.3. Esophagus. A. Transverse section through esophageal lumen of *Deontostoma californicum* showing portions of the esophageal glands, and concentered radial muscle attachments (drawn from TEM of Siddiqui and Viglierchio). B. Reconstruction of basal esophagus of *Xiphinema* showing platelike character of lumen cuticle and specialized apodemes for attachment of radial musculature (modified from Lopez-Abella et al.).

III. Cuticular Structures

External cuticular structures fall into two general categories: ornamentation or sensory organs. Ornamentation should not be interpreted as being nonfunctional. It generally functions to compensate for stress factors or to aid in the movement of the nematode. Sensory structures by definition enhance the nematode's awareness of the external environment. In some instances it is difficult to distinguish between ornamentation and sensory structures. Often these structures have dual functions, for example, in the warts of bunonematids, the probolae of cephalobids, the male supplementary organs of linhomoeids (Fig. 3.4), and in the various formations of the caudal alae of the male tail.

In general external ornamentation involves only the epicuticle and exocuticle. However, in some instances such as nippostrongyles and Chromadoria portions of the mesocuticle, as it is modified to form struts and structural pillars, may also be involved; these generally are referred to as punctations when viewed microscopically (Fig. 3.1). There are several types of cuticular modifications and markings, which are manifest as transverse incisures, longitudinal incisures or ridges, various inflations or alae, scales, punctations, and spines. Sensory structures on the other hand go beyond just simple markings and expansions. There are functional

Figure 3.4. A. Chromadoria:Monhysterida; cuplike preanal supplement of a male *Aponchium cylindricolle* with underlying gland (redrawn from Cobb). B. Rhabdita:Cephalobina; anterior end of *Acrobeles* showing long fringed inner circle of labial probolae and shorter external circle of cephalic probolae (original).

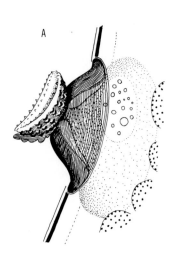
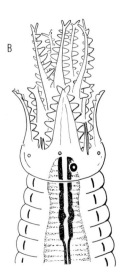

organs such as setae, papillae, amphids, phasmids, and deirids and super-ficial appendages as ambulatory setae, adhesion tubes, male supplementary organs, and such secondary sexual characteristics as suckers that involve the cuticle, hypodermal modification, muscles, and/or the nervous system.

A. Ornamentation

1. Transverse Striae

It now appears that transverse striae may be universal among Nemata. Earlier works, prior to the use of the electron microscope, report smooth cuticles for such forms as *Enoplus, Dorylaimus, Mononchus,* and *Tricho-dorus.* With the light microscope transverse striae in these forms are not visible; however, minute striae are visible with the electron microscope. When striae are deep and visible with the light microscope the interstice area is called an **annule** (Fig. 3.5A). Such striae **(incisures)** generally in-volve the epicuticle and only a part of the exocuticle. The minute transverse striae seen with the electron microscope usually involve the epicuticle and only a negligible portion of the exocuticle.

Annular formation can be so characteristic as to have significance in the family, generic, and species level. The degree of development can be so bizarre as to give the impression of overlapping tiles (*Ceramonema*)

Figure 3.5. External cuticular modifications. A. Tylenchida, transverse annula-tion interrupted by lateral longitudinal incisures. B. Tylenchida, transverse and longitudinal incisures in addition to lateral longitudinal field of incisures. C. Dorylaimida, longitudinal ridges. D. Ascaridida, Oxyuroidea, greatly expanded lateral longitudinal alae (original).

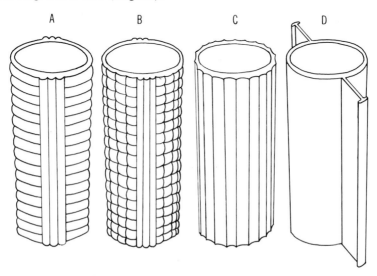

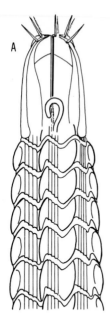
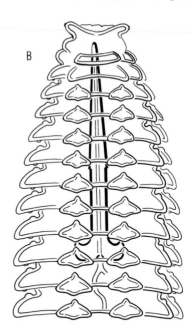

Figure 3.6. External cuticular body ornamentation. A. Chromadoria:*Cera-monema undulatum*; anterior end of female showing tilelike cuticular pattern. B. *Criconema octangulare*: anterior end of female showing coarse annulation with scalelike projections (modified from Cobb).

(Fig. 3.6A). In criconematids the annules are formed as heavy overlapping rings, with or without spines—a condition that allows the animal to move in a unique fashion for nematodes (Fig. 3.6B). Criconematids have the ability to extend and contract their length and thus move in a fashion similar to Kinorhyncha or Annelida. The normal serpentine movement in these nemas is greatly restricted. Other nemas such as the fish parasite *Spinitectus* have posteriorly directed spines on the posterior margin of the anterior annules. These may function to anchor the parasite in place while in the gut. Overlapping annulation is not always posteriorly directed. In *Euchromadora vulgaris* they are anteriorly directed in the forward part of the body and posteriorly directed on the posterior body. Therefore, annulation is more than simple ornamentation. Annules lend structural strength and may contribute greatly to lateral and longitudinal flexibility. Biologically, they aid the nematode in surviving a parasitic mode of life.

2. Longitudinal Striae

Longitudinal striae may take the form of either longitudinal incisures or ridges. When in the form of incisures around the entire circumference of the body they may interrupt the transverse striae, giving the external surface of the body the appearance of a corncob (Fig. 3.5B). Longitudinal markings more often take the form of ridges or alae, in the absence of transverse striae. Ridges usually encompass the whole circumference as in *Dorylaimus stagnalis* (Fig. 3.5C). They may either extend to a sharp apex or are rounded. The total number and spacing of these ridges vary according to species.

Other forms of longitudinal markings may occur around the body dorsally, ventrally, laterally, and/or submedially as spines on the transverse plates or annules. This formation is not uncommon among the Chromadoria as demonstrated by the Monoposthiinae. Longitudinal striae on individual annules may also occur as in *Ceramonema* (Fig. 3.6A). In males of some spirurids longitudinal ridges, often broken or crossed by transverse striae, are confined posteriorly to the ventral region of the body just anterior to the cloacal opening (Fig. 3.7).

Figure 3.7. Peloderan caudal alae. Spirurida:*Sterliadochona pedispicula*. A. Transverse section of male tail showing alae expansion, ventral ridges, and the structure of the caudal papillae within the alae. B. Male tail, ventral view showing ventral longitudinal ridges and the placement of caudal papillae (A,B redrawn from Maggenti and Paxman).

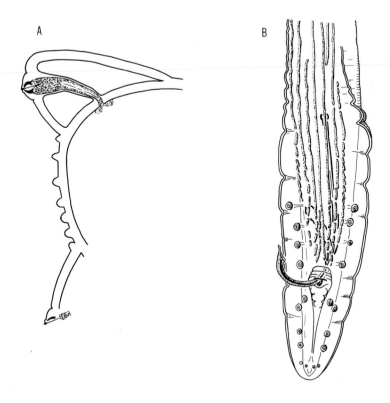

3. Longitudinal Alae

The most distinctive longitudinal feature among Chromadoria and Secernentea is the longitudinal **lateral alae** composed of one or more ridges separated by incisures (Fig. 3.5A, B). These longitudinal alae occur over the region of lateral hypodermal chords (Fig. 3.5A, B, D). Among Tylenchida, especially, the number of incisures and ridges can be of importance in generic and specific designations. Longitudinal alae are best developed among the Oxyuroidea where they extend beyond the body contour, often as much as a full body diameter (Fig. 3.5D). In Tylenchida they are only negligibly raised beyond the normal body contour.

4. Cervical Alae

On the anterior of some animal parasitic nematodes in Strongylida, Ascaridida, and Spirurida are rather wide lateral anterior alae. The expansions may be single, bifid, and trifid as in the thelazid *Physocephalus sexalatus*. Internally they often have supporting struts. As with most cuticular markings, they are formed of the epi- and exocuticle. One must be cautious of cervical (neck) alae because artifacts due to fixation often can be misinterpreted as alae, cervical expansions or collars.

5. Caudal Alae

These caudal finlike structures are found primarily among the Secernentea and are confined to the caudal region of males. Caudal alae are extremely rare among Adenophorea, and when present, they are very poorly developed, as in *Trichodorus*, *Oncholaimellus*, and *Anoplostoma*.

Caudal alae in males of Rhabditida and Strongylida are termed bursae. The term **bursa** means bell and was introduced by Dujardin in 1845 to designate the caudal expansions of some male strongylids (Fig. 3.8B, C) and dioctophymatids (Fig. 3.8F). It has since been applied by various workers to all **caudal alae.** There are no strong arguments against the usage of the term other than the fact that musculature is associated with some of the genital papillae (rays) that are found within the alae of male Strongylida. The fact remains, however, that all caudal alae are of similar construction, being formed laterally in the region of the anal opening (cloacal opening) as an expanded modification of the epi- and exocuticle, and they usually enclose within their confines one or more genital papillae (Fig. 3.7A). It is alleged that papillary secretions serve some adhesive function during copulation.

Descriptive terms for the varied shapes of caudal alae were introduced by Schneider in 1866, and are still useful in taxonomy. When the caudal alae are restricted to the two sides of the body and do not surround or meet posterior to the tail tip, they are called **leptoderan** (Fig. 3.8E); in

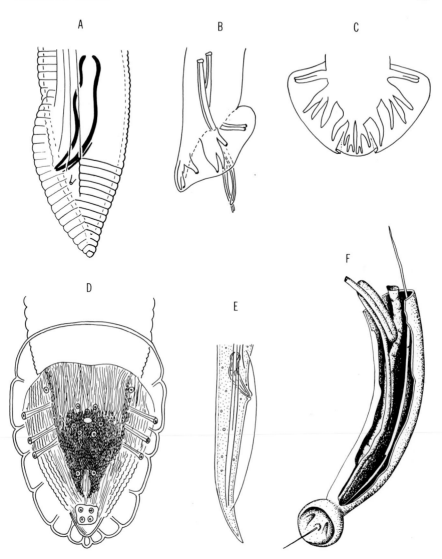

Figure 3.8. Types of caudal alae (bursae) occurring on male tails of Secernentea. A. Tylenchida:*Rotylenchus*; peloderan (redrawn from Thorne). B. Strongylida: *Chabertia*; peloderan with well-developed bursal rays, lateral view (redrawn from Yorke and Maplestone). C. Strongylida:*Delafondia*; peloderan, ventral view (redrawn from Loos). D. Spirurida:*Physaloptera turgida*; arakoderan, ventral view also shows perianal ridging common to spirurids (redrawn from Chitwood). E. Tylenchida:*Ditylenchus*; leptoderan alae, lateral view (redrawn from Thorne). F. Ascaridida:Dioctophymatoidea; arakoderan (redrawn from Goeze).

forms where the alae surround or meet at the tail tip the term **peloderan** (Fig. 3.8A, B, C) is applied. No special term has been proposed for caudal alae that meet anteriorly and posteriorly, that is, forming a complete oval as in *Rhabditis elegans*, *R. coarcta*, *Cylindridera icosiensis*, and *Physaloptera turgida*. For consistency and understanding I propose the term **arakoderan**, meaning bowl-skin for caudal alae that completely surround the cloacal area (Fig. 3.8D).

In general, caudal alae can be differentiated into three types: rhabditoid, strongyloid, and tylenchoid. The **rhabditoid bursa** is a narrow or wide alae containing seven or more pairs of genital papillae. Bursae among strongyloids are three lobed, often very elaborate, and the enclosed genital papillae are associated with musculature forming "rays" (Fig. 3.8B, C). **Tylenchoid bursae** are rather simple caudal fin expansions that contain no more than one pair of genital papillae (Fig. 3.8A); the **aphelenchoid bursae** contain three or more pairs of genital papillae.

It is speculated that the caudal alae act to clasp the female during copulation. This may be true in Strongylida where the "rays" are accompanied by muscle tissue. However, this does not seem plausible among the Tylenchida where the alae often contain neither papillae nor musculature. It should also be noted that there are almost as many, if not more, forms of Secernentea that do not possess caudal alae. Remember also that except for extremely rare instances caudal alae are absent among Adenophorea. Little phylogenetic significance can be given to the presence or absence of caudal alae except perhaps to speculate that among Secernentea their presence may represent a more ancestral condition as exemplified by Rhabditia. However, while characteristic of some large groups such as the Strongylida they may or may not be present among closely related genera in other groups, such as *Aphelenchus*, which have a bursa, and *Aphelenchoides*, which do not.

6. Spines and Setae

The two terms spines and setae have been employed as if they were synonymous; they are not. Each term has very important morphological and taxonomic significance and they should not be interchanged. **Spines** are noncellular cuticular protrusions, without muscular or nervous connections. In Adenophorea spines are rare, however; they are seen as cuticular ornamentation in the subclass Chromadoria, e.g., *Monoposthia*. Spines are much more common among the parasitic Secernentea, including the plant parasites such as *Criconema* (Fig. 3.6B). Spines are usually associated with annulation as either anterior or posterior extensions forming annular **collarettes**. They may be limited to the anterior region as in the fish parasite, *Spinitectus*, or on most annules as on *Criconema*. Sometimes a single spinated collar is present at the anterior end as in the genus *Heth* (Fig. 3.9A). Among spirurids posteriorly directed hooks may be well developed

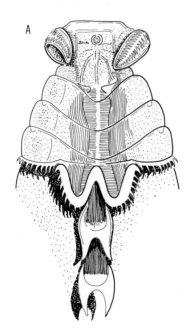
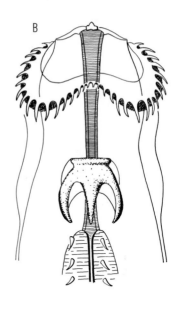

Figure 3.9. Cuticular spines. A. Oxyuroidea:*Heth*; anterior end of female with anterior fringed lobes followed by a lobed collar edged with spines; two massive body spines also illustrated (redrawn from Steiner). B. Seuratoidea:*Seuratia*; female anterior end showing spiny cordons; a massive trident body spine and two rows small body spines (redrawn from Seurat).

in transverse rows on the anterior end of the body, e.g., *Gnathostoma*. Spines are exceptionally well developed in the genus *Seuratia* (Fig. 3.9B). In this genus there occurs an anterior collarette, a large cervical trident spine, as well as longitudinal rows of body spines.

Scales are closely related to spines and are only distinguished from spines by their bluntness (Fig. 3.6B).

Setae, on the other hand, are sensory organs (sensilla) produced by specialized cells of the hypodermis. Setae, and their derivatives, may be very complex and serve a variety of functions. *Deontostoma* spp. possess both tactile setae and sublateral gland tubes (Fig. 3.10B). Desmoscolecida and Draconematidae have **tubiform setae** that function in locomotion; these are long hollow tubes associated with glands, muscles, and nerves.

A typical seta (Fig. 3.10A) is a slender to stout hairlike process composed primarily of exocuticle covered by epicuticle around a protoplasmic outgrowth of a single hypodermal cell called the sensillar gland or **trichogen**. The socket in which the seta or tube articulates is formed by another hypodermal cell called the **tormogen**; this cell is sometimes called the support or socket cell. Finally, associated with many, if not all setae, is a

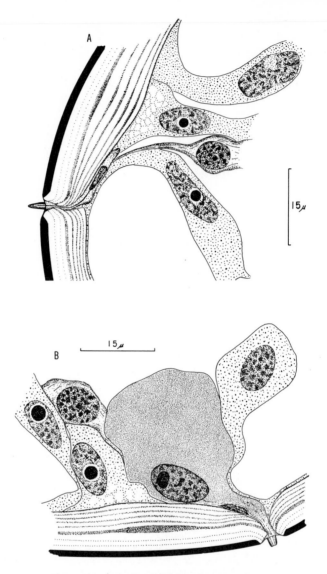

Figure 3.10. Enoplida:*Deontostoma californicum*; setal and hypodermal gland structure. A. Transverse section through sublateral somatic seta showing relationship of hypodermal cells and sensory neuron. B. Transverse section through a sublateral hypodermal gland (after Maggenti).

sensory nerve cell and sometimes glands. Even the labial and cephalic papillae are derivatives from setae. However, in papilliform sensilla the trichogen cell (hair forming cell) may be absent and only the tormogen (socket forming cell) appears to be retained along with sensory elements.

These will be discussed more fully under sensory organs of the nervous system.

7. Helmet or Capsule

At the anterior extremity there is a thickening of the inner layers of the external body cuticle (Fig. 3.6A). This may be infracuticle. In Enoplia certain groups are characterized as having anterior cuticle doubled, but it now appears that there is no doubling of cuticle, just additional depositions below the endocuticle or a modification of endocuticle. In addition the cheilostome may enfold and extend over the anterior esophagus forming an esophageal capsule. It is presumed that when the capsule or helmet is seen in Chromadoria, its formation and structure is similar to those capsules, cephalic and esophageal, seen in Enoplia.

8. Tail Shapes

Tail shape is not, strictly speaking, an element of cuticular ornamentation or modification. However, the changes in tail shape are used in taxonomy, and therefore, will be described here. The **tail** is that portion of the body posterior to the anus. Also of value in taxonomy are tail structures such as the caudal alae (Fig. 3.8).

The variety of tail shapes that may be encountered among Nemata are illustrated in Fig. 3.11. In addition the tail may be modified terminally by cuticular projections called **mucrones**; these may be simple or elaborate (Fig. 3.11M, N). Terminal mucrones should not be confused with the spinneret which is a cuticular modification associated with the caudal glands of Enoplia and Chromadoria (Fig. 2.9B).

Spinnerets themselves are variable; sometimes they are no more than a terminal pore. Usually the **spinneret** is a heavy walled cuticular cone (Fig. 3.12A), and is connected to three caudal glands. When caudal glands are present, there are always three (Fig. 3.12B). The literature sometimes lists four or more, but these cases have proved not to be caudal glands but coelomocytes or other body cavity cells. In some marine nematodes the glands are not located in the tail, but are found some distance anterior to the anus.

9. Preanal Sucker

Suckers are not a common feature of nematodes. They have never been known to occur on free-living nematodes; in fact, they are limited to the superfamily Seuratoidea, order Ascarida (Fig. 3.13). The function of suckers is unknown, but they appear to consist of a combination of glandular and muscular tissue. Structures on the tail of Drilonematoidea (earth-

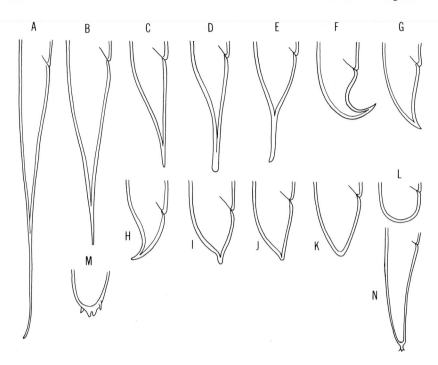

Figure 3.11. Variation of tail shapes. A. Filiform. B. Attenuated. C. Dorsally convex-conoid. D. Dorsally convex-conoid, then cylindroid. E. Uniformly convex-conoid, spicate. F. Conoid, ventrally arcuate. G. Conoid. H. Conoid, recurved. I. Digitate. J. Subdigitate. K. Bluntly convex-conoid. L. Hemispherical (A–L, redrawn from Thorne). M. Multidigitate. N. Mucronate (M,N originals).

worm parasites) and Rhabditoidea that utilize mice or phoretic (transport) hosts are not caudal suckers, but greatly enlarged phasmids (see Fig. 7.11).

B. Sensory Structures

1. Cephalic Sensilla: Phylogenetic Trends

Cephalic setae and papillae, which constitute the **cephalic sensilla**, recapitulate phylogenetic development probably more than any other organ or structure in the nematode body. Morphologically there is no reason to suspect that papillae did not evolve from setae. In fact, all available evidence concerning the cephalic sensory organs substantiates this conclusion. Furthermore, few would argue with the evolutionary theory that cephalization of structures demonstrates phylogenetic advancement. **Cephalization** is the accumulation of body structures—in this instance, sensory organs— at the anterior end. The principle is well developed in arthropods where

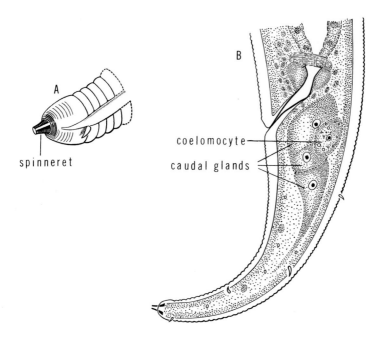

B

coelomocyte

caudal glands

A

spinneret

Figure 3.12. Chromadoria:Araeolaimida. A. Tail tip with cuticularized spin-neret. B. Female tail showing three caudal glands and a dorsal caudal coelomo-cyte which is sometimes mistaken as a fourth gland (A,B after Maggenti).

the cephalization of organs of locomotion evolved into the mouth parts of contemporary forms.

In addition to setae and papillae an additional pair of sensory organs, the amphids have undergone cephalization. It cannot be overemphasized that amphids are not "lateral papillae" that complete, as we shall see, the complement of the third whorl of cephalic sensory setae or papillae. These lateral organs, presumed chemoreceptors, were first named by Cobb in 1913. Goldschmidt in 1904 demonstrated, through studies on the nervous system, that the amphids are not related to the other sensory structures of the head. Many more recent studies have confirmed this.

Ancestral nemas probably had 16 sensory setae (sensilla) and one pair of amphids on the anterior region of their bodies (Figs. 3.14A, 3.15A). The 16 setae occurred between the level of the paired lateral amphids and the oral opening. Contemporary nematodes among Adenophorea do have the 16 sense organs plus the paired amphids; however, only among oxystominids are forms known in which all the sensory organs are seti-form and separated into three distinct whorls; amphids are never setiform. Cephalic sensory organs are often fused and reduced in number among Secernentea; this is especially true among parasitic forms where it is not uncommon to have only four papillae plus the two porelike amphids on

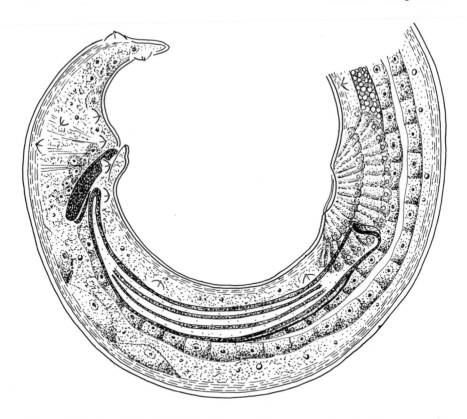

Figure 3.13. Ascaridida:*Bulbodacnitis ampullastoma*; male tail with a preanal sucker (redrawn from Maggenti).

the lips. However, the sensory receptor often remains embedded in the cuticle and no external aperture or structure is seen.

On presumed ancestral nemas, the sensory organs are distributed in three whorls on the lips and cephalic region. Oxystominoidea come closest to illustrating the ancestral distribution of organs. In these forms the oral opening is surrounded by six small lips: two subdorsal, two lateral, and two subventral. Each lip bears a single sensilla forming a circlet of either six papillae or six setae (Fig. 3.14A). These are known as the internal circle and are named by position as internodorsals (id), internolaterals (il), and internoventrals (iv). Immediately posterior to the lips the second circlet of setiform sense organs are located (Fig. 3.14A). These prominent setae are located dorsodorsally (dd), externolaterally (el), and ventroventrally (vv). The third circlet occurs on the neck or cervical region (Fig. 3.14A). This whorl has only four setae laterodorsally (ld) and lateroventrally (lv).

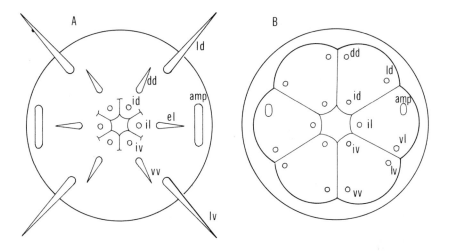

Figure 3.14. Face views of ancestral and derived nematodes. A. Adenophorea: Enoplida. B. Secernentea:Rhabditida. Abbreviations: id, internodorsal; il, internolateral; iv, internoventral; dd, dorsodorsal; el, externolateral; vv, ventroventral; ld, laterodorsal; lv, lateroventral; amp, amphid (original).

In the course of cephalization the two external circlets become one as in *Deontostoma* and are setiform (Fig. 3.15B). Along with this anterior movement of setae is movement of the amphid. The whorls are distinguishable not only by setae location, but by differential setal size. Generally those setae belonging to the first external row are shorter than the four from the second external row or third whorl. However, in some instances the setae of the first external row are longer than those of the second row, e.g., *Prismatolaimus* of the Tripyloidea, where six sensilla are long and four are short. A further step is seen in the genus *Tripyla* where the first external whorl of sensilla moves onto the lips in the form of very short stout setae; the second whorl remains just posterior to the lips and is also formed of short stout setae. This step is clearly shown in the genus *Plectus* (Fig. 3.15C). The amphids in *Tripyla* are almost at the level of these setae. The final development among the Enoplia is seen among the mononchs and dorylaims (Fig. 3.15D). All the cephalic sensory organs are papilliform and on the lips; the amphids remain postlabial, located at the constriction between the lips and body. In no known instance among Adenophorea are the amphids found on the lips.

Chromadoria manifest much the same steps of cephalization of the anterior organs that occur in Enoplia, with a few minor but interesting differences. The stepwise change from setae to papillae is perhaps more clearly demonstrated. In genera such as *Acanthonchus*, *Chromadora*,

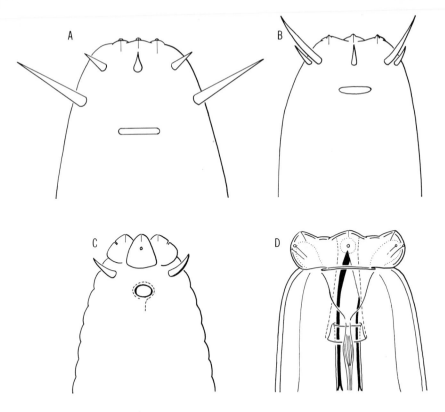

Figure 3.15. Schematics showing the sequence of events in cephalic sensilla cephalization. A. Enoplida:Oxystominoidea; sensilla in three whorls only one labial or circumoral. B. Enoplida:Enoploidea; in this step the two external whorls of sensilla have joined together. C. Chromadoria:Araeolaimida; here the first two whorls of sensilla have migrated onto the lips leaving the third whorl and amphids postlabial. D. Enoplia:Dorylaimida; all sensilla have moved onto the lips and only the amphids remain postlabial (original).

Grammanema, and others, the two external whorls are setiform and only slightly separated or combined into a single whorl, and the amphids are far anterior, being just posterior to the second external whorl. Unlike Enoplia, the next step in cephalization is not to form short, stout setae, that are almost papillae; rather the first external row moves onto the lips as papillae leaving the four setae of the second row postlabial along with the amphids (Fig. 3.15C). The final step, repeated in many groups of Chromadoria, is for all "setae" to become labial papillae with amphids remaining postlabial.

Secernentea show much less diversity or sequential cephalization. In all forms the anterior cephalic sensory organs, including the amphids, are on or in the lips (Fig. 3.14B). The cephalic sensory organs, now more accurately referred to as labial sensilla, are papilliform and the amphids porelike. Because the amphids are now labial the externolaterals of the ancestral first whorl are moved to a ventrolateral position while the amphids occupy a dorsolateral position. Fusion and reduction of labial papillae is common among Secernentea. It would appear that in some instances the internal circle of papillae is completely lost or moved to the inner lip surface, and further reduction is among papillae of the ancestral external circle, usually involving the dorsodorsals and the ventroventrals. The absence of any external evidence of the latter is common among ascarids and strongyles. With the loss of papillae from the lateral lips the amphids assume a more lateral position.

2. Amphids

It would be remiss to discuss the cuticular sensory organs and not include the amphids (Fig. 3.16). Amphid variability has taxonomic significance within groups, but beyond cephalization they show no sequential phylogenetic significance throughout Nemata. Phylogenies have been proposed on the basis of amphid external elaboration or shape, but these have little relationships to phylogenetic schemes based on the combinations of other morphological systems.

Figure 3.16. Amphid shapes. A. Oval. B. Slitlike. C. Ellipsoidal. D. Porelike. E. Horseshoe, sausage. F. Unispiral, shepherd's crook. G. Circular. H. Circular, broken. I. Multispiral (based on Schuurmans, Stekhoven, and DeConinck).

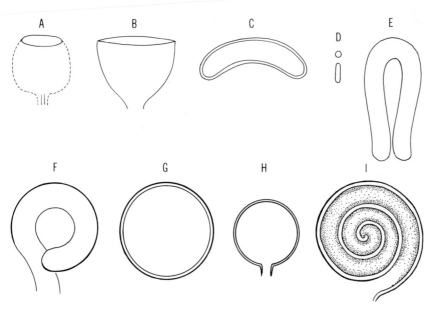

There are two schools of thought concerning the nature of an ancestral amphid. One school is based on the spiral amphid (Fig. 3.16I), and one on the simple transversally elongate oval (Fig. 3.16A). It should be noted that this discussion is based on the external appearance of the amphid and pouch and not on the internal morphology. Judging by current acceptance of Nemata classification and the phylogenies on which the higher classification is based, one is led to conclude that the oval amphid is the most ancestral and that the spiral amphid is derived. The amphid in advanced forms is labial and porelike. This is not difficult to accept when one considers both position and nature of the amphid.

Among Enoplia, as was seen with the cephalic setae and papillae, cephalization is orderly and sequential; so also with the amphids. These paired chemoreceptors are well back from the lip region in the Oxystomina. They appear elongate and pocketlike with a small circular or oval opening anteriorly. As these structures move toward the anterior the opening changes from oval to slitlike and the pocket becomes shortened. In Dorylaimida the amphid and pouch are similar in appearance to inverted stirrups (Fig. 3.16B). However, among parasitic forms they are often externally porelike (Fig. 3.16D).

Within the Chromadoria the anteriorly located, postlabial amphids show the greatest variability. In fact, almost all possible external manifestations can be seen: circular, oval, unispiral, multispiral, and slits (Fig. 3.16C, E–H). This has led some nematologists to believe that chromadorids are most representative of ancestral nemas. Such an assumption does, however, require that decephalization be accepted as a viable alternative to phylogenetic advancement. That is, we would have to accept that the amphids moved more posteriorly in Enoplia (Oxystomoidea) and that papillae moved away from the labial region to become setae. As will be discussed in other sections, other morphological reversals concerning cuticle, esophagi, reproductive system, and excretory system would also have to be accepted. However, understanding how the external appearance of amphids is produced gives an added dimension that supports the hypothesis that Chromadoria are not ancestral. Bear in mind that amphidial shape is an external manifestation and that their internal structure that connects them to the nervous system is remarkably similar (see Chapter 4, Section III).

Cobb in the early 1900s described spiral and multispiral amphids as channels or grooves impressed into the external cuticle. This notion of "impression" has persisted into modern literature. In a 1976 book on nematode classification and evolution Andrassy presents the view that in Chromadoria (= Torquentia) they are surface indentations "mirrors" with nerve endings against the surface.

We have already seen that among Enoplia the amphids are described as cyathiform, meaning that immediately following the external opening is a pocket. What must be understood is that the amphidial tube and pocket are the same. In a spiral or shepards crook amphid, the pocket (tube) is elongated and curved or twisted. The increased surface area may have physiological significance, but certainly not any morphological significance. Therefore, Cobb's impression of grooves and Andrassy's of indentations are wrong. The sensory element in all remains posteriorly located at the end of the amphidial tube. In multispiral forms the opening is central and the tube coiled counterclockwise. The tube penetrates into the body at the dorsal side of the spiral. The simplest yet most derived form of amphid is characteristic of Secernentea where the amphid is porelike (sometimes oval, cleftlike or slitlike) and located laterally or dorsolaterally on the lateral lips. Among some forms of Rhabditia and Diplogasteria the amphid in first stage larvae is postlabial and oval to reniform (kidney-shaped), substantiating that cephalization has occurred and that Secernentea evolved from nematodes with postlabial amphids.

3. Sense Organs

The sense organs of nematodes are traditionally placed into two categories: setae or papillae. These may be further characterized as labial, cephalic, cervical, somatic, or caudal, according to their location. A few are given additional recognition as amphids, deirids, or phasmids. These are all external sensory organs (exteroceptors), but there are also some internal sensory organs (interoceptors). The **interoceptors** are associated with cuticularly lined regions of the alimentary canal or reproductive system. Both types of sensory structures are designed for the reception of a stimulus and transmission of the stimulus to a nerve center. **Exteroceptors** respond to conditions in the external environment and interoceptors respond to internal conditions.

Sense organs of nemas are made up of cuticle, hypodermis, and sensory cells. Such simple receptor complexes are called **sensilla**. Nema sense organs cannot yet be identified as receptors for any particular group of stimuli; therefore, they are classed on an anatomical basis and functions are only assumed.

Sensory neurons in Nemata are of a type called primary sense cells, which receive stimuli either directly or through a distal process. If the distal processes from the neurocyte proceed to a specific organ, then it is of a class termed **Type I** (Fig. 3.17). If the processes are distributed not to a specific organ, but terminate on the inner surface of the integument, they are classed as **Type II** (Fig. 3.18). As primary sense cells their **axons** (sensory nerves) proceed to a nerve center, either in the central nervous system or esophagus, without interruption or connection to other neurons in their course.

The simplest form of sensory structure is the innervated hair (seta). Other sensory organs of the external body, excluding possibly the amphids, are modifications of this simple setiform sense organ. The external form of the sense organ determines the amount of associated hypodermal or

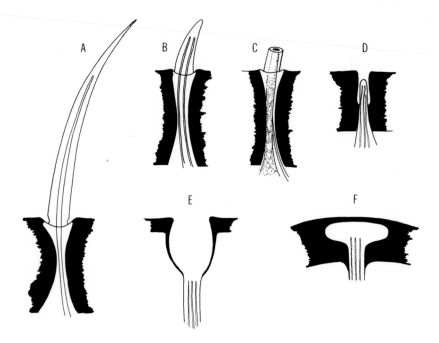

Figure 3.17. Some forms of sensilla found among Nemata. A. Sensilla trichodea. B. Sensilla basiconica. C. Sensilla basiconica, often associated with glands. D. Sensilla coeloconica. E. Sensilla ampullastoma. F. Sensilla insiticus (original).

other ectodermal cells, which are involved in any given receptor complex. As with other morphological features some of the terms for these cells and structures are borrowed from other invertebrate sciences without implication of homology. Therefore, the hypodermal cell that contributes to the formation of the seta is called the trichogen cell and the hypodermal cell that modifies the external cuticle to accommodate the sense organ is termed the tormogen cell. In some descriptions of sense organs these are not separately distinguished and are lumped as **support cells.**

On the basis of anatomical structure nematode sensilla, in this text, are segregated into five categories: Organs bearing typical elongate setae are called **sensilla trichodea** (Fig. 3.17A). Those with the external process reduced to a peg or cone are termed **sensilla basiconica** (Fig. 3.17B, C). When no external structure is evident, e.g., the sensory organ is porelike, then those that are in shallow pits are **sensilla coeloconica** (Fig. 3.17D), and those in deep pouches are **sensilla ampullacea** (Figs. 3.17E, 3.18B). If there is no evidence of an external structure or pore, but the ciliary process or modified cilia are embedded in the cuticle, then this highly specialized sense organ is called **sensilla insiticius** (Figs. 3.17F, 3.18A).

The structural anatomy of sensilla trichodea and sensilla basiconica is

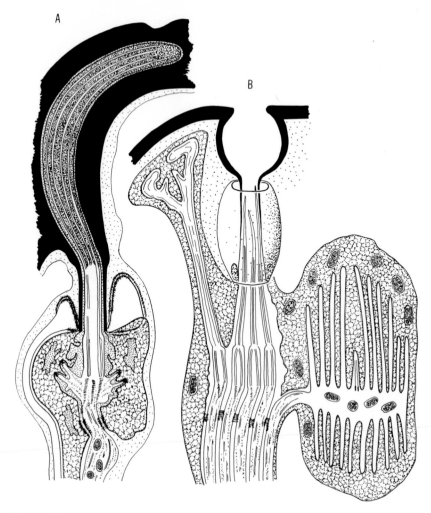

Figure 3.18. A. Sensilla insiticus; Tylenchida (modified from DeGrisse). B. Sensilla ampullastoma with an associated sheath sensilla, sensilla coleus (modified from Coomans).

very similar. The primary difference is in the length of external setae. If the seta is less than five times as long as broad at the base and general form is conelike, it is classed as sensilla basiconica. When the seta is elongate and hairlike, it is called sensilla trichodea.

In some Adenophorea and many Secernentea another type of sensilla, probably derived from a sensilla insiticius, has free endings under the cuticle. When associated with the amphid they are generally embedded in the support gland cells (Fig. 3.18B). This type of sensilla has been designated as a "sheath" sensilla. To be consistent with the nomenclature in this text, this form of sensilla is designated **sensilla coleus** (L. = sheath).

Not all sensilla are separate and simple. Mixed sensilla receptors have been reported in at least three genera: *Caenorhabditis*, *Capillaria*, and *Heterakis*. It is proposed that these may function as both mechanoreceptors and chemoreceptors.

IV. Hypodermis

In the majority of nematodes the hypodermis is a thin cellular layer underlying the cuticle. In a cross section of the nematode the hypodermis is seen to bulge out or protrude into the body cavity dorsally, laterally, and ventrally. These protrusions (chords) are most obvious in the lateral areas where they generally are composed of three or more rows of nucleated cell bodies. Ventrally there may be one or two rows of cells in the chord. From the concentrated areas of cell bodies thin sheets of cellular material extend between the cuticle and the muscle bands. It should be understood that these sheets are compressed extensions of the cell bodies that are seen laterally, dorsally, and ventrally in the chords. The dorsal hypodermal chord consists of discrete nucleated cells only in the region of the esophagus. From the esophagus posteriad any evidence of a dorsal chord shows no nuclei because the chord is formed by the meeting of hypodermal tissue extending from the lateral subdorsal cells. Laterally in the more generalized free-living marine nematodes, the chords are seen to consist of a cross section of three cell widths. Sometimes because of cellular displacement it appears that there may be even as many as four or five in a cross section of each lateral chord. Ventrally the chord is seen in cross section as being composed of two cell rows. The number of cells that constitute an individual chord along the length of the body varies according to the size and species of nematode. The larger the nematode the more cells will be seen to exist throughout the length of the chord.

1. Adenophorea

Variations do occur throughout Adenophorea. Anteriorly, in cross section, more cells are generally seen. For instance, Dorylaimida and some Enoplida in cross section often have as many as three rows of cells dorsally, three rows laterally, and sometimes even three rows of cells ventrally. However, this is not the general picture. The generalized description of the nematode hypodermis consists of the single row of cells anteriorly in the dorsal section, three rows of cells in the lateral chords, and one or two rows of cells in the ventral chord (Fig. 3.19A, B). The lateral cells, when

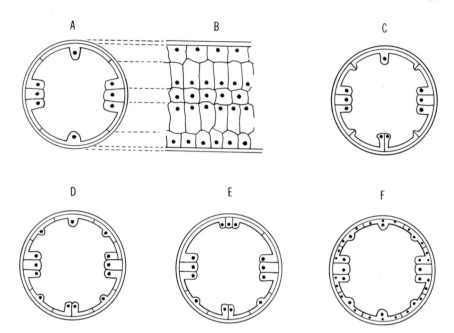

Figure 3.19. Arrangements of hypodermal cells in Nemata. A. Enoplida. B. Enoplida, lateral view. C. Chromadoria. D. Mermithida, *Mermis, Agamermis,* anterior region. E. Mermithida, anterior region. F. Secernentea, *Dioctophyma,* note multinucleate condition (original).

they occur three to a chord, are given topographical nomenclatorial designations so that we recognize one row as dorsolateral, one row directly lateral, and the third row as ventrolateral in position.

Generally speaking the same structure as seen in the free-living marine nematodes of the subclass Enoplia also persists into the subclass Chromadoria. One significant difference that appears in at least some Chromadorida, which cannot be claimed to be characteristic of all, is the presence of subdorsal and subventral hypodermal chords (Fig. 3.19C). These chords are anucleate and occur approximately halfway (submedial) between the dorsal chord and the lateral chord as minor protrusions into the body cavity. Such chords are seen in Desmoscolecida, Axonolaimida, Monhysterida, and among some Chromadorida.

Among the insect parasitic adenophoreans one change from the generalized plan is an increase in the number of chords. Anteriorly in Mermithids eight chords are visible (Fig. 3.19D, E): one dorsal, one ventral, two lateral, and four submedian. Anteriorly the four submedian chords are nucleated and each of the other chords is composed of one dorsal, three lateral and two ventral nucleated cells. Posteriorly the dorsal chord becomes more normal, that is, it consists of a single row of cells dorsally, three rows of cells laterally, two rows of cells ventrally, and four anucleate submedians. In other mermithids the subdorsals may move dorsally so that in the dorsal chord three rows of cells are seen (Fig. 3.19E): one is the basic dorsal cell and the other two are formed by subdorsals that have moved to a position alongside of this dorsal chord. However, in such forms there remain three cells in the lateral chord and two cells in the ventral chord.

Since the Adenophorea comprise the vast majority of nematodes one must assume that the manifestations of hypodermis in Adenophorea are more typical of Nemata and may represent a more ancestral state. As such, the hypodermis is composed of discrete uninucleate cells and therefore does not form a syncytium of any type. Some extreme modifications do occur as in the trichurids and capillarids, both of which are parasites. These nemas have a special modification called the **bacillary band** (Fig. 3.20). Bacillary bands occur among nemas placed in the superfamily Trichuroidea. The bacillary band may be limited to one hypodermal chord or all four chords may be modified. In *Trichuris* there is a single band located ventrally and limited to the esophageal region of the body; *Capillaria*, on the other hand, has bands in all four chordal regions. The bands involve modification of the cuticle as well as the hypodermal cells. The band is a complex of glandular and nonglandular cells (Fig. 3.20). The gland cells open to the exterior through complex pores. It has been postulated that these gland cells function in osmotic or ionic regulation.

Dioctophymatids again show an inconsistency in their placement in Adenophorea by various authors, which correlates with the differences in their cuticle. Their hypodermal characteristics are far more reminiscent of the Secernentea. It is true enough that posteriorly diotophymatids have eight chords, much as in mermithids; however, all eight chords are nucleated throughout the body while the chords in mermithids are not. In addition interchordal nuclei are found between the cell body and in its subcuticular extension (Fig. 3.19F). In dioctophymatids in the anterior region the typical four chord system is seen, one dorsal, two lateral (composed of three rows of cells), and one ventral chord. Posteriorly we find the mixed chordal and interchordal nucleated cells typical of many secernentean parasites. Interchordal nuclei are a multinucleate condition unknown among Adenophorea, except among dioctophymatids, which seem, therefore, to be misplaced taxonomically.

2. Secernentea

Among Secernentea, especially among parasitic forms, the inordinate increase in body size is not correlated with an increase in the number of individual cells forming the hypodermis as it is in Adenophorea. Rather increased body size is accommodated by the formation of multinucleate hypodermal cells; thus, an unbalanced nucleocytoplasmic ratio is avoided. This condition appears to occur within the Secernentea without regard for

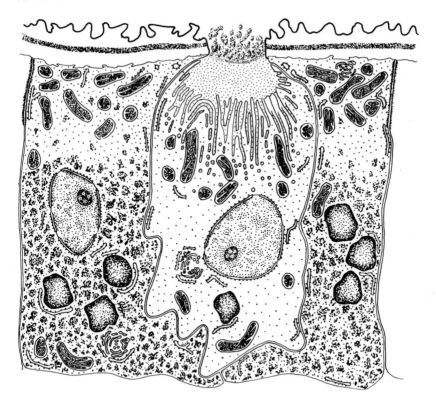

Figure 3.20. Transverse section of bacillary band in *Capillaria hepatica* showing a hypodermal gland cell and two nonglandular cells (modified from Wright).

taxonomic placement or phylogenetic considerations. That is, in any individual order or family of Secernentea the trend is to go from uninucleate cells to multinucelate formation without visible cell membranes separating the individual nuclei of the hypodermis.

In the interchordal zones, beneath the cuticle, cell membranes separating the individual cells of the hypodermis are extremely difficult to resolve. In most instances they are only visible with the aid of an electron microscope. However, it is to be emphasized that each cell is a discrete entity under the control of one nucleus and/or its daughters. A misconception has crept into modern nematological literature that most nematodes have a **syncytial hypodermis** resulting from the amalgamation of several cells; this is *not* true. Unfortunately, **syncytium** is an encompassing term with more than one definition and, therefore, has resulted in some confusion among nematologists. Among Secernentea when a "syncytial" hypodermis is present it is the result of **coenocytic** development, that is, a multinucleate condition of discrete cells resulting from repeated nuclear division unaccompanied by cell fission. The multinucleate hypodermis of nematodes does not result from the aggregation or fusion of cells, which is how a **lysigenoma** of plants is formed.

So far among the Secernentea, with the possible exception of the dioctophymatids, submedian chords are not described. Secernentea, therefore, almost always possess only the four generalized chordal areas as seen in other nematodes. In Secernentea there are greater differences in the cellular arrangement of the hypodermis than seen in Adenophorea and these changes or increased complexity are most easily seen in the lateral chords. The degrees of hypodermal change among Secernentea can be separated into six levels of increasing complexity that involve cellular division or nuclear division without cell membrane formation (coenocytes).

1. In the simplest form there are (longitudinally), in each of the three cell rows that comprise the lateral hypodermal chord, 12 to 14 cells whose nuclei are equal in size (Fig. 3.21A). In such forms, which includes the genus *Rhabditis*, there appears to be little or no increase in the number of

Figure 3.21. Nuclear arrangements in the hypodermal cells of Secernentea: Rhabditia. A. *Rhabditis*. B. Oxyuroidea and Trichostrongyloidea. C. Metastrongyloidea and Trichostrongyloidea. D. Ascaridida. E. Spiruria:Camallanida. F. Spirurida:Filarioidea (original).

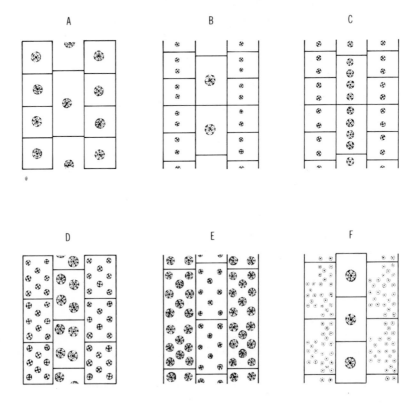

cells after hatching. If any differences in the size of nuclei occur in each of the rows, then the nuclei of the lateral row are somewhat smaller than those found in the dorsolateral and ventrolateral rows.

2. The Oxyuroidea and Trichrostrongyloidea show a slightly more complex condition. In the lateral hypodermal row there remain 12 to 14 cells; however, in the dorsolateral and ventrolateral rows there is an increase in the number of nuclei to 24 or 28. These may be binucleate cells rather than separate cells, but this has not been determined. The nuclei in the lateral row are larger than those in either the dorso- or ventrolateral rows (Fig. 3.21B). No changes are seen in the condition of the dorsal and ventral chords.

3. The metastrongyles and trichostrongyles show an increase in the number of cells in all three rows of the lateral hypodermal chords (Fig. 3.21C). The dorso- and ventrolateral chords are characterized by having twice the number of nuclei than the 24 to 28 nuclei seen in a generalized Secernentean nematode. In the lateral row there are four times the number of nuclei (48 to 56 nuclei visible), and these are larger than those in the dorso- and ventrolateral rows.

4. In some metastrongyles and some ascaroids an increase in the number of cells is evident. This does not appear to be an increment associated with the 12 to 14 cells seen in the basic form. Rather the number of cells appears to be associated more with the growth of the animal after hatching. However, there is again the notable difference in the size of the lateral nuclei, these being larger than those found in the dorso- and ventrolateral rows (Fig. 3.21D).

5. Among camallanoids and some ascaroids many more nuclei are present in the lateral rows of adults than are evident in the young. But in these instances the lateral nuclei are much smaller than those in the sublateral positions. There are also more nuclei in the dorso- and ventro- lateral rows than there are in the lateral row (Fig. 3.21E).

6. Uninucleate cells are seen only in the lateral row of cells, and the nuclei remain relatively large (Fig. 3.21F). Such is the case among some ascarids. The sublaterals, however, have a multinucleate condition with scattered nuclear nests. This is also seen in the Filarioidea. Nuclei are not confined to the region of the chord, but are found scattered throughout the hypodermal tissue in the interchordal regions and are also evident in the dorsal and ventral chords. Among the Spiruria thus far studied, such multinucleate chords are also noted.

The tendency to form a multinucleate cell (coenocyte) is noted even in the simplest secernentean forms. Therefore, one might in a general sense say that Adenophorea can be distinguished from Secernentea on the basis of the characteristics shown within the hypodermis. As such, Adenophorea maintains discrete uninucleate cells, whereas in Secernentea it is common to have distinctive multinucleate cells (coenocytes) with nuclei in the interchordal zones.

V. Excretory System

The excretory system is one of the most neglected organ systems in morphological studies of nematodes. Consequently there are conflicting reports and misunderstandings throughout the literature. There even is a lack of consistency in the nomenclature of the system. The "excretory" cell is variously called a specialized hypodermal cell, a renette, a sinus cell, or a ventral gland. Although all alleged excretory systems in Nemata are derived from hypodermal (= ectodermal) tissue, all processes of excretion are not necessarily hypodermal in origin.

An organized excretory system does not occur in all Adenophorea; however, one seems to be universally present in Secernentea. When an organized system is present, the one consistent observation throughout Nemata is that the external orifice (excretory pore) is always medio-ventral; its placement along the body is, however, variable.

A. Alternate Cellular Means of Excretion

Prior to the discussion of organized excretory systems some attention should be directed to other possible cellular systems of excretion available to nematodes. Deserving of consideration are the hypodermis and cuticle, hypodermal glands, coelomocytes, and the prerectum. Other possibilities are the tubular gland setae, and gelatinous secretions from caudal glands, rectal glands, or even the excretory cell.

Among adenophoreans, where no excretory cell has been reported, there often are numerous longitudinal **hypodermal glands**. These are often evident in species that do possess an identifiable excretory cell. This in itself does not exclude them from having an excretory function. In fact they may enhance the efficiency of excretion by functioning throughout the length of the body since the excretory cell is usually limited to the esophageal region and among adenophores there are no collecting tubules as there are in Secernentea.

Another structure that might have some possible excretory function is the **prerectum**, which is a morphologically distinct section of the intestine seen in many of the terrestrial adenophores. It may be that a soil exist-ence requires an additional elaboration over a single cell or even rows of hypodermal cells. The tubular systems developed in Secernentea lend some support to this speculation that the demands of a soil environment require adaptations uncalled for in the marine or freshwater environment.

Coelomocytes are cells situated in the body cavity that are not a part of the connective or isolation tissues of the pseudocoelom. Though most commonly reported in Secernentea, they do occur in Adenophorea. Some have a phagocytic function demonstrated by India ink injections into the pseudocoelom. The coelomocytes phagocytize (incorporate) the foreign ink particles. In Chromadoria and Secernentea the positions of the coelomocytes in the body are rather specific. Generally, two occur near the esophago-intestinal junction and at least one other pair occur more posteriorly. In some instances the anterior pair of coelomocytes are intimately associated with the excretory cell. This association can be seen in *Plectus* (Chromadoria) (Fig. 3.22C) and in strongyles (Rhabditia). In addition to these fixed position coelomocytes there are reportedly free migratory forms; this, however, should be confirmed.

When the coelomocytes are located in the tail region they are often confused with caudal glands. This may account for the frequent reports of more than three glands (Figs. 2.9B, 3.12B).

B. Excretory System, Glandular, with a Medioventral Orifice

1. Adenophorea

Among those Adenophorea with a recognized excretory cell only a non-canalicular system is known. In this class the entire excretory system consists of a simple single ventral gland cell without collecting tubules (Fig. 3.22A, B); however, in a few instances the gland may have extensions simulating tubules (Fig. 3.22D). There is one report that purports two gland cells in *Longidorus* (Dorylaimida:Longidoridae). This should be verified because not only is it a unique situation, but the report describes the glands as intimately associated with the first major ventral ganglion (retrovesicular?) of the ventral nerve cord. There is no evidence presented that assures us that the author was observing gland cells and not "glial" cells of the nervous system. The typical adenophorean system often has a proximal ampulla and the cell may have an elongated neck. Normally the duct is only cuticularly lined in the region of the pore. However, in some Chromadoria, i.e., *Plectus*, the duct is very elongated and cuticularly lined. Sometimes the single cell is lobed as in *Enoplus* (Fig. 3.22B), or possesses rather extensive lateral elongations such as in *Anonchus* (Chromadoria) (Fig. 3.22D). It has been proposed that these are ancestral to the lateral canal or canals of Secernentea. Beyond this speculation there has been no morphological evidence to support or reject the hypothesis. The only comment to be added is that if the hypothesis has a basis, the ancestral relationship would be more likely to *Anonchus* rather than Enoplida.

2. Secernentea

Secernentea all have a well-developed excretory system. Perhaps this has evolved because of their terrestrial existence. Certainly no similar system exists among marine and freshwater nemas. The known systems fall into four categories and these do not always have taxonomic or phylogenetic

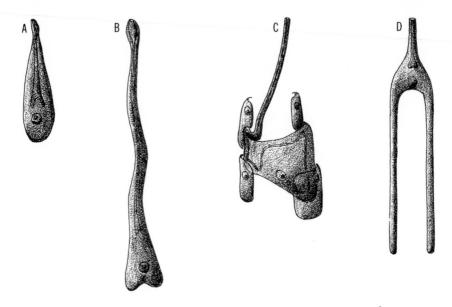

Figure 3.22. Excretory systems in Adenophorea. A. Chromadoria:*Chromadora*. B. Enoplia:*Phanodermopsis*. C. Chromadoria:*Plectus*. D. Chromadoria:*Anonchus* (A,B,D modified from Chitwood; C, redrawn from Maggenti).

significance. The first is called the **"oxyuroid"** system, which layed out has a basic H architecture with a medioventral excretory cell (sinus cell), associated with the crossbar (Fig. 3.23A, C). The second system is also an H system, but in addition to the medioventral excretory cell, there are two additional associated cells (Fig. 3.23B). A third modification, which is generally associated with some ascarids, spirurids, and free-living rhabditids, is an **inverted** U shape (Fig. 3.23D). The fourth system is **asymmetrical**, i.e., the collecting tubule is limited to one or the other lateral regions of the body (Fig. 3.23E, F).

a. Oxyuroid System

The simple H type (oxyuroid) system in free-living and some parasitic Rhabditia has several forms that are characterized according to the length of the cuticularly lined duct (Fig. 3.23A). Apparently the entire system is normally composed of three cells. Two of the nuclei of these cells are located near the orifice end of the duct and the third cell is the excretory cell proper. In some parasitic forms the tubules are reportedly multinucleate and each tubule has a ramifying lumen. The two cells associated with the terminal duct are apparently universal among Secernentea. These cells presumably represent the hypodermal cells that form the cuticularly lined duct.

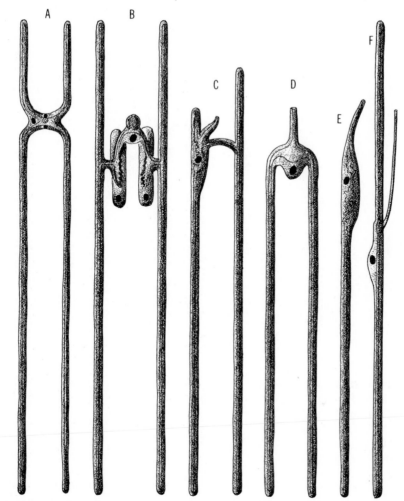

Figure 3.23. Excretory systems in Secernentea. A. Oxyuroid type. B. Rhabditoid type. C. Ascaroid type. D. Cephaloboid type. E. Anisakid type. F. Diplogasteria: Tylenchoid type. (B modified from Waddell; A,C–F modified from Chitwood).

There are four subcategories of this oxyuroid excretory system that are based on the excretory duct: (1) The cuticularly lined duct is long and tubular with one or two cells associated with the duct. (2) The duct is short, wide, and irregular until it merges with the excretory cell. This also is a three celled system with two duct cells and the single sinus cell. (3) The duct is elongated but vesicular. This system reportedly contains five nuclei, two associated with the terminal duct and three belonging to the excretory cell-canal system. (4) The terminal duct is short and vesiculate. The large diameter lateral canals come together at the sinus in the fashion of an X. Presumably, the vesiculate duct acts as a valve.

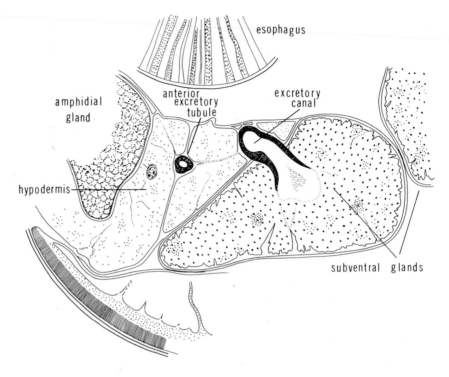

Figure 3.24. Transverse section through anterior region of *Stephanurus dentatus* showing connection between excretory canal and subventral coelomocytes (redrawn from Waddell).

b. Rhabditoid System

The second type of excretory system is also based on the H form, but is a combination of two discrete systems (Fig. 3.23B). The basic system has an elongate cuticularly lined terminal duct and an excretory (=sinus) cell connected with paired lateral canals. The second associated system consists of two subventral gland cells. The tubular system and excretory cell are developmentally and structurally distinct from the subventral cells. Waddell in 1968 showed that the canals and excretory cell are histologically similar and clearly dissimilar to the associated subventral cells (Fig. 3.24). The intimate association of these subventral cells with the excretory system is only evident in the fourth larval stage and adults. In younger larvae there is no connection between the tubular excretory system and the two subventral cells. In young larvae the two associated cells (=coelomocytes) lie free in the body cavity between the esophago-intestinal junction and the genital primordium. This is exactly the same position accorded to the anterior two coelomocytes seen in free-living rhabditis with the simple H (oxyuroid) excretory system.

The significance of Waddell's observations has not been readily appreciated because once again the findings were considered within the parameters of our traditional understanding of the excretory system. In part this is due to the nomenclature applied to the system. In Adenophorea there is a single excretory cell; in Secernentea the basic system consists of the terminal duct and the excretory cell with its collecting tubules. Unfortunately because of the two subventral cells in the so-called rhabditoid system the nomenclature was changed. The single excretory cell with its tubular extensions was renamed the sinus cell and the subventral cells (coelomocytes) were thought to be duplicated "excretory" cells. It should now be apparent that the rhabditoid, perhaps more appropriately the "strongyloid," system is a combination system derived from two separate morphological entities, which are in ontogeny and through histology still clearly identifiable. The concept of this system should not be applied to other derived systems as occur in Tylenchida. Apparently this is a unique development among strongyles and their ancestral relatives, and is to be seen no where else among Secernentea. As such, this system should be viewed as a highly developed modification of the basic H system, having significance to the groups in which it occurs, but not to the general phylogeny of Secernentea.

Two other modifications of the excretory system remain to be discussed: the inverted U and the asymmetrical system. The inverted U or "tuning fork" occurs among ascarids, spirurids, and cephalobs (Fig. 3.23D). It is identical to the H system but lacks the anterior lateral collecting tubules. The asymmetrical system has two forms: one occurs in anisakids, Ascaridida and the other is typical of Tylenchida and Aphelenchida (possibly it is characteristic of Diplogasteria). The anisakid form (Fig. 3.23E) is derived from the modified H of ascarids (Fig. 3.23C), i.e., only the left side remains. The right lateral canal and the connecting bridge, as well as the short anterior canal of the left tubule, are lost. This asymmetry of anisakids correlates with the phylogeny presented throughout this text, because it is proposed that the anisakids (parasites of marine vertebrates) evolved from the ascarid parasites of terrestrial vertebrates. Accepting this asymmetrical excretory system as derived from an H system supports this hypothesis.

The tylenchid excretory system is also asymmetrical but differs from the anisakid form by being truly one-half of the H system (Fig. 3.23F). From the excretory cell, a collecting tubule extends both anteriorly and posteriorly. There is no consistency as to which side of the body is serviced by the system. The canal may be in either of the lateral regions. This variability extends to members of the same species.

In members of the Criconematoidea the excretory cell may become greatly enlarged without any evidence of collecting tubules. In these situations the gland exudes a gelatinous substance that forms a matrix into which eggs are layed. Citrus nematode, *Tylenchulus semipenetrans*, which belongs to this superfamily, has the orifice of the excretory pore on the posterior half of the body and the cell itself, which secretes the gelatinous matrix, occupies 30% of the body volume.

c. Excretion

Excretion is not the exclusive responsibility of the excretory system per se. The cuticle, for example, is not impermeable and it is believed that CO_2 is excreted by transfer through the body wall. Apparently even the major nitrogenous waste product, ammonia, is eliminated through the body wall and anus. Only a small percentage of nitrogenous waste is in the form of urea, but the percentage increases when, through osmotic stress, ammonia accumulates in the nema's internal tissues. The excretory products of nematodes are numerous and include, in addition to the above, uric acid, amino acids, amines, peptides, and fatty acids; however, the excretory routes of these substances are unknown.

That the excretory cell and associated tubules can pick up foreign substances from the internal body and excrete them to the exterior has been demonstrated with fluorescent substances. It has also been suggested that the lateral "collecting" tubules may function in a manner similar to the Malpighian tubules of insects. It is certain that more than one function is attributable to the excretory system, such as the secretion of the gelatinous matrix, and that all excretion is not dependent on an organized excretory organ. The excretory cell has also been incriminated in the molting process.

VI. Molting

In their development to adulthood, nematodes undergo four molts after **eclosion** (hatching). Among Adenophorea it is almost universal for the first stage larva to emerge from the egg. However, it is almost a characteristic of Secernentea for the first stage larva to molt one time within the egg, so that it is the second stage larva that emerges.

The term **larva** is entrenched in nematological literature and therefore, is used throughout this text. However, it has been often pointed out that the more accurate designation would be **juveniles** or nymphs. The entomological term larva refers to forms that undergo complete metamorphosis, that is, as young they possess no features of the adult, e.g., caterpillars and maggots. Juveniles and nymphs, on the other hand, resemble the adult throughout development; however, the characteristics of sexual maturity are absent, thus the use of these terms by some nematologists.

Molting, though repeatedly observed, has been studied in only a few nematodes. To date two processes are recognized. In one, the entire cuticle with all layers intact is shed, a process known in Adenophorea and some Secernentea, especially those with endocuticle and mesocuticle. In some Secernentea, particularly among the parasitic forms in Diplogasteria, where only the epicuticle and exocuticle are present, the exocuticle is dissolved during the process and only the epicuticle is left to be shed.

The processes involved in initiation, activation, and final accomplishment are unknown. If the process were similar to that in arthropods then the sequence of events would be (1) neurosecretory materials triggering (2) some tissue (generally glandular) to produce a hormone, which (3) causes the epidermis or hypodermis to produce an enzyme capable of (4) initiating the molting of the cuticle by dissolution or loosening. The process of laying down a new cuticle would proceed concurrently. Few of these steps have been verified for nematodes and much of the evidence is circumstantial.

Neurosecretory cells are known to occur in nematode nervous systems. They have been reported in the ganglia of the nerve ring as well as in the ventral nerve cord. The triggering mechanism and its source of origin initiating the process of molting are yet to be determined.

There is no evidence, currently, to support the second step, hormonal production. In insects this function is assumed by the prothoracic gland after stimulation by neurosecretions. If the sequences were similar, then the structure or structures (glands) in the nematode performing this vital function remain to be identified. There are within the nematode body a number of cells whose function is yet unknown. Notable among these are the coelomocytes occurring at strategic points, e.g., in the esophageal region, vulval region, and caudal region in many Chromadoria and Secernentea. Coelomocytes are known in Adenophorea, but are not always in fixed positions. In addition, there are the so-called hypodermal glands.

Waddell showed that "coelomocytes" in the region of the excretory cell only become associated with the excretory system in the late larval and adult nematode. This system is described in the literature as an excretory system with two subventral glands. This observation could be highly significant. For example, the thoracic gland that produces the "molting" hormone in insects is not present in the adult. It is for this reason that insects do not molt as adults, since the hormone producing organ is no longer present. Therefore, a similar step occurring in nematodes would require cells that (1) are present in larval stages, but absent in the adult, or (2) change their characteristics from larva to adult. There is yet no evidence for hormone producing glands, but Waddell's observation on the excretory system and coelomocyte association could be interpreted to offer an interesting alternative in the connection of the coelomocytes of the adult *Stephanurus* to the excretory system, but not in early larval stages.

The third step in the sequence is the hormonally triggered production of an enzyme capable of dissolving or loosening the cuticle. A potential enzyme, leucine aminopeptidase, has been reported in the excretory cell of the fish parasite *Phocanema decipiens*. The presence of the enzyme within the excretory system does not indicate the production site. Since many nematodes do not have a recognizable excretory cell, other tissues must also be capable of serving this function. The presumed function of the excretory cell is to remove or collect a variety of substances from the body cavity fluids. The molting enzyme could be expected to be widespread throughout the body and merely accumulated in the excretory cell. In insects the presence of the molting enzyme in tissues completely disassociated with molting is well documented. In arthropods the enzyme is produced by the epidermal cells (= hypodermis of nematodes); there is no evidence for a different mechanism for nematodes.

As the molting process begins, there are notable changes in the nematode hypodermis. There is a noticeable increase in the number of mitochondria and a thickening of the hypodermis underlying the somatic musculature. These features are correlated with enlargement of the nucleus and nucleoli. These are taken to be indications of increased cellular activity, which may be associated with secretion of new cuticle.

The fourth step, the actual loosening or dissolution of specific cuticular layers in conjunction with the formation of a new cuticle, does occur. These two processes are illustrated in Fig. 3.25 and 3.26. This portion of the sequence has similarities to observations among insects and other arthropods. Prior to the separation or dissolution of any layers the epicuticle of the newly forming cuticle is layed down (Figs. 3.25C, 3.26D). This could mean that the molting fluid in nematodes is secreted in an inactive form remaining until the new epicuticle is laid down. Entomologists explain this as a mechanism to prevent the activity of the molting fluid from interfering with the deposition of the new cuticle. After epicuticle formation is completed the enzyme is activated to dissolve or loosen the old cuticle.

It is apparent that dissolution of the exocuticle occurs among the plant parasitic nematodes (Fig. 3.25D–F). Whether resorption of this material really occurs as reported needs further verification.

New cuticle may continue to form and condense long after the shedding of the old cuticle. In *Ascaris* cuticular growth apparently continues throughout the nematode's life. This is not unexpected since as an adult this nema grows from a few millimeters in length to over 15 cm.

An additional interesting observation has been made during the molting process of *Phocanema*. At about the time when the nematode is beginning to secrete the new cuticle there is an alleged secretory activity at the junction of the muscle cell and hypodermis.

It has been further noted that the cuticle often loosens first in the region of the excretory pore. This is taken as further proof of the involvement of the excretory cell. However, if the neurosecretion comes from the

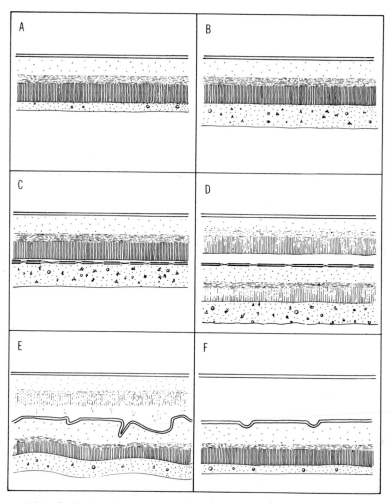

Figure 3.25. Cuticular changes occurring in Tylenchida:*Meloidogyne javanica* during molting. A. Normal cuticle before the initiation of molting. B. Hypodermis thickening signals the initiation of the molt. C. A new epicuticle starts to form between the hypodermis and the old cuticle. D. Exocuticle is deposited and a separation of the old and new cuticle is evident. E. Exocuticle of the old skin is dissolved and possibly resorbed; new cuticle becomes convoluted to allow for growth. F. Molt completed, new cuticle formed and only the epicuticle of the old skin remains to be shed (modified from Bird and Rogers).

ganglia of the nerve ring or the retrovesicular ganglion, then diffusion rates alone would explain why this region of the body would react first and, therefore, would not necessarily incriminate the excretory cell as the source of the molting fluid.

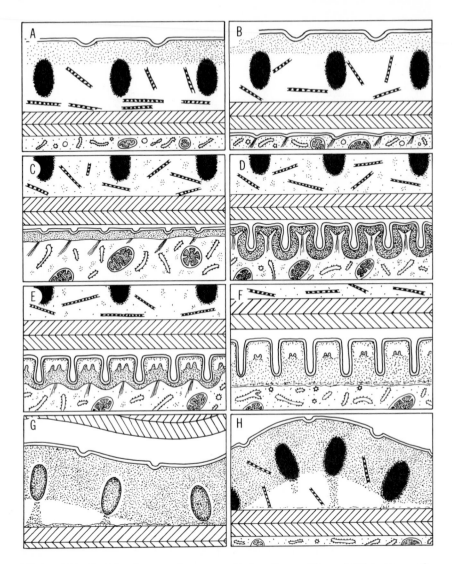

Figure 3.26. Cuticle changes that occur during the molting of *Nippostrongylus brasiliensis*. A. Normal cuticle before the initiation of the molt. B. Old cuticle begins to separate from the thickened hypodermis and the new epicuticle begins to form. C. Annular convolution and deposition of the exocuticle is initiated. D. Convolution continues. E. Cuticular struts, M-shaped structures are becoming evident. F. Convolution severe during the first deposition of the fiber layer, mesocuticle. G. Struts and fiber layers begin to take on definition. H. The new cuticle is complete and the old cuticle, all layers intact, is shed without evidence of any dissolution (modified from Lee).

Obviously a great deal of research remains to be done. The only phase of the process that appears clear so far is the secretion of the cuticle by the hypodermis. To be verified or discovered are the roles of neuro-secretory cells, the source of the hormone that triggers the production of the molting fluid, and the tissue source that produces the fluid. In short, we still know little about molting in nematodes.

Chapter 4
Internal Morphology

I. Somatic Musculature

The cells of the somatic musculature of nematodes are, with few exceptions, longitudinally oriented and spindle-shaped. Each cell is composed of two regions: a noncontractile portion that gives rise to the myoneural process and a contractile portion. The noncontractile region can be further described as the body of the cell containing the sarcoplasm and nucleus. From the body of the cell a process extends to the neural junction, a condition unique to nematodes (Fig. 4.1A). The contractile portion consists of oblique contractile elements variously arranged within the muscle fibers. The muscle bands are composed of two or more fibers (cells) that are grouped into four sectors in any given body cross section: two subdorsal and two subventral. These muscle band quadrants are separated from each other by the four major hypodermal chords (Fig. 2.8).

A system of nomenclature to describe the somatic musculature of nematodes was proposed more than one hundred years ago by Schneider. The system categorizes nematode musculature by the cross-sectional position of the contractile elements within individual cells and by the number of muscle cell rows in a given quadrant. Three cell types are named by contractile fiber position: platymyarian, coelomyarian, and circomyarian. **Platymyarian** cells have the contractile elements limited to that portion of the cell lying closest to the hypodermis. No elements extend up the sides of the cell toward the body cavity (Fig. 4.2C). In **coelomyarian** cells the elements extend to varying degrees up the cell sides toward the pseudocoelom (Fig. 4.2A, B). In some instances the sacroplasm is almost enclosed by the contractile elements. When the sarcoplasm is encircled by the contractile elements the muscle cell is said to be **circomyarian**, a term first introduced into nematological literature by Chitwood (Fig. 4.2D).

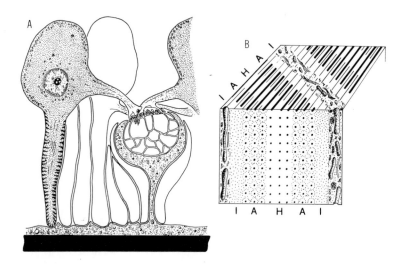

Figure 4.1. A. Schematic transverse section of muscle cells and their armlike extensions that effect the myoneural junction. The nerve cord is embedded in troughlike hypodermis. B. Schematic of a muscle fiber showing striation patterns in two planes. In actuality both filaments and striations are probably slightly oblique. This obliqueness is greatly exaggerated in this diagram. The angle of striations in reality is less than 6° (A,B modified from Rosenbluth).

This type of musculature is most often found among the specialized muscles of the body, such as, somato-intestinal, vulva, and spicular muscles. It is not uncommon for coelomyarian muscle to be circomyarian distally; however, the term is generally reserved for the specialized muscles.

Two terms are used to describe the number of cells in a sector: **meromyarian** and **polymyarian**. A third term was included in the original proposal: **holomyarian**. This term was discarded when Bütschli in 1873 showed that, by definition, no examples exist among nematodes. In order to be holomyarian a nematode must have two or fewer hypodermal chords in cross section. Originally, Schneider placed trichurids and mermithids in this category, but it was later shown that four chords exist anteriorly, even though only two chords may be seen in other regions. The term is not useful and has only historical interest.

Strict definitions cannot be applied to the terms meromyarian and polymyarian. However, generally when there are four or fewer rows of cells in each muscle band (quadrant) the condition is called meromyarian and the individual cells are platymyarian. Six or more cells in a quadrant is accepted as characteristic of a polymyarian condition and it is normal for each cell to be of the coelomyarian type. The transitional condition is five cells to a sector, which makes a categorical assignment difficult because the individual cells are generally shallow coelomyarian.

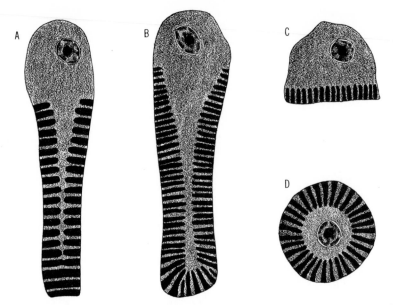

Figure 4.2. Traditional diagrams of nematode muscle cells in transverse section. A. Coelomyarian, *Deontostoma*; showing contiguous contractile elements at the base of the cell. B. Coelomyarian, *Ascaris*; showing the contractile elements as discrete entities around the cell base. C. Platymyarian. D. Circomyarian (originals).

Various authors have tried to attach an evolutionary significance to these categories of cell type and number. In recent years we are being led to the conclusion that there is no phylogenetic significance to be attached to these categories and characteristics, primarily because they are names without a sound morphological basis. For example, these characteristics occur in widely divergent groups. Enoplida, Dorylaimida and ascarids all have coelomyarian, polymyarian representatives. At one time the platymyarian condition was presumed ancestral; however, it occurs in such derived groups as the animal parasites of Trichostrongyloidea. Currently few nematologists believe these to be ancestral or primitive. In fact, the trend is to seek early evolutionary development among Enoplia and these are generally coelomyarian and polymyarian.

The number of cell rows in a quadrant and the individual cell pattern of contractile fiber distribution is related to the animal's activity rather than its placement in the evolutionary scale. When the anatomy of a single muscle cell is examined one realizes there is no great morphological significance or special mystique about platymyarian, coelomyarian, or even circomyarian muscle cells.

These terms, though useful as points of reference, are, because of electron microscope studies, becoming less and less useful for categorizing muscle types. For instance, the muscle cells of *Deontostoma* (Enoplida), *Xiphinema* (Dorylaimida), and *Ascaris* (Ascaridida) are all listed as coelomyarian. It is true that all have contractile elements extending up the lateral portions of the cell away from the hypodermis, but the form these elements assume is different in each case. In *Deontostoma* the contractile elements form two discrete rows apically, but near the base of the cell they extend as continuous bands from side to side (Fig. 4.2A). *Xiphinema* does not have the contractile element bundles isolated from each other by sarcolemma, sarcoplasmic reticulum, or sarcoplasma; the "bundles" are incompletely separated by fingers of sarcoplasmic reticulum (Fig. 4.3). *Ascaris*, on the other hand, has discrete elements apically and basally (Fig. 4.2B). The same is true of platymyarian musculature. In forms like *Desmoscolex* the contractile fibers are discrete units separated by sarcoplasmic reticulum and sarcoplasma. However, in Tylenchida the contractile elements are not clearly separated, but have fingers of sarcoplasmic reticulum penetrating the contractile fibers irregularly.

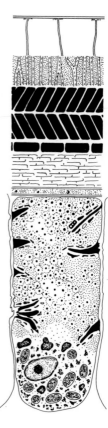

Figure 4.3. Portion of a transverse section through *Longidorus* showing the cuticle underlying hypodermis and a transverse section through a muscle cell showing fingerlike extensions of the sarcolemma (modified from Aboul-eid).

When individual cells of the three types are dissected open and laid flat, a similarity of fine structure is seen. In each instance the contractile element ridges run slightly oblique, approximately 6°, to the longitudinal axis of the cell. In a platymyarian cell the sarcoplasm is above the contractile elements as opposed to coelomyarian cells where the lateral upfolding includes the contractile elements. Circomyarian musculature is merely a further development where the contractile portion of the cell is completely rolled and the contractile elements meet apically and often axially, forming an elongated helix pattern. Therefore, the various cell forms are the result of lateral compression, crowding, and/or elongation. This is an over-simplification, but does visually clarify how each type is constructed. In fact, it may be this simple, that the accommodation of several cells in a sector requires lateral compression and this is why phylogenetic significance cannot be attached to the form seen in any given species.

That there are no great differences among the cell types seen is further confirmed by their microscopic anatomy. Each contractile element in longitudinal section contains five bands: a central band of thick (complex) filaments (approximately 200 Å) called the "H" band, bordered on each side by bands containing both thick and thin filaments designated as "A" bands (60 Å) and finally two outer bands of thin filaments, called "I" bands (Fig. 4.1B). These bands are comparable to the H, A, and I of vertebrate and other invertebrate striated muscle. In some species, e.g., *Ascaris* the thick and thin filaments of the A band may form regular patterns where each thick filament is surrounded by a circlet of 10–12 thin filaments. Contractile element ridges are separated from each other by Z zones, areas that are reportedly composed of tubules, vesicles, and fibrous material surrounded by sacroplasmic reticulum.

Presumably the **sarcoplasmic reticulum** may be involved in muscle relaxation. The zones supposedly take up calcium which produces a "relaxing factor." Circumstantial evidence based on the more common occurrence of sarcoplasmic reticulum in the muscles of free-living nemas is used to support this theory. Another hypothesis is that the substance of the sarcoplasmic reticulum is from "isolation tissue" forming a kind of basal membrane produced in *Ascaris* by cells in the region of the esophagus. Still another hypothesis is that infolding aids in the distribution of metabolites and also aids in distribution of nerve pulses.

The above account of somatic musculature is simplified for clarity. Many aspects of nematode musculature remain unexplored. The apparent lack of a well-defined Z zone presents a problem to contractile element operation. In one frontal section electron micrograph of an enoplid I have seen a Z zone that appears similar to that observed in crayfish. Our concept of nema muscle is greatly influenced by micrograph interpretations that are, for the most part, based on cross sections, which can be misleading. More studies need to be done with frontal and sagittal sections.

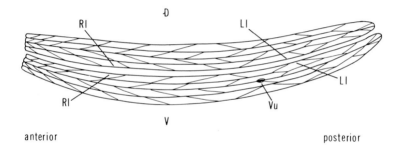

Figure 4.4. Diagram of the muscle cell arrangement in *Decrusia additicta*. The nematode is split through the lateral line. The lower sector represents the ventral portion of the body and the upper section the dorsal sector of the body. Posterior is right, anterior to the left. Abbreviations: V, ventromedian line; D, dorsolateral line; L, left lateral line; R, right lateral line; Vu, vulva (redrawn from Ohmori and Ohbayashi).

A. Muscle Cell Orientation

It is true that nema somatic musculature is oriented to the longitudinal axis of the body; however, this does not mean that they align like logs directly parallel to the longitudinal axis. As one observes nemas, it becomes evident that the muscles align anywhere from directly parallel to some 45° off the parallel. When parallel the cells are staggered like bricks, so that a spiral is produced around the body (Fig. 4.4). This is why simple explanations of nema movement are so misleading.

There is no question but that nemas with sinusoidal movement must obey the laws of physical forces for such movement. These laws do not explain the muscular contractions and relaxations necessary for the movement. The fact is few nemas are restricted to such a mode of movement. The range of twisting, bending, and rotation is almost unlimited. This is especially true of the anterior region and is best appreciated among the predaceous *Tripyla*. In a dish, one can observe the nervous probing of the head region as if the nema is testing the environment for a potential victim.

Criconemella spp. are allowed their unusual annelid locomotion because their somatic muscles are oblique to the longitudinal axis (Fig. 4.5). Contraction shortens the body and relaxation extends the body. Draconematids are allowed to creep in a fashion similar to inchworms because of their musculature and ambulatory setae.

Nematode movement and locomotion in their natural environment cannot be truly appreciated by observing them in water in a petri dish where no purchase is given and movement is restricted to a single plane. A better appreciation of nematode gracefulness and flexibility is gained

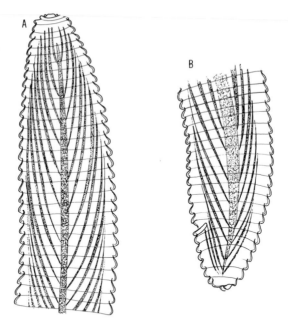

Figure 4.5. Muscle arrangement in *Criconemella xenoplax*, showing oblique orientation from the dorsomedial and ventromedial lines to the lateromedial lines. A. Anterior end. B. Posterior end (original).

when they are observed in a medium such as agar. Even so, the full wonder of these animals is only to be glimpsed.

II. Alimentary Canal

The alimentary canal is divisible into three regions: the stomodeum, the mesenteron, and the proctodeum. The **stomodeum** begins at the oral opening and includes the "mouth cavity" and esophagus. The **mesenteron**, formed from the embryonic endoderm, is the midgut or intestine proper. Following the midgut is the **proctodeum** or rectum, which ends at the anal or cloacal opening. Both the stomodeum and the proctodeum are recognized by their cuticular lining.

A. Stomodeum

The stomodeum is probably the most complex and least understood organ of the nematode body. The tissues and structures within this organ originate from two embryological tissues: the mesoderm and ectoderm. The main structure, the esophagus, arises during embryology as a primary invagination of ectodermal tissue, which combines with another embryological tissue, the mesoderm (forerunner of muscle), to form the esophagus, which is a combination of ectoderm (hypodermis, cuticle, glands, and nerves) and mesoderm (muscle). The most anterior "mouth" cavity results from a secondary invagination of the external ectoderm. As a result, two types of cuticle are present in the stomodeum: external cuticle, which lines the anterior "mouth" cavity, and internal cuticle, which lines the posterior "mouth" cavity and the esophagus. The two cuticles may be distinguished by differential staining (Mallory's triple stain); however, the structuring of the internal cuticle is reminiscent of the external epicuticle and exocuticle.

1. Lips

The lips surrounding the oral opening are not strictly part of the stomodeum because they are modifications of the external integument. However, they are a functional unit of the stoma. That is, their number and shape are determined by the form and functions of the stoma. The sensory structures associated with the anterior of nematodes and their cephalization onto the lips has already been discussed (Chapter 3); however, the lips (labial region) need some elucidation. The anterior extremity of many nematodes is hexaradiate and each sector is called a lip. The term lip is applied whether the sectors are movable or not. When the complete complement of six lips are present, the pairs are designated as subdorsal, lateral, and subventral, and each has one papilla of the internal circle. In parasitic nematodes the number of lips and labial sensilla are often reduced. The order Ascaridida is characterized by three lips, and the Spirurida by two lips, called **pseudolabia**. Lips may also be replaced by very ornate structures, often fringed and/or branched, as is characteristic of the Cephaloboidea, where the modified lips, termed **probolae** (labial and cephalic), are used to separate species, genera, and families (Fig. 3.4B).

The primitive state would seem to be represented by six lobes, each bearing a papilla or seta in the case of a protonematode. This is surmised by the hexaradiate pattern of the nerves supplying the cephalic and labial sense organs. Some authors have pointed to the nine-radiate symmetry of the arcade cells as representing the primitive evidence for a three lip condition. The **arcade cells** are the modified hypodermal cells that form the lips, and thus on the face of it the argument seems valid. However, the nine-radiate condition of the arcade cells is divisible into a hexaradiate formation and a triradiate system, where one cell is dorsal and two are submedian. In the arguments for three or six lips, no consideration is given to the formation of the oral cavity or cheilostome. Therefore, six arcade cells may form the lips, and three the cheilostome, which basically is divisible into three rhabdions or plates, one dorsal and two submedian. This is consistent with the position of the three extra arcade cells. Others have related the lip symmetry to the triradiate condition of the esophagus. The lips are not formed from esophageal primordeum, and therefore, the triradiate symmetry of the esophagus has nothing to do with the lips. It is assumed here that the six lip condition is basic.

2. Stoma

The stoma of nematodes is often referred to as the mouth cavity, buccal capsule, buccal cavity, and, of course, stoma. These terms are not founded on histological morphology or embryology and it was not the intent at the time of their incorporation into the language of nematologists that such should be implied. However, their usefulness in general discussions cannot be denied. In actuality the stoma of most nematodes is formed of two parts: anteriorly it is the product of external cuticle formation, and posteriorly it is a formation of the anterior esophagus (Fig. 4.6). The anterior section, called the **cheilostome**, is lined by external cuticle and is formed during embryological development as a secondary invagination. The second part of the stoma, called the **esophastome**, is a modified section of the anterior esophagus formed at the time of the primary invagination during gastrulation. The two parts of the stoma are, in general, easily recognized regardless of the nematode being studied. That portion not surrounded by esophageal tissue is the cheilostome and that part surrounded by esophageal tissue is the esophastome. The so-called "astomatous" nematodes lack the cheilostome, but retain an unexpanded esophastome.

Figure 4.6. Stomatal division in Nemata (*Butlerius* head redrawn from Goodey).

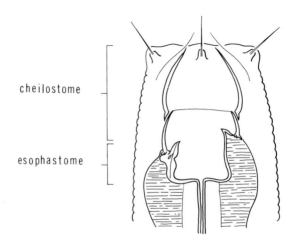

cheilostome

esophastome

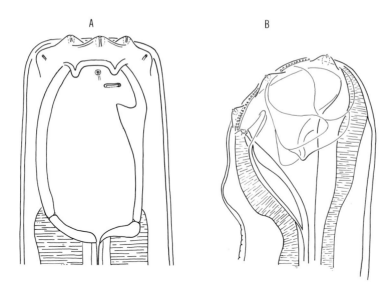

Figure 4.7. Globose stomas of differing embryological origins. A. Mononchida: *Mononchus*; stoma principally cheilostome, only the base is surrounded by esophageal tissue (head redrawn from Cobb). B. Seuratoidea:*Bulbodacnitis*; stoma almost entirely surrounded by esophageal tissue (head redrawn from Maggenti).

This uncomplicated system of recognition and nomenclature was first proposed by Inglis. With this system we can compare homologous parts among complex stomas or among stomas that are only superficially similar. For instance, large vaselike or globe stomas are not uncommon in widely divergent taxonomic groups. Even when viewed under the microscope these stomas look alike and are often described similarly. However, the value of the Inglis system is easily appreciated by examining two nematodes with globe stomas: *Mononchus*, a predaceous nematode, and *Bulbodacnitis*, a fish parasite (Fig. 4.7). The stoma of *Mononchus* is almost entirely cheilostome (Fig. 4.7A), the esophastome being represented only by basal plates covering the anterior esophagus. While in *Bulbodacnitis* only the lip lining is cheilostome, the remainder of the globe stoma ("jaws") is esophastome (Fig. 4.7B).

In the subclass Rhabditia the cheilostome and esophastome are often composed of a series of rings or plates. The plates and the region they enclose were first named by Steiner. The cheilostome and attendant plates (cheilorhabdions) remain unchanged in the Inglis system. However, the esophastome in the Steiner system is divided into two regions: the **pro-stome** and **telostome**. The prostome is further subdivided in three parts: **prostome** (prorhabdions), **mesostome** (mesorhabdions), and **metastome**

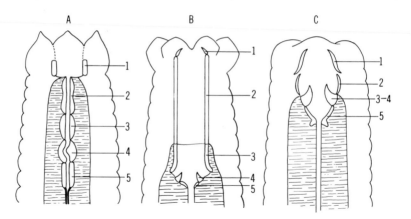

Figure 4.8. Diagrams comparing rhabdionic nomenclature to embryological origin. This figure illustrates the lack of homology when using stomal plates. A. Cephalobid stoma, only the first set of rhabdions (1) are cheilostomal, all others (2–5) are esophageal in origin. B. Rhabditid stoma, rhabdions 1 and 2 cheilostomal; rhabdions 3–5 esophageal in origin. C. Diplogasterid stoma, rhabdions 1–4 cheilostomal and only 5 is of esophageal origin. Rhabdionic nomenclature: 1, cheilorhabdion; 2, prorhabdion; 3, mesorhabdion; 4, metarhabdion; 5, telorhabdion (original).

(metarhabdions). There is disagreement as to whether the "glottoid" apparatus is a modification of the metarhabdions of the telorhabdions. The **glottoid** apparatus is a toothed projection at the base of the esophastome and its characteristic shape and number of teeth is utilized in species and generic identifications.

The identification of additional subdivisions of the esophastome is useful among select groups within Rhabditida. However, the system does not serve the purpose of allowing comparisons of homologous stomatal parts even among Rhabditia let alone other groups (Fig. 4.8). The esophastome plates in Rhabditida cannot convincingly, or with confidence, be homologized with the stoma in Diplogasterida or the spear of a tylenchid plant parasite or a predaceous or plant parasitic adenophorean. The comparison of cheilostomes and esophastomes, on the other hand, is useful and can be done throughout Nemata.

The stoma also illustrates the adage that ontogeny recapitulates phylogeny. In certain groups, such as Strongylida, the capacious adult stoma is not seen in the first and second stage larvae. These larval stages possess a stoma like free-living members of Rhabditida. Similar changes in stoma from larva to adult are seen in other groups. The larval form is believed to correspond to the group's ancestral origin. This does not seem unreasonable since the concept is consistent with observations in other animal groups.

a. Stomatal Armature

Movable and nonmovable stomatal structures (armature) such as teeth and stylets occur throughout Nemata. Movable structures in the nema stoma, with rare exceptions, are thought to be products of the cheilostome; and their movement is affected by the anterior musculature of the esophagus and may incorporate some of the esophastome, e.g., the odontophore of the dorylaim spear. Numerous descriptive terms are applied to these parts of the stoma. Large cheilostomal teeth are called **odontia** (Fig. 4.7A), while large esophastomal teeth are referred to as **onchia** (Fig. 4.9B). These terms should be used regardless of the fact that onchia means barb. The important fact to keep in mind is that applying these terms, as Cobb first intended, allows us to distinguish their origin; thus giving us more information than simply "stoma armed with large tooth." It is much more accurate and serves the needs of taxonomy better when we consistently apply odontia to cheilostomal teeth and onchia to esophastomal teeth. In describing a stoma with teeth, it is **dentate** if the teeth are moderate to large. If the teeth are numerous and small, the stoma is **denticulate.** Movable teeth, apparently always cheilostomal, are called **fossoria** or **mandibles** (Figs. 4.9C, 4.10A).

Protrusible spears occur in several unrelated groups of nematodes; however, they are a characteristic feature in three orders: Dorylaimida (Adenophorea), Tylenchida, and Aphelenchida (Secernentea). In Tylenchida and Aphelenchida the protrusible axial spear is called a "**stomatostyle.**" The descriptive names in older literature for the corresponding structure in those Dorylaimida possessing an axial spear are **onchiostyle** or **odontostyle**. The literature concerning dorylaims applies the term onchiostyle to the anterior portion of the spear found in the cheilostome and onchiophore

Figure 4.9. A. Oxystominoidea:*Amphidelus*; stoma almost entirely esophastome; sometimes this form is called astomatous. B. Tripyloidea:*Tripyla*; stoma esophastomal with an onchia. C. Enoploidea:*Enoplus*; cheilostome armed with three jaws (original).

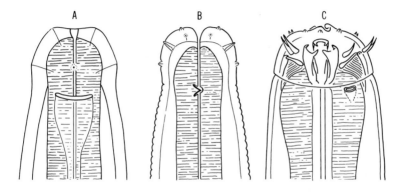

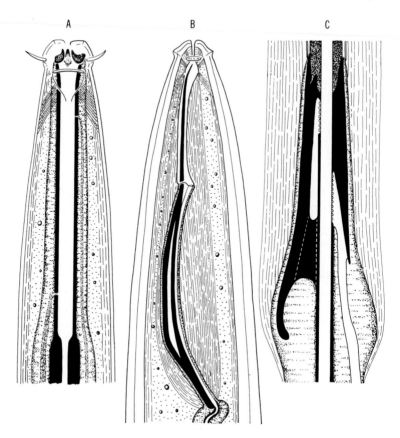

Figure 4.10. A. Ironoidea:*Ironus*; elongate cylindrical portion of stoma, esophastome; anterior movable teeth, cheilostomal (modified from Van der Heiden). B. Diphtherophoroidea:*Trichodorus*; the basal portion of the tooth is esophastomal; the anterior slender portion (extending from where the esophageal lumen meets the cheilostome) is cheilostomal in origin (modified from Allen). C. Esophastomal flanges (odontophore) of *Xiphinema* showing the internal sensory (gustatory) organs (modified from Robertson).

to the "spear extensions" which are modified esophastome. Though these terms are those proposed by some workers in the group and are commonly used, it is unfortunate because they are applied in the opposite manner of that used for onchia and odontia in the rest of Nemata. Therefore, throughout this text when discussing the spear of dorylaims the terminology will be used in a fashion consistent within Nemata. Therefore, the anterior spear will be referred to as **odontostyle** (= onchiostyle) and the posterior extensions will be termed the **odontophore** (= onchiophore) (Fig. 4.11A). This will allow phylogenetic proposals and homologies to be made without confusion in nomenclature.

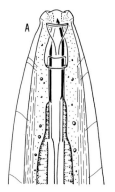 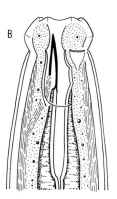

Figure 4.11. A. Dorylaimida:*Dorylaimus*; shows an axial spear. The odonto-style (cheilostomal) is situated over the odontophore (esophastomal) in manner that forms a hollow spear (redrawn from Thorne). B. Dorylaimida:*Nygolaimus*; shows a mural tooth (cheilostomal) not placed over the esophastome so that a hollow spear is not formed (redrawn from Thorne and Swanger).

b. Principle Types of Stomas

The stoma is more reflective of the ecological niche of a nematode than any other structure in the body. Because stoma form is so influenced by feeding habits within Nemata, shape in itself has little phylogenetic significance. Similar stomas show up in widely divergent groups and dissimilar stomas are seen in closely related groups.

Variability in stoma shape is, however, one of our most valuable taxonomic tools. Stoma morphology is valuable and has long been used in the identification of nematodes. However, the amount of phylogenetic significance that may be given to its gross form is subject to question because gross form is a manifestation of biological habit or niche. Among cell feeders the stoma is characteristically provided with a hollow stylet or protrusible tooth which functions to allow ingestion of cell contents. This type of stomatal armature is also common among predaceous nematodes. Most predators and "carnivorous" nematodes, however, possess well-developed stomas armed with large teeth, which may or may not be movable jaws or "mandibles." Bacterial and microbial feeders are usually recognized by their cylindrical unarmed stomas.

b1. Astomatous   The term astomatous should be used cautiously. It is misleading because a true astomatous condition exists in only a few forms of parasitic nematodes. When it occurs it is generally seen among males, though some larval stages also lack a developed feeding apparatus. The term, unfortunately, has come to be synonymous with the absence or rudimentary condition of the cheilostome and a coincident unexpanded esophastome (Fig. 4.9A). This condition is also described as "having the stoma collapsed." The presence of only an esophastome is not correlated with a lack of functionality. In fact, it is not uncommon in such a stomatal manifestation for the esophastome to be armed with one or more teeth (Fig. 4.9B).

The limitation of the stoma to only the esophastome is recurrent throughout Nemata and is common in the subclasses Enoplia and Rhabditia. The delimitation of the esophastome and its placement of armature is usually marked by a change in esophageal tissue. It may well be that the primitive nematode stoma was principally represented by the esophastome. This in fact was proposed by Filipjev in 1934. The Oxystominoidea fit this concept because they not only have the stoma limited to the esophastome but, as reported under Cephalic Sensory Structures (Fig. 3.15A), they illustrate the ancestral placement of sensory structures posterior to the lip region. The possession of just the esophastome in these basic Enoplia also complements the embryological observations that the stomodeum (esophastome plus esophagus) is a primary invagination. The presence of the cheilostome, which results from a secondary invagination, is likely to be a later phylogenetic development.

All other stomatal manifestations that follow include both the cheilostome and esophastome.

b2. Stoma without Armature   Expanded unarmed stomas are a common feature found in all groups of nematodes except Dorylaimida, Tylenchida, and Aphelenchida. Three general shapes are most typical: stoma cylindrical over its entire length, stoma globular (Fig. 4.7B), and stoma expanded anteriorly and somewhat or completely constricted posteriorly (Fig. 2.9A). The expanded portion often corresponds to the cheilostome and the constricted portion often corresponds to the esophastome. However, this is not always true and should be determined on an individual basis. In some nematodes the expanded region involves both the cheilostome and anterior esophastome.

It is generally assumed that unarmed, subglobular, cylindrical, or partially expanded stomas are indicative of microbial feeders. Often the esophagus has internal structures that aid in the grinding or slitting of the microbe, thus releasing the internal contents. There are many forms of nemas with unarmed cylindrical stomas that reveal no such esophageal modification and whose feeding habits are unknown.

It was noted earlier that among Rhabditia the esophastome may be subdivided into plates, but the function of these plates is unknown. Closer examination may reveal a masticatory function, especially among cephalobs where the plates have specific shapes, often appearing toothlike (Fig. 4.8A). In fact, when fully understood the latter type of stoma may be inappropriately described as unarmed.

When the anterior stoma is expanded and the posterior collapsed the anterior portion is called a vestibule. The expanded portion is generally cheilostome but may also involve the anterior esophastome.

b3. Stoma with Immovable Armature   Stomas in this category are prevalent among the predaceous nematodes (Fig. 4.7A). However, large globular stomas with armature (teeth, denticles, plates) are also seen among animal parasites of the intestinal tract. With predaceous nematodes the teeth function to slit the cuticle of the prey, thus making the internal tissues readily available for ingestion and digestion. Cuticular structures are not easily digested and predatory habits are often confirmed by observing spicules, spears, stomas, or even whole nemas in the intestinal tract. Animal parasites, particularly those like hookworm, use the **stomatal plates** and teeth to rupture blood vessels and to maintain their position in the intestinal tract (Fig. 8.6).

The stomatal armature may be either cheilostomal or esophastomal. Whenever cutting plates are present they are cheilostomal. Among marine nematodes one or more of the esophageal glands may open directly through the stomal teeth of the esophastome (onchia). In such instances it is believed that the secretions injected act to predigest the internal contents of the prey. There may also be some paralytic action associated with these glands. A predigestive function would be most useful in those cases where the prey is not ingested, but only the contents of the body are sucked out. Paralysis would prevent thrashing that could free the victim from the stomatal grasp of the predator.

b4. Stoma with Movable Armature   All movable armature, whether mandibular, fossoria, odontia, or spears, are thought to be derived from cheilostomal structures. However, the nature and formation of the armature of dorylaims needs clarification. The esophastome may be modified and attached to the cheilostomal armature to create a total functional unit that moves. Though movable armature is found throughout Nemata it is more characteristic of some groups than others. Mandibles (**jaws** or **gnathi**) are found principally among Enoplia (Fig. 4.9C); **fossoria** (outwardly moving teeth) are most characteristic of Diplogasterida; protrusible odontia generally denote members of the Dorylaimida. It is also among Dorylaimida (Adenophorea) and all Tylenchida and Aphelenchida (Secernentea) that the **hollow axial spear** is found.

b4a. Mandibular   With few exceptions a mandibular stoma is confined to free-living marine nematodes in Adenophorea. The stoma in these nemas is usually modified into three opposing mandibles: one dorsal and two subventral (Fig. 4.9C). Each mandible is composed of two parts, one cheilostomal and one esophastomal. There may or may not be any recognizable separation or suture between these two cuticular components. The cheilostomal mandible may be so fused to the onchial plates of esophastomal origin as to appear as one structure. The mandibles, therefore, are a single functional organ representing a specialization of the cheilorhabdions and esophastomal plates. Among Chromadoria the number of mandibles may be reduced to two by a reduction of the dorsal mandible and the lateral adjustment of the two subventrals, bringing them into opposition.

The musculature that operates the mandibles individually and as a unit is modified from the anterior musculature of the esophagus. The origin of the muscle is within the esophagus and the insertion is on the esophastomal plate. This arrangement appears to be constant throughout Nemata wherever movable armature exists.

b4b. Fossoria, Mural Teeth, and Axial Spears   Outwardly moving opposable teeth that are products of the cheilostome walls (cheilorhabdions) are called fossoria. In movement fossoria move in a scissorlike fashion, but the teeth cut on both the outward and inward strokes. Fossoria when protracted extend beyond the oral opening. The movement of fossoria are under the control of esophageal musculature.

Stomas with fossoria are found in both Adenophorea (Fig. 4.10A) and Secernentea (Fig. 4.8C). It is thought that their independent development, in the two classes of Nemata, is an important factor in the evolution of the axial spear in both Adenophorea (Dorylaimida) and Secernentea (Tylenchida, Aphelenchida). However, in the two classes the formation of fossoria, though both are thought to be cheilostomal, is different. In Secernentea (Diplogasterida) the fossoria are formed in place, that is, they are produced as teeth within the limits of the cheilostome, near the junction with the esophastome. In Adenophorea, on the other hand, the fossoria or teeth may be produced by special cells embedded in the anterior tissues of the esophagus. During larval molts these teeth migrate to take their place in the cheilostome. This is an important point because in Dorylaimida movable mural teeth and the anterior portion of the axial spear are produced within the confines of esophageal tissue. The only way this can be explained, if the fossoria, mural teeth and odontostyles are cheilostomal, is for modified external hypodermal cells to have become embedded within the anterior esophagus. The available information does not allow any conclusion to be made. There is circumstantial evidence from differential staining that these migratory fossoria in Dorylaimida are of cheilostomal origin and only secondarily associated with esophageal tissue. When stained with Mallory's triple stain the mural teeth and anterior spear (odontostyle) of Dorylaims react the same as the external epicuticle. This may mean that these structures are also keratinlike in chemical constitution. If they were products of the esophagus or esophastome then they would be expected to have the chemistry of esophageal cuticle, which they apparently do not. In addition the cell in which they are formed has a long duct whose opening or beginning is at the junction of the cheilostome and esophastome.

The movable **mural tooth** that occurs in Nygolaimidae (Dorylaimida) then could represent the development of a single fossor (Fig. 4.11B). Its formation is within the esophagus and its musculature is the same as described above except that protrusion from the stoma involves movement of the esophastome. This is an important step relative to the development of the hollow protrusible axial spear.

The axial spear of dorylaims is composed of two parts: the odontostyle and odontophore (Fig. 4.11A). The hollow odontostyle, which has a dorsal seam extending almost its entire length, is formed as is the mural tooth, by specialized hypodermal tissue in the anterior esophagus. The odontophore or spear extensions equate to the esophastome of other Nemata. The odontophore may be symmetrical or modified with flanges or knobs. In any event it is to this part that the contractile portion of the protractor muscles are attached (origin). The noncontractile portion of the muscle is in the anterior esophagus.

The cheilostome of a dorylaim with its hollow axial odontostyle is made up of three components: cheilorhabdions, sheath, and odontostyle. The anterior thick-walled cheilostomal region is connected to a membraneous sleeve (guiding sheath) that attaches to the base of the spear (odontostyle) where it joins the esophastome (odontophore). The vestibular area thus formed is the true stoma (Fig. 4.11A). This thin membraneous sheath is variously folded in different genera and species; the refracted light from the folds are referred to in the literature as "guide rings." Such a guide ring is most easily seen where the sheath is joined to the heavy-walled stoma. However, the true function is to permit stomatal flexibility, which allows the spear to be protracted.

Within the walls of the esophastome (odontophore) of Longidoridae (plant virus vector) there are three sinuses. The largest of the three is ventral, the two subdorsals are narrow and short. Each sinus contains nerve processes ending in ciliary bodies appressed to the thin indented cuticle near the food canal wall (Fig. 4.10C). These sensory organs are believed to have a gustatory function. How widely such structures may occur in Nemata is unknown.

The protractor muscles of the spear are once again the modified musculature of the anterior esophagus. Other muscles that appear to act as retractors and spear stabilizers are probably somatic (body wall) muscles modified from the specialized somato-esophageal muscles common throughout Enoplia (Fig. 4.10B).

A second type of "stylet" is also found among Dorylaimida, specifically in the Trichodoroida, which also vectors plant viruses. In this group the "stylet" is not hollow; it acts as an elongated mural tooth proximally attached to the dorsal wall of the esophastome (Fig. 4.10B). The tooth is a development by reduction of the dorsal half of the complex odontostyle-odontophore found in other members of the suborder Diphtherophorina. This tooth is referred to as an "odontostyle." Even though this is misleading it has become entrenched in the literature. In many ways the structure is similar to other odontostyle-odontophores. The odontostyle is formed posteriorly in the dorsal sector of the esophagus and migrates to the anterior limit of the esophastome where it becomes attached to the esophastome's heavy dorsal wall called the odontophore in this case. Dorsally the cheilostomal cuticle makes a simple fold before attaching to the odontostyle where it joins the odontophore (esophastome). This marks the extent of the original cheilostome (Fig. 4.10B). On the subventral walls the cheilostomal cuticle also makes a simple fold before joining the esophastome (thin-walled subventrally). Among the less advanced Diphtherophorina the subventral walls are thickened and act as an axial spear. However, in Trichodoroidea these walls are greatly reduced and no longer can be protruded. Thus, instead of a hollow axial spear these nematodes have a continuous food channel into which the dorsal "stylet" or "tooth" projects.

The available observations indicate that from fossoria bearing adenophoreans such as *Ironus* (Fig. 4.10A). three separate lines of stomatal armature development have occurred among dorylaims: (1) the true mural tooth, where one fossor became dominant (Fig. 4.11B); (2) the axial spear, where the tooth is hollow and locates directly on the food channel (Fig. 4.11A); (3) the dorsal "stylet," developed from the modified dorsal cheilorhabdion and dorsal wall of the esophastome with a reduction of the subventral walls, which functions in a fashion similar to a spear (Fig. 4.10B). The musculature that operates the protraction of the stylet is esophageal. Modified somatic musculature retracts the stylet and acts to stabilize the cheilostome and operate the lips (Fig. 4.10B).

The final spear form to be discussed occurs in the class Secernentea, orders Tylenchida and Aphelenchida. It is the most characteristic feature of plant parasitic nematodes; however, it is also associated with fungus feeders and secernentean parasites of insects. The precursor of the "tylench" spear is to be found among the Diplogasteria with fossoria.

The cheilostome and its modifications are more complex in Tylenchida and Aphelenchida than in other forms so far studied. In these the cheilostome includes the cephalic framework, the stomatal cavity (vestibule), and spear cone (modified from fossoria) (Fig. 4.12). The **cephalic framework** is reminiscent of the helmet of Enoplia and Chromadoria. It is an elaboration of the cheilostomal wall that extends over the modified anterior esophagus to the body wall. The spear muscles are the modified anterior muscles of the esophagus. The framework is much like an upturned basket that supports the labial region and gives partial purchase (apodeme) for the insertion of protractor spear muscles. The stomatal cavity or vestibule is thin walled, folded as a collar where it attaches to the spear cone at the junction with the esophastome (Fig. 4.12). This region is referred to in the literature as the "guiding apparatus." The **spear cone** is the portion of the stylet protruded through the oral opening and through which cell juices are ingested. The cone is hollow with a subterminal ventral opening.

Figure 4.12. A three-dimensional reconstruction of the anterior extremity of Tylenchida. This illustration shows the cheilostome as consisting of the basketlike cephalic framework and the tubelike extension attached to the base of the spear cone. The spear cone is also cheilostomal in origin. These parts are contributed to by external hypodermis seen lining the right side of the central tube. The base of the spear (shaft and knobs) are esophastomal in origin as are the protractor muscles, one of which is shown in section. The protractor muscles are conversions of the three anterior esophageal muscles (modified and reconstructed from Baldwin and Hirschmann).

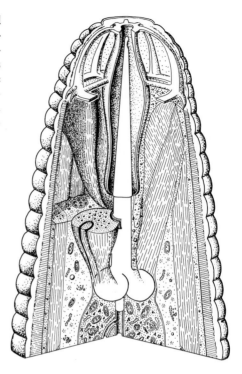

The esophastome is composed of an elongate tubular **shaft** that terminates, normally, in three **knobs**, one dorsal and two subventral. The three knobs function as **apodemes** for spear muscle origin. The three muscles are modified from anterior esophageal muscles. The contractile portion of these muscles extends from the knobs (origin) to the cephalic framework and body wall (insertion). The noncontractile portion of these muscles remains part of the anterior esophagus. In at least one observation deposits of β-glycogen, indicative of high muscular activity, are associated at the region of insertion. Prior to insertion on the framework and body wall each muscle branches so that at insertion it appears as if there are six or more protractor muscles operating the spear. No retractor muscles or labial support muscles (the framework contributes to this latter function) have been described in Tylenchida and Aphelenchida as existed in Dorylaimida.

3. Esophagus

Nemic esophagi, though variable in shape and reflective of feeding habits, have proved to be useful taxonomic and phylogenetic tools. In the construction of higher classification based on phylogenetic concepts the esophagus is probably the most important organ in the nematode body. Its phylogenetic significance is based not only on structural diversity, but on the internal morphology and relationship of the various tissues contributing to its makeup.

a. General Morphology

The esophagus of nematodes is made up of four tissues: epidermis, nerves, muscles, and glands. Therefore, the esophagus is the most complex organ in the nematode body. Varying degrees of cell constancy exist among nemic esophagi. Reconstruction from contemporary species leads to the conclusion that among ancestral nematodes the esophagus was composed of 36 muscles, 18 epithelial and/or marginal cells, plus some 40 neurons and 5 glands.

Nemic esophagi have certain features in common because of the rather constant relationship of the internal tissues to each other. All esophagi can be divided into two parts: the **corpus** (anterior region) and **postcorpus** (posterior region where the esophageal glands are located) (Fig. 4.13). Even in cylindrical esophagi the regions are identifiable by the change in tissues and location of the glands. In forms derived from the ancestral cylindrical esophagus these two regions are often further subdivisible. The corpus can be divided into a **procorpus** and **metacorpus** (with or without valves) (Fig. 4.13D). Between the corpus and posterior bulb in derived esophagi there is a narrow isthmus devoid of nuclei and usually surrounded by the circumesophageal commissure (nerve ring). The **isthmus** is the anterior region of the postcorpus (Fig. 4.13C, D). The postcorpus, com-

Figure 4.13. Diagram comparing homologous regions of nemic esophagi. A. Enoplida. B. Dorylaimida. C. Rhabditida. D. Tylenchida (original).

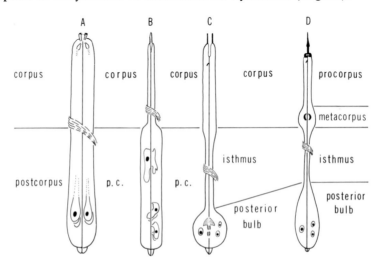

monly called the **terminal bulb**, is extremely variable in shape and contained valves; however, it always houses the esophageal glands, of which there may be five (ancestral) or three (derived). Among Adenophorea, Enoplia characteristically have five glands and Chromadoria three glands. However, in Chromadoria and Secernentea five glands may be derived from three gland species by duplication of the subventrals. Externally the esophagus is surrounded by a basement membrane that isolates it from the body cavity and other internal organs.

Internally, the esophagus is triradiate (cross section), that is, divided into three sectors: one dorsal and two subventral (Fig. 3.3A). The cuticularly lined lumen of the esophagus divides the sectors and is also triradiate. One ray (arm) of the lumen points ventrally and the other two point subdorsally (Fig. 3.3A). The cuticular lining of the radii (arms of the lumen) are often modified for muscle attachment. These apodemes may be very elaborate when seen in cross section or simple thickenings on each ray where muscles are concentered (Fig. 3.3A). The apex of each ray may be expanded in the corpus into longitudinal tubes. Among Mononchida when the apex of the rays are expanded in the posterior region of the postcorpus the esophagus is said to be **tuberculate**.

Until recently much of our understanding of the nematode esophagus was based on nuclear identification rather than cellular organization. In fact, it was believed that the esophagus was a syncytium without cell membranes between the various tissues, but studies with the electron microscope have shown that this is not true. The cells of each tissue type are discrete; however, some are binucleate and the glands in parasitic forms may be multinucleate.

Since cell integrity was not easily visible with the light microscope, tissues were recognized by their nuclei and the names given to them persist in the literature. Chitwood, in turn, devised a system for mapping the distribution of the various nuclear types in the esophagus. Such maps are extremely useful in understanding tissue distribution and derivation. For instance, among Enoplia with cylindrical esophagi the maps show that in forms displaying ancestral characteristics the tissues are randomly distributed (Fig. 4.14A), whereas in derived forms, even though the esophagus appears cylindrical without noticeable divisions, the nuclei are aggregated anteriorly and posteriorly in the corpus and postcorpus, respectively (Fig. 4.14B). Indicating tissue distribution is similarly aggregated for division of labor of the corpus and postcorpus.

Some confusion has arisen because a nuclear nomenclature rather than a tissue nomenclature has been used to describe the internal structure of nemic esophagi. The recognized nuclear types are radial, marginal, nerve, and glands. **Radial** nuclei belong to the esophageal muscles. The name implies that the muscle is radial; however, the muscle cell is longitudinal and its internal fibrils are radial or oblique. This lack of understanding has impeded our appreciation of the workings of nematode esophagi. Many

A   TRIPYLA

	R.Sv.			D			L.Sv.	
V	Sv	DI	Ld	D	Ld	DI	Sv	V
		n_3 m_3	n_1			m_1	n_2	m_2
		n_6	r_6 n_4 r_1			n_5		
r_4		r_5				r_2		r_3
		n_8 r_{11}	r_{12}		r_7	r_8	n_7	
r_{10}								r_9
		n_{11}	m_6		m_4	n_{10}		
		n_{13}		n_{14}		n_{12}	m_5	
		r_{17}	r_{18}		r_{13}	r_{14}		
r_{16}								r_{15}
		n_{15}		n_9		n_{14}		
		n_{17} n_{20}		n_{10}		n_{19} n_{16}		
		n_{23} r_{28}			r_{18}	r_{20} n_{22}		
		n_{25}	r_{24}			n_{24}		
		n_{27}		n_{21}		n_{26}		
r_{22}	m_9					m_7		r_{21}
	n_{28} r_{29}	r_{30}		r_{25}	r_{26}			
m_8 n_{29}								
r_{25}		n_{24} n_{31}	n_{29} n_{30}	n_{33} n_{37} r_{27}				
g_3			n_{32}			g_2		
				n_{30}				
r_{34} n_{41} m_{12}			m_{10}			r_{33} n_{40}		
m_{11}								
	g_3	r_{30}		r_{31}			g_4	
			r_{34} n_{44} n_{42} n_{43} r_{32}					

B   ANONCHUS

	R.Sv.			D			L.Sv.	
V	Sv	DI	Ld	D	Ld	DI	Sv	V
r_4	n_3 r_5	r_6		r_1		r_2	n_2	r_3
r_{10}								r_9
		r_{11} r_{12}		n_1 r_7	r_8			
r_{16} n_6				n_4 r_{13}	r_{14} n_5 r_{15}			
		r_{17} r_{18}						
n_8					n_7			
n_{11} r_{23}			n_9		r_{20} n_{10}			
n_{13}					n_{12} r_{21}			
r_{22}		r_{24} n_{11}						
n_{16}				r_{19}		n_{15}		
n_{19} r_{29}			n_{17}		r_{26} n_{18}			
r_{28}		r_{30}	r_{25}			r_{27}		
					n_{20}			
n	n				r_{32}			
		r_{33}	r_{31}					
					n_{25}	g_3		
g_2		g_1 r_{34} n_{29}	n_{23} r_{35}					
r_{26} r_{36} n_{24} n_{30} n_{28}								

Figure 4.14. Nuclear maps. A. One-part esophagus with nuclei rather randomly scattered (*Tripyla*). B. Nuclei aggregated anteriorly and posteriorly, cleared area indicates the isthmus (*Anonchus*). Abbreviations: n, nerve nuclei; m, marginal nuclei; r, radial nuclei; g, glands (redrawn from Chitwood).

workers have the impression that the fibrils seen in a series of cross sections all act independently. In reality, they represent the fibrils of individual cells that may extend for some distance along the length of the esophagus between the basement membrane and esophageal lumen. The contractile fibrils act as a unit within the sarcomere and not as independent muscles as is often assumed. Esophageal musculature differs from somatic musculature in that it does not possess the muscle arms that extend to the nerve cords to synapse. In the esophagus neurons run within muscle sectors and most synapses occur along the muscle. The two muscle classes, esophageal and somatic, are further distinguished by the occurrence of myosin heavy chains in esophageal muscle.

Marginal nuclei belong to cells that are located near the apex of the radial arms of the esophageal lumen. The function of these cells remains unconfirmed. It is believed that they contribute to the formation of the "tendonlike" fibers that extend from the apex of the radial arm to basement membranes enclosing the esophagus. There is no agreement as to whether these fibers are contractile or noncontractile. Some support the hypothesis that these cells are active in secreting the cuticular lining of the esophagus. It seems likely that they may perform both functions. Another theory for the laying down of esophageal cuticle involves the esophageal muscles. It has been proposed that the muscles are myoepithelial and that they secrete the esophageal lining. I do not adhere to this hypothesis because it seems so unnecessary in the presence of marginal cells that appear to have characteristics of cells originating from ectoderm. Since the hypodermis of the external cuticle performs more than one function, it seems unreasonable to assume a very limited function for esophageal "hypodermis" or marginal cells and at the same time assume a unique function for the muscle cells.

The esophagus of nematodes is an extremely variable and complex organ and explanation of how it works, based on engineering formulae, are interesting academically, but may have little to do with real events.

b. Esophageal (Esophago-Sympathetic) Nervous System

The central nervous system (circumesophageal commissure or nerve ring) and the esophageal nervous system are connected by neuron fibers that run from the central nervous system (CNS) to the esophagus in the lateral papillary nerves. These connecting nerves depart from the papillary nerve anteriorly and enter the esophagus in the region between the esophastome and cheilostome. Within the esophagus three major trunks are discernible, axially located in each sector, one dorsal and one in each subventral sector. Throughout the length of the esophagus these longitudinal trunks are interconnected by commissures. In general, there is a commissure at the base of the corpus, and two in the postcorpus, one anteriorly and one near the base. The large commissure and associated neurons located in the corpus are sometimes referred to as the esophageal enteric nerve ring.

Some six neuron types are identified within the esophagus: motor neurons, interneurons (probably proprioceptors) neurosecretory motor neurons, motor interneurons, marginal cell neurons, and neurons with mechanoceptive-like endings. Most of the esophageal neurons are unbranched, unipolar or bipolar cells.

Motor neurons innervate the muscles of the corpus and postcorpus. The anterior esophageal musculature that operates the stomatal armature (fossoria, jaws, spears) among various nematodes is also innervated from these neurons.

Interneurons are associated with the lining of the lumen and the basement membrane and are presumed to be proprioceptive (internal sense organs). Though not confirmed, the gustatory organs in the base (odonto-

phore) of the spear among Dorylaimida may be innervated by proprioceptive interneurons (Fig. 4.10C).

c. Types of Esophagi

Systems for categorizing nemic esophagi are based on the number of
differentiated regions visible in totomounts. Three basic forms are recognized: a one-part esophagus, which is cylindrical to conoid without distinguishing regions externally; a two-part esophagus recognized by the
narrow corpus followed by an expanded postcorpus; and the three-part
esophagus, with three distinct regions, a corpus, and a postcorpus which
is divided into a narrow isthmus and an expanded posterior bulb (Fig.
4.13). Depending on the species, these basic regions may be further
subdivided.

c1. Enoploid (One-Part)   This form of esophagus is typical of the order
Enoplida and the general conformation corresponds to the one-part esophagus that is grossly either cylindrical or conoid (Fig. 4.13A). Five esophageal glands are characteristic of the subclass Enoplia, being reduced in
number in only a few families of Dorylaimida, for example, Longidoridae.
Of the five glands in the order Enoplida one is dorsal and four are in the
subventral sectors. The three major glands (one dorsal and two subventral),
though large and easily seen, are unequal in size; the remaining subventrals
are sometimes smaller and more difficult to distinguish. In any event, this
type of esophagus is considered ancestral and typical of the order Enoplida.
The orifices of the three major glands in Enoplida open either directly into
the stoma or immediately posterior to it. In some representatives the orifices
are farther back, but the three anterior most glands of the postcorpus enter
anterior to the nerve ring.

Due to the distribution and form of muscular attachments the external
contour of the conoid esophagi may appear smooth in *Oxystomina*, crenate
in *Phanodermopsis* (Fig. 4.15A), or multibulbar in *Bobella* (Fig. 4.15B).
The development of muscle attachments on the cuticular lining of the
esophagus appears to have little phylogenetic significance, even though
the presence or absence and degree of development are of taxonomic
importance. The cylindrical esophagi often approach a two-part type; however, because of their symmetrical outline the regions are only readily
distinguished internally.

It is interesting and of some phylogenetic significance that among Enoplia
it is only two superfamilies (Enoploidea and Oncholaimoidea) of the
order Enoplida that possess pigment spots (ocelli), which allegedly have
some light sensory function.

Esophago-intestinal valves are variable being either triradiate or dorsoventrally flattened and comprised of as few as eight cells or as many as 100.

The order Mononchida also has a one-part esophagus; however, the five

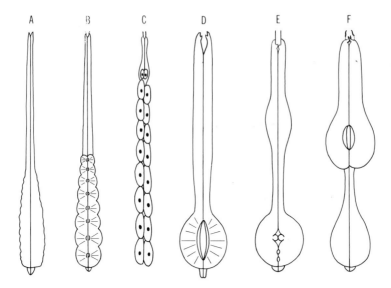

Figure 4.15. Types of esophagi. A. Crenate, *Phanodermopsis*. B. Multibulbar, *Bolbella*. C. Stichosome, *Agamermis*. D. Two-part esophagus with terminal bulb, *Microlaimus*. E. Three-part esophagus with terminal valved bulb, *Rhabditis*. F. Three-part esophagus with valved metacorpus, *Diplogaster* (original).

gland orifices are posteriorly located in the region of the gland body. They are not reported anterior to the nerve ring. The esophagus musculature is connected to well-developed cuticular apodemes situated at the base of each esophageal ray. In mononchs the esophago-intestinal valve is triradiate and contains fewer than 30 cells.

c2. Dorylaimoid (Two-Part)   The two parts of a dorylaimoid esophagus are externally visible (Fig. 4.13B). The cylindrical corpus is slender and may or may not be set off from the expanded, often elongate, postcorpus by a constriction. The lumen of the corpus is subtriangular, often appearing circular in cross section. Twenty-four muscles are present in the corpus. The six anteriormost operate the stomatal armature. As previously noted the stomatal armature is generated by a cell in the left subventral sector of the procorpus in forms having the axial spear or mural tooth. The number of esophageal glands in the postcorpus varies from three to five, five being far more common. In all known instances the gland orifices are located near the gland body and do not open anterior to the nerve ring. The postcorpus may be represented by a short pyriform region almost devoid of musculature. In forms with a muscular postcorpus the apodemes for the attachment of radial musculature of the esophagus are well developed. The esophago-intestinal valve is generally dorsoventrally flattened and contains less than 30 cells.

c3. Spiruroid Superficially esophagi among spirurids (Secernentea) look much like those which are common to dorylaims (Adenophorea). However, histological studies show that the internal tissue organization is not homologous. In dorylaims the expanded posterior portion is postcorpus, whereas in spirurids the expanded posterior region is made up of a part of the corpus (metacorpus) and postcorpus. This is recognized because the narrow anterior region (procorpus) contains 12 muscles and the swollen region which contains the multinucleate esophageal glands has 24 muscles instead of 18 in each region as occurs in the related ascarids and camallanids.

c4. Mermithoid This mermithoid esophagus is always associated with parasites of animals either invertebrate or vertebrate. Perhaps the morphological term stichosome is more characteristic than a nonmorphological designation as mermithoid. The **stichosome** is a series of large gland cells (stichocytes) attached to the posterior region of the esophagus (Fig. 4.15C). These are considered subsidiary or secondary esophageal glands, probably representing multiple duplication from the posterior pair of subventral glands. In larval forms three large unicellular glands are present in addition to the stichosome. The stichocytes of the postcorpus have orifices opening into the esophageal lumen in their immediate vicinity. Two types of stichosome esophagi occur among parasitic nemas. In mermithids (insect parasites) there is a row of stichocytes on either side of the esophagus. The anterior esophagus in both forms contains radial musculature either well developed or degenerate. The esophago-intestinal valve is dorsoventrally flattened. In vertebrate parasites (Trichocephalida) the stichosome may be limited to one side of the esophagus.

c5. Chromadoroid (Three-Part) The chromadoroid esophagus, in totomounts, is sometimes divisible into corpus, isthmus, and posterior bulb. When the gross outline of three sections is not easily discerned, the divisions can be recognized by the change in esophageal tissue (muscles) between the corpus and isthmus (Figs. 4.13C, 4.15D), remembering the isthmus is a division of the postcorpus and is recognized by its lack of nuclei. The chromadoroid esophagus is characterized also by the presence of three or five uninucleate glands in the postcorpus. The dorsal gland opens at the base of the stoma and the subventral glands open at the base of the corpus. The corpus is undivided, and therefore, a procorpus and metacorpus are seldom apparent except in a minor swelling in the metacorporal region. The arms of the corporal lumen often are terminated in well-developed tubes. Seldom are apodemes for muscle attachment evident. The postcorpus is bulbar and muscular. The chromadoroid esophagus contains 24 to 36 muscles equally divided between the corpus and postcorpus. The postcorpus may or may not be valved. When a valve is present its operation differs from the rhabditoid three-part esophagus. In chromadoroids the valve is heavily cuticularized and operates as a bellows or as three denticulate opposing plates. The esophago-intestinal valve is dorso-ventrally flattened and composed of approximately 20 cells or less. Sometimes specially differentiated intestinal cells are associated with the esophago-intestinal valve forming a distinct organ called a ventricular column.

c6. Rhabditoid   Esophagi classified as rhabditoid have a cylindrical muscular corpus, which can be further subdivided into a procorpus and metacorpus (Fig. 4.15E). The corpus is separated by a distinct isthmus that leads to the muscular valved postcorpus containing the esophageal glands. The procorpus is recognized as that anterior region where the arms of the esophageal lumen end in well-developed tubes; in the metacorpus the terminal tubes are small. In general, 12 muscles are found in the corpus: six in the procorpus and six in the metacorpus. The postcorpus also contains 12 muscles. The orifices of the esophageal glands are similar to the chromadoroid esophagus. The dorsal gland opens near the base of the stoma and the subventral glands open at the base of the metacorpus. The distinguishing characteristics of this type of esophagus are the muscular development of the metacorpus (almost bulblike) and the form of the valve in the postcorpus. In the rhabditoid postcorpus the lumen expands into a trilobed reservoir, into which project three muscular lobes, one dorsal and two subventral. Each lobe is lined by cuticle, often denticulate. Contraction of the postcorporal musculature rotates the three lobes posteriorly. This action creates a reservoir posterior to the lobes. Food drawn into this reservoir is forced posteriorly into the intestine by the subsequent relaxation of the lobe muscles, which rotates the lobes anteriorly. There is a secondary reservoir posterior to the lobes that also aids in the movement of food into the intestine through the triradiate esophago-intestinal valve.

c7. Diplogasteroid   Included in the diplogasteroid category is the so-called tylenchoid esophagus. These are combined here because structurally the differences are negligible. Grossly the diplogasteroid esophagus is divisible into the corpus, isthmus, and posterior bulb; and the corpus further divisible into a procorpus and metacorpus (Figs. 4.13D, 4.15F). The number of esophageal glands in the postcorpus is variable from three to five, rarely more. Such esophagi are distinguishable by the muscularity and valved condition of the metacorporal region. The valved metacorpus is the chief pump of the esophagus as opposed to the other esophagi previously described in which the chief pumping organ was in the postcorpus. Phylogenetically from diplogasterids to tylenchids there is a progressive degeneration of the musculature of the postcorporal region; however, the nuclei for muscle and other tissues remain. Therefore, we know that the postcorporal region of chromadoroids, rhabditids, diplogasterids, tylenchs, and aphelenchs are homologous. The muscular degeneration in the post-corpus of Diplogasteria is correlated with an increase in the size and form of the esophageal glands. The glands may become so enlarged in tylenchs and aphelenchs that they overlap the anterior portion of the intestine considerably. This is a development often repeated in individual families. The assumption made is that those with "overlapping" glands are more derived than forms without overlapping glands. Among these esophagi a triradiate esophago-intestinal valve is most common.

The orders Tylenchida and Aphelenchida can be distinguished from each other in part by the differences in their esophagi. In both orders the esophagus is divisible into the corpus (procorpus-metacorpus) and post-corpus. The isthmus of the postcorpus is present in Tylenchida even when the postcorpus glands overlap the intestine. Additionally, the tylench esophagus is distinguished by the placement of the dorsal esophageal gland orifice. It is always in the anterior procorpus close to the base of the stylet.

Aphelenchida only has the typical corpus, isthmus, and posterior bulb (glands), apparent in the family Paraphelenchidae. The other families are distinguished by having the glands not just overlapping the intestine, but taking the form of an appendage coming directly from the metacorpus bulb (Aphelenchoidoidea) or as a distinct appendage of the isthmus (Aphelenchidae). In the latter there is no blend of the isthmus and posterior glands. The isthmus is columnar and the glands connect to the column in a cecalike fashion. The dorsal esophageal gland in Aphelenchida also differs by opening into the metacorpus just anterior to the valve (Figs. 6.23, 6.24, 6.25). In both orders the subventral glands open posterior to the valve in the metacorpus.

B. Mesenteron (Intestine)

1. General Morphology

The tubular **mesenteron**, commonly called the intestine, is composed of a single layer of endodermal epithelium (Fig. 4.16). Three subdivisions called the ventricular region, midgut, and prerectum are sometimes recognized. These regions are rarely discernible in totomount studies, except for the prerectum among Dorylaimida, which is distinguished by being separated from the anterior intestine by a conspicuous change in cell character. This difference in cellular structure can even be seen under a dissecting microscope; however, the **ventricular** region, which extends from the junction with the esophagus to some undetermined distance posterior, is distinguished rather arbitrarily by the cellular inclusions and number of cells in the intestinal tube circumference. The cells in this region are often more packed with cell inclusions and insoluble spherocrystals and are more numerous in a cross section than in the general intestine. Among the Tylenchida and Aphelenchida subdivisions of the intestine are not recognized.

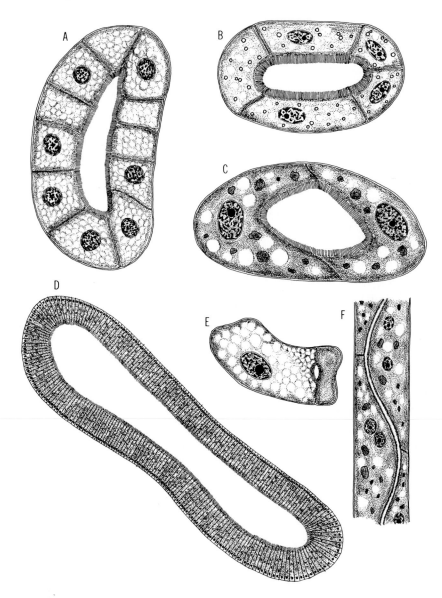

Figure 4.16. Histological sections of nematode intestines. A. Enoplia, *Deontostoma.* B. Chromadoria, *Axonolaimus.* C. Rhabditia, *Rhabditis.* D. Rhabditia, *Ascaris.* E. Diplogasteria, *Ditylenchus.* F. Diplogasteria, *Ditylenchus* (longitudinal section) (modified from Chitwood).

The intestine in some parasitic nemas is disconnected from the stomodeum and proctodeum and though nonfunctional in the usual sense it becomes a food storage organ called a "trophosome."

The ventricular region of Secernentea other than Diplogasteria is some-

A   B   C   D   E   F

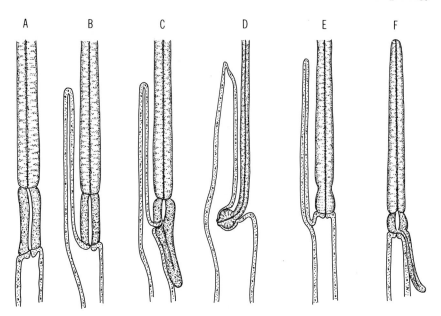

Figure 4.17. Intestinal and esophageal diverticula among Ascaridida. A. *Anisakis.* B. *Porrocaecum.* C. *Contracaecum.* D. *Dujardinia.* E. *Angusticaecum.* F. *Raphidascaris* (modified from Yamaguti).

times distinguished by the development of a **cecum** or **diverticulum** (pouch) that may be directed anteriorly or posteriorly (Fig. 4.17). When two ceca are described on the intestine the anterior one is an extension of the intestine and the posterior "cecum" is an extension of the posterior esophagus. A true ventricular cecum of the intestine may be either anteriorly or posteriorly directed. Most commonly it is anteriorly directed. Though caeca occur among free-living nematodes in Rhabditia they are more commonly found in the parasitic ascarids and spirurids. The function of a caecum is generally unknown; however, as a modified anterior pouch it plays an important role in the life cycle of insect "parasites" among the neoaplectanids. In these nematodes an interior cecum is filled with bacteria which are released into the insect's hemocoel when the nema penetrates the insect gut. The bacteria are pathogenic to the insect, but serve as a source of food for the nematode which cannot survive directly on the insect's tissues.

The intestine internally is bordered by microvilli. The length of the microvilli often can be used to distinguish the three regions of the intestine. Externally the intestine may be bounded (Fig. 4.18A) by a basal lamella. The origin of the basal lamella is unknown; however, it acts in the nature of a sheath and has an affinity for collagen stains. As such this mesenterial sheath appears to be a mesodermal product. In most nematodes the sheath is not sufficiently developed to be called a basal lamella. Rather

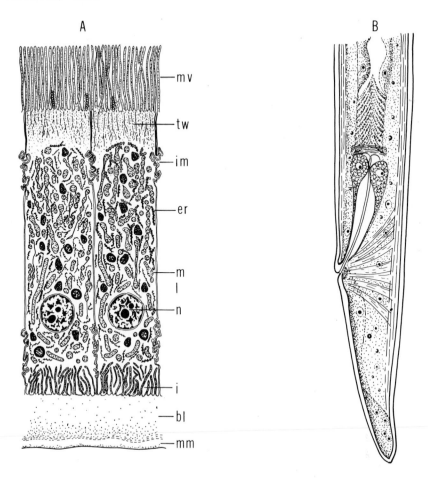

Figure 4.18. Ultrastructure diagram of intestinal cells of *Ascaris*. Abbreviations: mv, microvilli; tw, terminal web; im, infolding of cell membrane; er, endoplasmic reticulum; m, mitochondria; l, lipid inclusions; n, nucleus; i, infoldings of plasma membrane; bl, basal lamella; mm, mesenterial membrane (redrawn from Kessel et al.). B. Female tail of *Bulbodacnitis* showing intestino-rectal valve (redrawn from Maggenti).

an extremely thin membrane isolates the intestine from the body cavity; this membrane is called the pseudocoelomic membrane. Beneath the basal lamella or pseudocoelomic membrane muscle fibers may occur. Normally these fibers, modified from somatic muscles, are confined to the posterior region of the gut and do not form a continuous layer. However, in some nematodes the muscle fibers form a coarse meshwork giving the appearance of a separate muscle layer. These muscles aid in the movement of food through the intestine, and may be sufficiently numerous and well developed so as to create a peristalsis of the intestine.

The intestinal cell as it occurs in uninucleate ascarids can serve as a model for Nemata even though most Secernentea do not have uninucleate cells. With the exception of the mermithid insect parasites (which have a degenerate digestive system called a trophosome) all known Adenophorea have uninucleate intestinal cells.

The internal surface of an intestinal cell is covered by microvilli (finger-like projections of the cell surface) of varying length (Fig. 4.18). The continuous layer of microvilli lining the lumen of the intestine is referred to in the literature as the bacillary layer or brush border (Fig. 4.16). The microvilli, regardless of length, are all about 1 mμ diameter. Within each microvillus is an electron dense core of longitudinal filaments surrounded by an electron lucid region. Below the microvilli there may be a network of electron dense fibrils known as the **terminal web** or plasma cap. The terminal web is perforated by cytoplasmic connections between the apical layer and the cytoplasm of the remainder of the cell. When there are extensions from the terminal web into the core of the microvilli the apical cytoplasmic layer is distinguished as the subbacillary layer. However, in many nematodes only the cytoplasm of the cell proper is seen below the microvilli; neither the terminal web nor the apical cytoplasmic layer is present.

Intestinal cell cytoplasm is rich in mitochondria, Golgi complexes, endoplasmic reticulum, ribosomes, glycogen granules, lipid droplets, and lamellar bodies. Lamellar bodies probably correspond to the spherocrystals seen with light microscopy. Although their function is unknown, it has been speculated that they may be involved in the destruction and elimination of waste material. Bird has proposed that these bodies may be associated with the synthesis of the mucopolysaccharide or mucoprotein externally associated with microvilli. When associated with microvilli this material is called the glycocalyx or filamentous coat.

Nuclear placement is influenced by cell shape: in cuboidal cells the nucleus is centrally located while in columnar cells it is usually in a basal position. When a terminal web is present mitochondria accumulate apically, vesicles of rough endoplasmic reticulum are central, and lamellar bodies, glycogen deposits, the nucleus, and other subcellular components are located basally. In cuboidal cells where the terminal web is lacking the cellular components are randomly distributed. In either instance the main storage material in the intestine is lipids. In plant parasites lipids have been reported to account for 30% of the total dry weight of the animal.

The intestine is separated from the body cavity, as previously stated, by a basement membrane or basal lamella that does not appear to be an intestinal cell deposition. Its collagenlike composition lends credence to a mesodermal origin.

When intestinal cells differ in character throughout the intestinal epithelium the condition is termed **heterocytous**. When no difference or specialized cells are seen the intestine is said to be **homocytous**. If all the cells in the intestinal epithelium are of equal height in a given cross section, the intestine is **isocytous**. Sometimes there are definite differences in the height of the cells and this condition is called **anisocyty**.

The number of cells in a given cross section of nematode intestine varies from two to over 100. Larval forms and some adult forms may have fewer than 20 cells in the entire intestine, whereas the total number of cells in adults in Ascaridida may exceed 1,000,000 (Fig. 4.16D). In general, cell numbers increase from larva to adult. As the number of cells in the intestine increases the shape also changes. When only a few cells are present the cells are longitudinally elongate and rectangular (Fig. 4.16B, C, E). As the number of cells exceeds 100, all cells become hexagonal and either cuboidal or columnar.

These complexities of size, shape, and number are perhaps best understood when viewed in the light of embryological development. Martini, near the turn of the century, noted that when the definitive first stage larva was formed, the egg had undergone 10 cleavages. At this point, theoretically, the larval intestine should contain 128 cells. This means that at the end of three egg cleavages the endodermal stem cell is recognized and seven subsequent cleavages of this cell would result in 128 cells. In fact, however, this is seldom achieved because of lag times and lack of synchronization in cleavages. Therefore, many nemas contain less than 128 cells in their entire intestine and to such a condition the term oligocytous is applied. In this process a theoretical first stage larva would contain 1024 cells of which 128 would be intestinal and 64 somatic muscles. The remainder of the cells would constitute the hypodermis, esophagus, nervous system, excretory system, and germinal primordia.

Using this theoretical approach nematodes have been divided into two groups: those that do not achieve the theoretical number (**oligocytous**) and those forms in which the endoderm exceeds 10 cleavages, that is, in theory have undergone 11 cleavages (256 cells). In the latter instance the intestine is designated as **polycytous**. It has been further observed that after 16 cleavages (8192 cells) a change in shape and cross-sectional number occurs. Between 256 and 8192 (polycytous) there are 20 to 50 cells in a given circumference and the shape is cuboidal (Fig. 4.16A). Intestines exceeding 8192 cells (17 cleavages or more) are called **myriocytous** and in a given circumference the number of cells exceeds 100 and each is columnar in shape (Fig. 4.16D).

The nature of nematode intestinal cells can be further complicated by the occurrence of polynucleation. Uninucleation is the more common condition and is considered ancestral. Among the entire class Adenophorea polynucleation is reported to occur only among the mermithid parasites of insects. However, the condition is common among the parasitic Secernentea and alleged to be nearly universal among Tylenchida.

In polynucleate nematodes the number of nuclei ranges from two (Fig. 4.16F) to over 500. Such cells are coenocytic, having developed by nuclear division without cell membrane formation, rather than by cell aggregation as is sometimes alleged. As I cautioned in the discussion of the hypodermis one must be very careful when using the term syncytia. It becomes more and more apparent that syncytia resulting from cell aggregation does not occur in nematodes. If it does occur, it is extremely rare and should be carefully confirmed. For example, if cell aggregation accounts for the polynucleate condition of the intestine or any other tissue then the number of nuclei in the cells of a given intestine should be variable, since it is unlikely that an equivalent number of cells would aggregate to form an intestine of some 20 cells with each cell containing equal numbers of nuclei.

However, not only does the number of cells in the intestine follow a geometric progression indicative of cell cleavages, but so also does the number of nuclei in a given cell. For instance, the number of nuclei closely conforms to theoretical cleavages (2, 4, 8, 16, 32, etc.), even among the Strongyles with over 500 nuclei (theoretical 512 nuclei). If cell aggregation occurred, then nuclear numbers should conform to an arithmetic progression and so far this does not seem to be the case. Currently for example, none is reported to have nuclear counts arithmetically between 300 and 500 (theoretical 256–512); arithmetically any number between these two should occur if cell aggregation were involved.

The cross-sectional shape of the lumen to some extent reflects the number of cells in the intestine. Oligocytous intestines generally have a cylindrical or rounded lumen; polycytous intestines characteristically have subpolygonal lumens; and myriocytous intestines have a lumen that is highly irregular in shape, being multifolded, or when empty, a flattened tube.

Currently little emphasis is placed on the taxonomic significance of the intestinal variability. However, understanding intestinal variation could be significant. As an example, three species of horse strongyles can be distinguished by the number of intestinal cells in the third stage larva: *Strongylus equinus* has 16 intestinal cells; *S. edentatus* has 20 cells; and *S. vulgaris* has 32 cells. Some authors place these species in three genera: *Strongylus*, *Alfortia*, and *Delafondia*, respectively.

The intestine, as we have seen, is more than a simple tube that is more or less straight, which accommodates itself to the available space in the body cavity.

2. Somato-Intestinal Muscles

These muscles extend from the body wall to the intestine. Their form and number vary from group to group. In some, four submedian longitudinal rows of transverse muscles occur from the body wall to the intestine. These muscles are extensive and in at least one group of nematodes (Diocto-phymatida) they move gut contents in peristaltic waves. More often they are limited to a few scattered pairs of subdorsal and subventral muscles found posteriorly near the intestino-rectal valve.

C. Proctodeum

The **proctodeum** or **rectum** is the posterior ectodermal part of the alimentary canal (Fig. 3.2C). Like the stomodeum it results from an invagination of the ectoderm during embryonic development. It is a simple tube more or less flattened, subtriangular or irregular in cross section, constituting merely a conduit from the intestine (mesenteron) to the ventromedian generally subterminal anus. The anterior end of the proctoderm is marked by the presense of an uninucleate circomyarian **sphincter muscle** which surrounds the posterior extremity of the intestine.

The epithelial cells of the rectum are flattened or cuboidal. These epithelial cells are believed to secrete the cuticular lining of the rectum (Fig. 4.18B). The cuticular lining is continuous with the external body cuticle and is shed at molting time. Though the cuticle is connected to the external cuticle, it is most similar in appearance and staining reactions to the epicuticle layer.

The opening from the posterior intestine into the rectum is often guarded by an intestino-rectal valve or **pylorus.** This valve is formed by projection of the posterior intestinal epithelium either as a cellular reflexure or cellular extension (Fig. 4.18B). The structure of the intestino-rectal valve is not unlike that of the esophago-intestinal valve.

In some nematodes, especially among Secernentea, there are ectodermal glands that open into the rectum near its anterior extremity. In Rhabditia, females commonly have three rectal glands and males six glands. Though rectal glands are rarely reported in the order Tylenchida, the females of root-knot nematode (*Meloidogyne*) have six rectal or anal glands (Fig. 6.17A). The function of the anal glands of the root-knot nematode is the production of the gelatinous matrix in which eggs are layed. The function of rectal glands in other nematodes is unknown. Among Adenophorea it is only in the subclass Chromadoria that rectal glands are known, and then only among the family Plectidae (Fig. 3.2C).

The musculature associated with the proctodeum is limited to the sphincter already mentioned, and two other sets directly attached to the rectum. The muscle sets attached directly to the rectum are called **depressor ani** and **dilator ani**. The latter, dilator ani, are not common. When they do occur, they extend from the ventral body wall to the posterior surface of the anus. Paired ventral preanal muscles connected to the anterior lip of the anus have also been described. Universal, however, is the depressor ani muscle (Fig. 4.18B). This H-shaped muscle consists of two vertical groups of fibers between the dorsal wall of the rectum, the posterior lip of the anus, and dorsolateral side of the body. The nucleus of this muscle is situated in the horizontal band of sarcoplasm between the two groups of fibers.

In females of root-knot nematode the anus and the rectal glands have no connection with the alimentary canal. The extrusion of the gelatinous matrix produced by the rectal glands is aided by the contractions of the depressor ani muscle. The muscle in other nematodes aids in defecation. Because the rectum has no circular muscles, defecation is aided by internal pressure supplied presumably by the somato-intestinal muscles. Upon contraction these muscles would dilate the posterior intestine (prerectum); upon relaxation of the sphincter, the bolus would move into the rectum. Anterior movement of the bolus is prevented by the pyloric valve. Subsequently, the sphincter contracts and the depressor ani contracts; the bolus then moves from the distended rectum by the body pressure, which forces the waste bolus out upon depressor ani relaxation. In males the rectum joins the reproductive system to form a cloaca. The complexity of the male proctodeum is discussed in Chapter 5.

III. Nervous System

It should now be evident that nematodes are complex animals composed of a number of different tissues and organs, and as with all living organisms; their protoplasm is excitable or irritable. Yet of itself this mass is inert unless stimulated to action. Each tissue and organ of the nematode has its own role to perform, but these independent roles must operate in harmony if definite results are to be accomplished. Coordinated responsive action is a necessity if the nematode is to make advantageous adjustments to its immediate environment. These functions of stimulation, coordination, and responsive actions are the responsibility of the nervous system.

The nervous system perceives stimuli from the environment and transmits them to the internal tissues in which the latent energy for activity is stored. It is the nervous tissue that controls and directs the results of this liberated energy. Stimuli for activity which originate from the external environment are transmitted by way of nervous tissue to internal tissues; however, the distribution of internal stimuli may be accomplished by the nervous system or by hormones. Little is known about hormones of nematodes, but the structure of the nervous system is fairly well understood in many respects.

The functional unit of the nervous system is the neuron. A **neuron** consists of the cell body (**neurocyte**) where the nucleus is located; an elongated branch, which is the nerve fiber (**axon**), and aborizations or branching fibrils (**dendrites**) which conduct impulses toward the neurocyte. Nerve impulses travel inward to the neurocyte by way of the dendrites and away from the neurocyte along the axon. An aggregation of neurocytes constitutes a **ganglion**.

Neurons are classed into three basic types: sensory or afferent neurons; motor or efferent neurons; and adjustor or associative (internuncial) neurons. **Sensory neurons** conduct impulses from the receptor to or toward the central nervous system; and **motor neurons** conduct impulses to the effectors. **Association neurons** join sensory and motor neurons; thus a single stimulus from the receptor may excite more than one effector.

The nervous system of nematodes differs from most other invertebrates in that it is not limited to nerve cords that are all ventral with nerves passing from the ventral chain of ganglia and cords to the various organs of the body. Even though the main nerve cord is ventral, nematodes have three other nerve cords extending longitudinally through the body: one dorsal and two lateral. In addition processes from the somatic muscles extend to the dorsal and ventral nerve cords, rather than nerve processes extending to the muscles as in other invertebrates (Fig. 4.1A). The total system is divisible into four parts: the central, the peripheral, the recto-sympathetic, and the esophago-sympathetic system. The esophageal nervous system has already been discussed.

A. Central Nervous System

The central nervous system of nematodes consists of ganglionated nerve tissue or the "brain" connected with the **nerve ring** (circumesophageal commissure) and the ventral ganglion chain, which is called the ventral nerve. The nerve ring generally surrounds the esophagus at its midpoint, in that region between the corpus and postcorpus. This region of the esophagus is sometimes recognized as the isthmus. The nerve ring is obliquely oriented in the nema body with the dorsal side most anteriad.

1. Ganglia

The ganglia comprising the so-called brain of nematodes lies anterior and posterior to the nerve ring (Fig. 4.19). Anterior to the nerve ring there are six small, radially placed, cephalic papillary ganglia (cpg). The dorsal ganglion (dg), two large lateral ganglia (lg), and the ventral ganglion (vg) are located posterior to the nerve ring.

Anterior to the nerve ring and attached to it are the six small cephalic papillary ganglia: two subdorsal, two lateral, and two subventral. Each ganglia is composed of 3–7 bipolar sensory neurons whose axons pass from the posterior side of each ganglion directly into the nerve ring. The dendrites of these neurons form the six cephalic papillary nerves (cpn), which end distally in the three whorls of cephalic sensilla.

Posterior to the nerve ring the major ganglia are concentrated; dorsally there is a small dorsal ganglion composed of two neurons, either bipolar or unipolar. There are also two small subdorsal ganglia (sdg), each consisting of two neurons. Laterally there are two large complex masses of

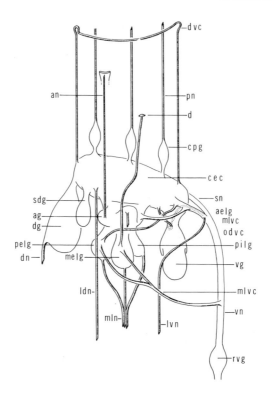

Figure 4.19. Generalized anterior central nervous system showing the right side of the nerve ring and associated ganglia, nerves, and commissures. Abbreviations: dvc, dorsoventral commissure; an, amphidial nerve; pn, papillary nerve; d, deirid; cpg, cephalic papillary ganglia; cec, circumesophageal commissure; cn, subventral nerve; sdg, subdorsal ganglion; ag, amphidial ganglion; aelg, anterior externolateral ganglion; mlvc, major lateroventral commissure; dg, dorsal ganglion; odvc, oblique dorsoventral commissure; pelg, posterior externolateral ganglion; dn, dorsal nerve; melg, median externolateral ganglion; pilg, posterior internolateral ganglion; vg, ventral ganglia; mlvc, minor lateroventral commissure; ldn, laterodorsal nerve; mln, mediolateral nerve; lvn, lateroventral nerve; vn, ventral nerve; rvg, retrovesicular ganglion (original).

21–42 nerve cells which are collectively referred to as the lateral ganglia. The lateral ganglia may be subdivided into six lesser ganglia: the internolateral ganglia which lie directly against the nerve ring and connect with it; the amphidial ganglia (ag), composed of eight or more neurons; the posterior internolateral ganglia (pilg); the anterior externolateral ganglia (aelg); and the median (melg) and posterior externolateral ganglia (pelg). The major part of the lateral ganglia connect with the nerve ring by way of the major lateroventral commissures (mlvc) which proceed through the massive subventral nerve trunks (svn). The posterior and median externolateral ganglia connect to the ventral nerve through the minor lateroventral commissures (milvc). The last of the ganglia associated with the nerve ring is the bilobed ventral ganglion (vg) containing 16 to 33 neurons.

2. Ventral Nerve

From the ventral side of the nerve ring arise two subventral nerve trunks uniting to form the ventral nerve (vn), which is a chain of ganglia. The most anterior and largest of the ventral nerve ganglia is the retrovesicular ganglion (rvg) located posteriad to where the minor lateroventral commissures enter the ventral nerve. This ganglion is reportedly an association center consisting of about 12 bipolar neurons; in some forms the postretrovesicular ganglion merges with the retrovesicular ganglion, in which case the ganglion contains about 20 neurons. The ventral nerve continues posteriorly passing to the right of the excretory pore (when present) and to the right of the vulva. Though the neurocytes in the vulvar region are few and scattered, some authors consider these loose aggregations as prevulvar and postvulvar ganglia.

In the rectal region the ventral nerve gives off paired rectal commissures (rc), which unite dorsal to the rectum (Fig. 4.20). Laterally they contain paired laterorectal ganglia (lrg). The dorsorectal ganglion (drg) is located where the commissures unite dorsal to the rectum. From this

Figure 4.20. Central nervous system, caudal region. Abbreviations: dn, dorsal nerve; dllc, dorsolateral lumbar commissure; drg, dorsorectal ganglion; lrg, laterorectal ganglion; mcn, median caudal nerve; ln, lateral nerve; rc, rectal commissure; alg, anolumbar ganglion; lcn, laterocaudal nerve; alc, anolumbar commissure; rc, rectal commissure; gpc, genitopapillary commissure; vn, ventral nerve (original).

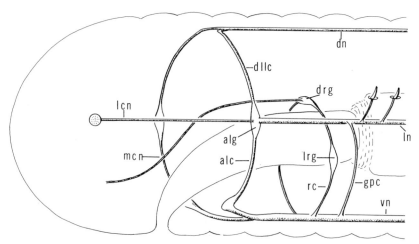

dorsorectal ganglion extends the median caudal nerve (mcn). Because the commissures and ganglia are extensions of the ventral nerve the system is sometimes referred to as the rectal sympathetic system. Posterior to the emergence of the rectal commissures from the ventral nerve are the most posteriad ganglia, the bifid preanal, anolumbar ganglia (alg), from which extend the anolumbar commissures (alc) that extend to the lateral nerves in the region of the lumbar ganglia from which the laterocaudal nerves (lcn) proceed posteriad in the tail. These lumbar ganglia are connected to the dorsal nerve (dn) by the paired dorsolateral lumbar commissures (dllc).

B. Peripheral Nervous System

The peripheral nervous system includes the cephalic papillary nerves, the amphidial nerves, the genital papillary nerves, the nerves innervating deirids, and the somatic nerves.

1. Cephalic Papillary Nerves

The six cephalic papillary nerves have their sensory ending (dendritic fibers) in the 16 cephalic sensilla. In ancestral forms the sensilla are in three whorls (see Chapter 3, Section III), an internal circumoral circlet of six sensilla and two posterior or external circlets of six and four sensilla. In derived forms there are usually two whorls, the second being formed from the combined external circlets on the lips. The circumoral circlet may or may not be present, but the external circlet of 10 sensilla is almost always present. In some instances there may be fusing of the sensilla, but in any event, the pattern of innervation from the cephalic papillary nerves remains remarkably constant. Even when the sensillum is absent the nerve ending of the ancestral condition remains (Fig. 3.18A). The six cephalic papillary nerves, two subdorsal, two lateral, and two subventral, have direct axonic connections with the nerve ring (Fig. 4.19). The dendrites of the subdorsal papillary nerves terminate in the dorsodorsal and laterodorsal sensilla of the external circle and in the internodorsal sensilla of the internal circle (circumoral) (Fig. 3.14). The sensory endings of the subventral papillary nerve end in the ventroventral and lateroventral sensilla of the external circle and the internoventral sensilla of the internal circle (circumoral). The sensory endings of the subventral papillary nerves end in the ventroventral and lateroventral sensilla of the external circle and the internoventral sensilla of the internal circle. The lateral papillary nerves terminate anteriad in the ventrolateral sensilla of the external circle and internolateral sensilla of the internal circle. The lateral papillary nerves also contain the nerves that connect the central nervous system with the esophago-sympathetic system.

2..Amphidial Nerves

The amphidial nerves (an) from the anterior chemoreceptors have an indirect connection with the nerve ring. The 10 or so bipolar neurocytes of the amphidial nerves are clustered posterior to the nerve ring in the amphidial ganglia located in the paired large lateral ganglia. These bipolar cell bodies are much larger than the bipolar cell bodies of the papillary nerves. The dendritic nerve fibers from the amphids to the amphidial ganglia run in close proximity to the lateral papillary nerves, but are easily distinguished by their greater diameter. The axonic fibers join into a bundle near the center of each respective lateral (amphidial) ganglion and emerge in the lateroventral commissures; near the ventral line they turn inward through the ventral nerve trunks to the ventral ganglion where they form synaptic contacts.

3. Genital Papillary Nerves

Paired preanal papillae are innervated by bipolar neurons located in the ventrolateral nerve (Fig. 4.20). The dendritic fibers from these papillae join the ventrolateral nerves from the sensory neurocyte; the axionic process proceeds to the ventral nerve via the genito-papillary commissure (gpc). Ventromedian papillae are innervated by bipolar neurons connected directly to the posterior end of the ventral nerve. Postanal or caudal papillary nerves have neurocytes in the lumbar ganglia, and all the postanal sensory organs, including the phasmids of Secernentea, are innervated by processes from the laterocaudal nerves (Fig. 4.20).

4. Deirids

These paired papillae in some Secernentea and Chromadoria are commonly called **cervical papillae**. The sensory cell of each is located in the medial externolateral ganglia (Fig. 4.19). These lateral sensory sensilla are located anteriorly, usually on the so-called neck region.

5. Somatic Nerves

The subsidiary nerves extending from the nerve ring, transversing in the lateral and dorsal hypodermal chords, are referred to as the somatic nerves. From the dorsal ganglion of the nerve ring extends the dorsal nerve (dn). It proceeds posteriad to the postanal region where it bifurcates forming the dorsolateral lumbar commissures, which join the dorsal nerve with the lumbar ganglia (Fig. 4.20). The dorsal nerve is without ganglia along its length, except among Spiruria where neurocytes have been reported. The dorsal nerve is considered to be motor because it has direct connection with innervation processes from the somatic muscles.

Two pairs of nerves, the laterodorsals (ldn) and lateroventrals (lvn), have their origin from corresponding regions of the nerve ring where the neurocytes are located (Fig. 4.19). From the nerve ring the laterodorsal nerve proceeds through the hypodermis to the submedian lateral hypodermal chord region where it assumes a longitudinal posteriad course. The lateroventral nerve takes an indirect route before joining the lateral nerves posteriad. The dendrite proceeds from the nerve ring to the neurocyte (bipolar) in the ventral ganglion. The axonic fibers proceed to the ventral nerve by way of the major lateroventral commissures (Fig. 4.19). The lateroventral nerve emerges from the ventral nerve and extends through the hypodermis to the lateral hypodermal chords; from here they extend posteriad as did the laterodorsal nerves.

The mediolateral nerve (mln) consists of the nerves extending from the median and posterior externolateral ganglia and the posterior internolateral ganglia (Fig. 4.19). They proceed posteriad to the lumbar ganglia (Fig. 4.20). The paired lateral caudal nerves continue posteriad from these ganglia. The mediolateral lateral nerve is both sensory and motor in function.

6. Hemizonid

Along the length of the nematode body occur several nerve commissures that connect the longitudinal nerves of the body to each other. Some commissures are symmetrical and connect the lateral nerves with the main ventral nerve; others are asymmetrical and are limited to one side of the body. These asymmetrical commissures may connect the dorsal nerve with the lateral nerves or the lateral and ventral nerves. The best-recognized asymmetrical commissure occurs in the anterior body and connects the dorsal and ventral regions of the nerve ring (Fig. 4.19).

Some of these commissures are prominent in totomounts and appear as refractive bodies just under the cuticle dorsally and ventrally. Their prominence has been given taxonomic significance in several nema groups. In these groups they have been called **cephalids** and **hemicephalids**, and most commonly recognized is the **hemizonid** which generally occurs near the excretory pore in Secernentea. The hemizonid is formed by the subventral commissures that connect the nerve ring to the ventral nerve cord. In longitudinal section the nerves of the commissure are seen to lie just under the cuticle and separated from it by the hypodermis (Fig. 4.21).

C. Peripheral Nerve Net

Among many forms of Adenophorea in addition to the central and peripheral nervous system is a peripheral nerve net (Fig. 4.22A). The system is most easily demonstrated among marine Enoplia and Chromadoria. The neural net forms a meshwork from anterior to posterior. A notable feature

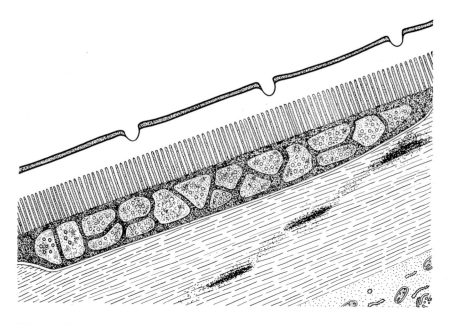

Figure 4.21. Longitudinal section through the hemizonid (drawn from a TEM photograph of Smith).

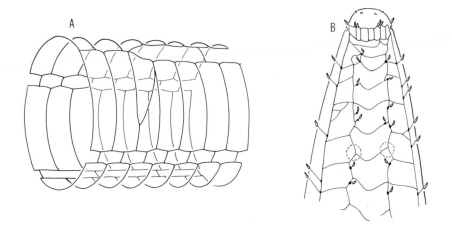

Figure 4.22. Peripheral nerve net in *Deontostoma.* A. Three-dimensional reconstruction of the nerve net in the neck region (redrawn from Croll and Maggenti). B. Ventral view, cephalic and cervical plexus.

is the linking of the somatic setae and papillae, which, in turn, are individually connected to sensory neurons of the peripheral nervous system. The transverse commissures of the net occur at rather constant intervals of 25–35 μm. Anteriorly the system forms a radially symmetrical plexus (Fig.

4.22B); in the vulval region of the female and in the caudal region of males the system forms zygomorphic plexuses. It is postulated that the network coordinates impulses from seta to seta and thence to the central nervous system, thus compensating for the paucity of sensory organs over the body.

In Secernentea a network has not been described; however, free nerve endings are known to occur under the cuticle at variable points along the body. Whether these free endings represent a modification or a compensation for a nerve net is unknown. In any event, they still could function to provide nemas with awareness of their external environment.

Chapter 5
Reproductive System

I. Introduction

The nematode's reproductive organs, in both sexes, are always tubular and may consist of one, two or multiple tubules (gonads) (Figs. 5.1–5.3, 5.6). The system is a complex of tissues derived from mesoderm and ectoderm. The primary constituent of the tubules and the germ cells is mesoderm. Ectoderm forms the secondary structures produced from invaginations of the body wall. The mesodermal parts of the genitalia house the germ cells, and provide for their nutrition and development. The ectodermal structures aid in the union between sexes and in the transfer of sperm or deposition of eggs.

In the majority of species the males and females are very similar in appearance except for the characters of the reproductive system and size, females generally being somewhat larger than males. There are instances of marked sexual dimorphism in both free-living and parasitic nematodes. **Sexual dimorphism** takes several forms; the mature female may become swollen and saclike while the male remains vermiform. Both sexes may remain vermiform, but the male lacks any form of feeding apparatus; or the male may remain normal, but the female degenerates to a reproductive sac by prolapse of the reproductive organs, which then grow independent of the female. Less spectacular forms of dimorphism may be seen in number, size, and form of body spines or in amphid size.

Nematode species, with few exceptions, have both males and females. In some species only the female is known and then reproduction is generally accomplished by parthenogenesis, which may be either meiotic or mitotic. Hermaphroditic individuals are rare and many of the reported cases presume hermaphroditism because of a lack of males. In only a few cases has hermaphroditism been verified by the observation of both male and

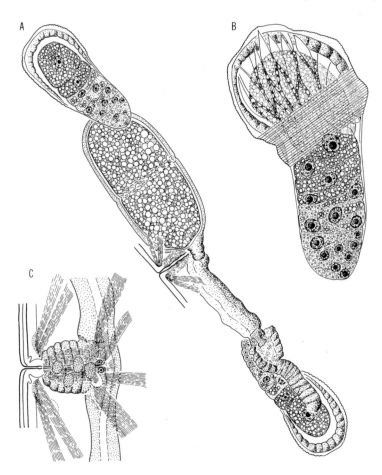

Figure 5.1. Female reproduction system in Dorylaimida:*Discolaimus.* A. Complete reproduction system. B. Ovary showing spindle-shaped epithelial cells and striated connectivelike tissue. C. Vulva and vagina showing associated musculature (modified from Coomans).

female organs, *Heterogonema ovomasculis* (Fig. 7.7) or the presence of male and female gametes (*Caenorhabditis?*).

There are many records of abnormal individuals that as intersexes exhibit characteristics of both males and females. These are not hermaphrodites, but generally are abnormal females. Usually the female reproductive system is complete and functional, but posteriorly the female may have spicules. Rarely, is the intersex a male with female characteristics. This condition should not be confused with sex reversal such as occurs among a number of species of root-knot nematodes (*Meloidogyne*). Second stage larval females of root-knot nematode under environmental stress reverse

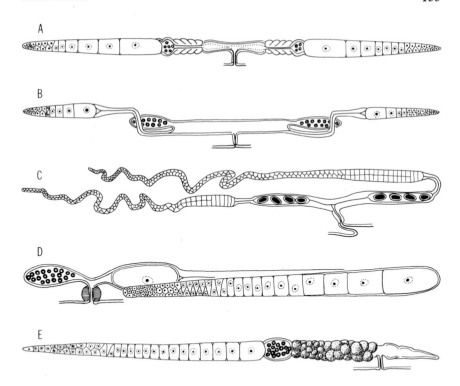

Figure 5.2. Variations in amphidelphic female reproductive systems. A. Ovaries outstretched. B. Specialized spermatheca between oviduct and uterus. C. Anterior ovary outstretched with flexures; posterior ovary reflexed and anteriorly directed, also with flexures. D. Postpudendum reproductive system. E. Antepudendum reproductive system (A–E redrawn from Chitwood).

sex and develop into males. These males are recognized by the possession of two testes, an extremely rare occurrence among Secernentea.

Most nematodes are **amphigonus** (males and females separate) and **oviparous (exotokia),** i.e., the eggs develop outside the body of the nematode. Eggs are generally fertilized when passing by the sperm receptacle, which may be at the posterior end of the ovary (Fig. 5.2A), or at the junction of the oviduct and uterus (Fig. 5.2B). Spermatozoa may be stored in a **spermatheca** or other receptacle at the time of mating; however, many nemas mate several times and may or may not store sperm in the reproductive tract. Monovarial forms often use the vestigial uterine sac of the degenerated second gonad as a seminal receptacle (Fig. 5.2D).

Though **exotoky** is the common phenomenon among nematodes, **endotoky** (egg development within the body) does occur. The type of development within the females varies such that both viviparous and ovoviviparous are used to describe variable development. **Viviparous** is

when a well-defined egg shell is not seen during embryological development and the larvae are retained in the uterus until deposited. This occurs
in two important parasites of humans: trichinosis and the guinea worm.
Ovoviviparity generally occurs among normal oviparous species and is
often associated with senility of the female. When these eggs hatch within
the body of the female, the released larvae often feed on the decaying
carcass of the mother. It is incorrect to term this phenomenon *endotokia
matricida* as is commonly done. This term was originally proposed by
Seurat to denote the retention of eggs inside the female body as occurs in
Heteroderidae and some Oxyuridae.

A. Hermaphroditism or Self-Fertilization

There are many reports, most unverified, of **hermaphroditism** among
nematodes. Most such reports are based on the presence of a spermatheca
in the female reproductive system when males are unknown in the species.
Such reasoning is unacceptable because when investigated in depth, two
factors generally become apparent. Males may not be completely lacking,
but rare and often under proper conditions abundant. Second, reproduction in such reported hermaphrodites turns out to be parthenogenetic. The
best evidence of hermaphroditism is with the mermithid insect parasite
Heterogonema ovomascularis, where males produce not only sperm, but
viable eggs. Females in this species are normal and egg development occurs
by cross-fertilization with the hermaphroditic males (Fig. 7.7).

The other observed instance of hemaphroditism is reported among
rhabdits in the genus *Caenorhabditis*. In these the female is the hermaphro-
dite and allegedly produces both male and female gametes in the same
gonad, but this report needs verification.

B. Sex Determination

The most common mechanism for sex determination ascribed for nematodes is that of an XO-XX mechanism. This is particularly true of Secernentea as few studies have been conducted in Adenophorea. In this
mechanism the male has at least one less chromosome than the female and
therefore produces two types of gametes. However, the XY-XX mechanism
is also reported among free-living soil nematodes. In many nemas sex
chromosomes are difficult to distinguish and it is presumed that they must
resemble autosomes on morphological and behavioral bases during nuclear
division.

Epigenetic factors may also control the production of males. Crowding is
one such factor and is seen among mermithid parasites of insects. When
1–3 parasites are present in the insect, only female nematodes are pro-

duced; when 4–14 parasites are present, both males and females develop; when more than 14 nemas are present in the insect, only males are produced. Temperature sometimes influences the production of males. When temperature regimes are changed *Caenorhabditis* produce males, which otherwise may never be seen. Root-knot nematodes, especially *M. incognita*, produce few males when environmental conditions are favorable. However, when crowded or where there is a deterioration of the root system high populations of males are produced; these are primarily produced by sex reversal of "female" larvae.

Triantaphyllou reports that in root-knot nematodes, males are produced from fertilized eggs and females from unfertilized eggs. This report leaves a question as to where the male comes from if fertilized eggs are required, since many populations proceed for several generations without males. Males are generally few in number, short-lived, and would be unlikely to sustain themselves or function over long periods. It may be that fertilization is first supplied by sex reversed females; this should be investigated.

A mechanism not reported in Nemata that should be investigated is one that occurs among invertebrates below and above nematodes in the evolutionary scale, for example, rotifers and insects. One mechanism available to these animals is the opposite to that reported for *Meloidogyne*, i.e., in certain insects (honeybee) and rotifers, males are the product of unfertilized haploid eggs (arrhenotoky) and females are produced by mitotic parthenogenesis or cross-fertilization with haploid males. The possibility of such mechanisms being operative among nematodes warrants consideration.

II. Female Reproductive System

The essential parts of the female system are the tubular ovary, oviduct, uterus, and vagina, which opens to the exterior through the vulva or gonopore (Fig. 5.1). In addition to these basic parts, there may be a seminal receptacle (spermatheca) or accessory glands that contribute to shell formation or produce mucoids to either adhere the eggs together or provide protection against water loss.

A. Terminology

The variability in female reproductive organs, i.e., their number, and direction motivated Seurat over a period of years, 1913–1920, to propose a nomenclature for the types observed. In recent years some of the terms and definitions of Seurat have been misused. In order to clarify the situation the original nomenclature and definitions of Seurat are presented here:

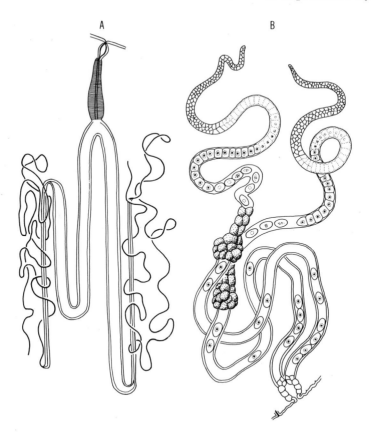

Figure 5.3. A. Opisthodelphic reproductive system, *Ascaris*. B. Prodelphic reproductive system, *Meloidogyne* (A modified from Chitwood; B modified from Hirschmann).

Amphidelphic: uteri opposed (Figs. 5.1, 5.2)
Opisthodelphic: uteri parallel and posteriorly directed (Fig. 5.3A)
Prodelphic: uteri parallel and anteriorly directed (Fig. 5.3B)

Of prime importance to this nomenclature is the position and direction of the uteri and not the ovary. In recent years these terms have incorrectly been used as synonymous with the ovary. As such prodelphic has been used to mean one ovary anteriorly directed and opisthodelphic to describe one ovary posteriorly directed. Seurat wisely did not apply terms to the ovary because of the many varieties of form and direction that occur. For example:

1. Ovaries outstretched (Fig. 5.2A)
2. Ovaries reflexed (Fig. 5.1)

3. Ovary with flexures (Fig. 5.2C)
4. Ovaries reflexed and with additional flexures (Fig. 5.2C)
5. Anterior ovary outstretched anteriorly; posterior ovary reflexed and also directed anteriorly (Fig. 5.2C)
6. Anterior ovary absent (Fig. 5.2D)
7. Posterior ovary absent (Fig. 5.2E)

Most nematodes, whether they possess one ovary (**monovarial**), two ovaries (**diovarial**), or several ovaries (**polyovarial**) are amphidelphic. Generally among monovarial forms the second reduced or vestigial uterus is still evident and is called the postuterine or prouterine sac, depending on whether it is either a remnant of the posterior or anterior genital tube (Fig. 5.2D, E). Species that lack the vestigial uterus are still considered amphidelphic, because closely related species show the ancestral amphidelphic condition. Furthermore, the prodelphic and opisthodelphic conditions are limited to a few parasitic groups, for example, among animal parasites *Ascaris* is opisthodelphic (Fig. 5.3A), and among the plant parasites the species of Heteroderidae are prodelphic (Fig. 5.3B).

Two common conditions among amphidelphic forms remain without adequate terminology: (1) one genital tube anteriorly directed and (2) one genital tube posteriorly directed. We cannot arbitrarily use Seurat's "prodelphic" and "opisthodelphic" designation, because they, by definition, are limited to parallel uteri. It would be unwise to base the nomenclature on the ovary because it may be with flexures, reflexed or outstretched. Therefore, the terminology proposed is dependent on the main direction the genital tube follows away from the vulva. Two terms are being proposed for these conditions that are so common among both free-living and parasitic nematodes. Both terms are to be applied to ancestrally amphidelphic species:

Antepudendum: meaning the genital tube proceeds anterior from the vulva (Fig. 5.2E)
Postpudendum: Meaning the genital tube proceeds posteriorly from the vulva (Fig. 5.2D)

Four other terms are applied to the female's genital tube and again they make reference to the uteri: **monodelphic**, one uterus; **didelphic**, two uteri; **tetradelphic**, four uteri; and **polydelphic**, more than four uteri.

B. Ovary

The ovary is that portion of the reproductive tract that contains the germinal cells covered by an epithelial sac. The outer epithelial sheath is mesodermal in origin. The sheath consists of a single layer of elongate, flat, spindle-shaped cells (Fig. 5.1B). At the blind end the ovary is terminated by a large cap or apical cell. Presumably this cell forms the epithelial sheath covering the ovarial terminus. This cell is not to be confused with the underlying terminal proliferative cell that gives rise to the germinal cells.

The ovary can be divided into two regions: (1) the germinal zone and (2) the growth zone. The **germinal zone** is always relatively short and is the region of rapid division. The cells contained by this zone are relatively small. In most nematodes the **growth zone** constitutes the greater part of the ovary. This is especially true among parasitic forms, but is by no means limited to them.

Two types of ovary are reported among Nemata: **telogonic** and **hologonic**. Hologonic ovaries are restricted to a few parasitic groups including the Trichuroidea and Dioctophymatoidea. In a hologonic ovary the germinal cells are proliferated from a series of germinal areas extending the length of the ovary either on one or both sides. This should be confirmed with modern techniques. The majority of nematodes have telogonic ovaries in which new germ cells originate at the blind end from one cell.

In female ascarids the germ cells are aggregated around a central protoplasmic core called the **rachis** (Fig. 5.4A). It has been shown that the developing oocytes are connected to the rachis by cytoplasmic bridges containing microtubules (Fig. 5.4B). The function of the rachis remains unclear. It is speculated that it may have some nutritional function. The rachis is also reported in some oxyurids.

When the ovary is turned back at the junction with the oviduct the gonad is described as reflexed (Figs. 5.1A, 5.2D); bends within the length of the ovary itself are called flexures (Fig. 5.2C).

Figure 5.4. A. Transverse section of *Ascaris* ovary showing sector of rachis in the growth zone (modified from Hirschmann). B. Oocyte connection to the rachis by a cytoplasmic bridge containing microtubules (original, drawn from a TEM photograph by Bird).

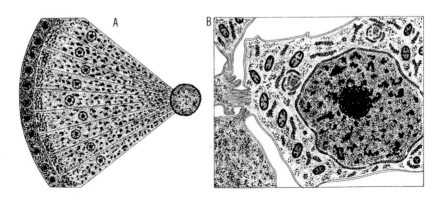

C. Oviduct

When distinguishable, the oviduct consists of a narrow tube of high columnar epithelium (Figs. 5.1A, 5.2). There may or may not be a sphincter muscle at the junction of the ovary and oviduct. Some nematodes appear to have a muscle sheath covering the oviduct. It is not uncommon to have the region nearest the ovary converted to a seminal receptacle or spermatheca (Figs. 5.1A, 5.2A, E). In oxyurids the oviduct has the function of laying down the eggshell. Normally in Nemata eggshell deposition occurs in the proximal region of the uterus.

D. Uterus

The uterus is the most complex region within the reproductive tract. For the most part the uterus has a squamous epithelium covered by circular and oblique muscles, which vary in degree of development within groups as well as between widely divergent groups. The proximal region often is composed of tall columnar secretory cells known as the **crustaformeria** or **quadricolumella**. These cells are believed to secrete the outermost egg membrane known as the uterine layer. The distal region of uterus often acts as a seminal receptacle and may be so modified as to form an external pouch. Specialized spermathecal pouches occur in several nematodes (Fig. 5.2B). In *Ascaris* the cells lining the seminal receptacle region are tufted along their luminal side. These cells have been reported to be phagocytic on unused or aging spermatozoa.

Uterine musculature acts to move the egg along through peristaltic waves. When an ovijector is formed in the uterus, the epithelium becomes thicker and the musculature more highly developed.

E. Vagina

Unlike the other portions of the reproductive tract, which are germinal or mesodermal, the vagina is of ectodermal origin (Fig. 5.1C). The **vagina vera** is always recognized by its cuticular lining, which is continuous with the external body cuticle (Fig. 5.1C). The cuticular lining seems most closely related to epi and exocuticle. In some nemas there is a noncuticularly lined tube, similar in construction, interjected between the vagina and the uterus. This nonectodermal tube is called the **vagina uterina** (Fig. 5.3A).

The vagina opens to the exterior through a slitlike vulva. The vagina in only two instances is reported not to open through the vulva, but through a cloaca, like the males. This occurs in *Lauratonema* and *Rondonia*.

Vaginal musculature is continuous with that of the uterus, but is often more greatly developed. Because the muscle layer may be several layers thick, the vagina often has a laminated appearance (Fig. 5.1C). The non-contractile sarcoplasmic portions of the muscles commonly assume a bladderlike shape and have erroneously been interpreted as glands. In several free-living nemas there is, in addition to the normal vaginal musculature, a large sphincter muscle located near the vulva.

The vulva is provided with its own musculature. Two types are known: (1) *dilator vulvae* and (2) *constrictor vulvae*. It is possible that the *constrictor vulvae* is actually the large sphincter of the vagina.

The *dilator vulvae* muscles are better understood. The placement of the muscle cell bodies and their nuclei at the body wall indicates that the dilators are modified from cells of ordinary somatic muscles and are not specialized muscles such as those associated with the uterus and vagina. Generally there are four fiber bands anterior to the vulva and four bands posterior to the vulva.

F. Comparative Morphology

Free-living species of Adenophorea and Secernentea possessing two ovaries generally have the transverse vulva near equatorial. When only one genital tube is present, then the vulva is shifted anteriorly or posteriorly in Adenophorea, whereas in Secernentea it generally remains either equatorial (cephalobs) or is shifted posteriorly. Among parasitic Secernentea the vulva may be far anterior (spirurids) or it may be terminal (Heteroderidae).

The most significant contrast between free-living Adenophorea and Secernentea is the general tendency among adenophores to produce fewer eggs. This is evident in the shortened growth zone in which there are generally fewer than 20 developing oocytes (Fig. 5.1B). There are exceptions among the monhysterids (Chromadoria), which have a long outstretched ovary with a lengthy area for oocyte development. These forms usually have in excess of 25 developing oocytes. Most Secernentea have elongated ovaries, which may or may not be reflexed. In these females there may be more than 100 developing oocytes (Fig. 5.2E).

The plant parasitic species in both classes of Nemata show little or no differences from free-living species. Animal parasites, however, often show great complicity of the ovary in size, number, and convolution. The ovaries of *Ascaris* if layed out would measure around one meter; the sperm whale parasite, *Placentanema gigantissima*, has 32 ovaries which can produce nearly 80 $\times$ 10^6 eggs. These ovarial variabilities receive further attention in Chapters 7 and 8.

In *Xiphinema* special attention has been given to a muscular region of the uterus located near the eggshell glands. This area is called the Z organ. It appears to be no more than a muscular region to aid eggs on their way down the uterus. The so-called Z particles sometimes seen in the region appear to be artifacts, having no relationship to the muscular region itself.

With the female reproductive system of some species of Oncholaimidae there occurs a most unusual system of organs known as "**Demanian vessels**." This is a complex system, variable in form (Fig. 5.5), but generally consisting of a complicated system of double efferent tubes, which connect the posterior intestine through an **osmosium** with the uterus or uteri. These "tubes" are confluent at a glandular region called the **uvette**. From the uvette the ducts lead to one or more exit pores in the body wall. The function of this unique system among Nemata is unestablished. However, Rachor's work would indicate that the Demanian system is a complex seminal receptacle. The observation that sperm is present within the system was made some 60 years ago by Filipjev. It is proposed by Rachor that the osmosium transports nutrients from the intestine for the nourishment of the sperm. The osmosium is a combination of modified intestinal tissue

Figure 5.5. Diagrams of the basic types of Demanian systems in Oncholaimina. A. *Viscosia*. B. *Meyersia*. C. *Adoncholaimus*. D. *Metoncholaimus*. E. *Oncholaimus* (redrawn from Rachor).

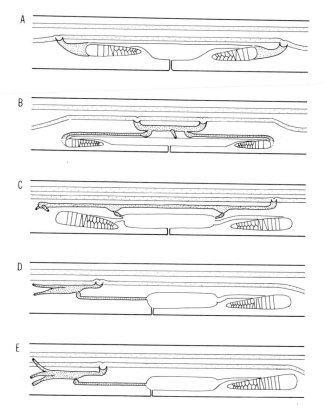

that protrudes into the tissue of the Demanian organ, which is of gonadal origin. No explanation is offered for the efferent tubes that open through the body wall.

III. Male Reproductive System

A. General Morphology

The general morphology of the male reproductive system is divided into three or four regions. In some the vas efferens is not recognized, so only three basic parts (testis, seminal vesicle, and vas deferens) are seen. The testis may be either **hologonic** or **telogonic**. The latter, being by far the more common, is considered the basic form. Hologonic testes are only known to occur in some parasitic groups as are listed for the female ovary.

The common telogonic testis is covered by an epithelium continuous with that of the seminal vesicle (or vas deferens), similar to that which in females covers the ovary, and is continuous with that of the oviduct. The germ tissue within the epithelial sac is subdivisible into a germinal zone and growth zone. In Adenophorea the cells in the growth zone are first seen as a four cell cord that soon changes into a double row, and finally a single chain of cells. In Secernentea the growth zone is first in the form of a cord of six or more cells attached centrally to a protoplasmic rachis. In some nemas there is a distinguishable region between the testis and the seminal vesicle called the vas efferens. When present, it is recognized by its high cuboidal or columnar epithelium. Generally following the testis is the seminal vesicle, which is the storage region for sperm.

Fortunately, no confusion comparable to females surrounds the nomenclature of the male reproductive system. In general, adenophorean males have two testes **(diorchic)** that are opposed (Fig. 5.6A, B); only rarely are they both anteriorly directed as in the genus *Anticoma* (Enoplia: Leptosomatidae). One testis **(monorchic)** is the normal condition among Secernentea (Fig. 5.6C). The most widely recognized exception occurs among those species of Heteroderidae that have undergone sex reversal. These sex reversed males have paired anteriorly directed testes, and if named the condition is called **proorchic**.

In addition to the presence of one or two testes, there are other characteristics that generally separate the hierarchy groups among Nemata. Enoplia, so far as known, have heavy and extensively muscled ejaculatory duct extending to the seminal vesicle (Fig. 5.6A). Chromadoria characteristically have greatly reduced musculature associated with the ductus ejaculatoris (Fig. 5.6B). They are, therefore, similar to Secernentea, which have almost no muscle around the ejaculatory duct (Fig. 5.6C).

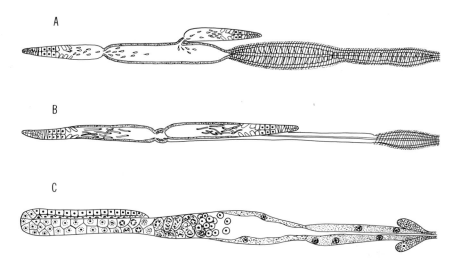

Figure 5.6. Male reproductive systems. A. Enoplia:*Enoplus*; seminal vesicle and vas deferens heavily muscled. B. Chromadoria:*Tobrilus*; musculature limited vas deferens. C. Secernentea:*Rhabditis*; musculature lacking but gland present (A–C redrawn from Chitwood).

B. Secondary Male Sex Organs

Included as secondary sex organs are the cloaca, spicular pouch, spicules, gubernaculum, copulatory muscles, and supplements.

1. Cloaca

The cloaca in males is characteristic of Nemata and is one character that separates them from related groups. Only two females are known to possess a cloaca, and these are to be considered unique situations of no phylogenetic significance to nematodes (see Section II). In all known nemas, save the Trichuroidea, the cloaca is formed by the ventral confluence of the vas deferens with the rectum. In Trichuroidea the vas deferens enters the rectum dorsolaterally. There is a slight difference in the form of the cloaca between Adenophorea and Secernentea. In adenophores the vas deferens joins the rectum posteriad so that both a cloaca and a distinguishable rectum exist (Fig. 5.7A). This may be an indication that in nematode progenitors both males and females possessed separate gonopores. Secernentean cloacas are formed by the entrance of the vas deferens into the hindgut either at or just posterior to the intestino-rectal value; thus no distinguishable rectum exists (Fig. 5.7B).

The cloaca is a complex structure with pouches that accommodates the

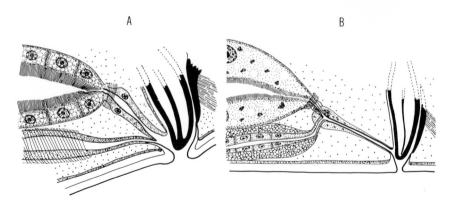

Figure 5.7. Male cloaca, structure difference between Adenophorea and Secernentea. A. Adenophorea; vas deferens and rectum join to form cloaca near anal opening. Therefore, rectum still intact. B. Secernentea; vas deferens joins alimentary canal near terminus of intestine; rectum not distinct (original).

spicules and other accessory structures like the gubernaculum, and in strongyles the telamon. A group of cells called the spicula primordia form the spicular pouch and spicules. These cells are located on the dorsal wall of the cloaca. The spicula primordia undergo proliferation that is both a quasi-evagination to form the pouch and an invagination that forms the spicules. The pouch is lined with cuticle that is continuous with that of the cloaca. The cuticular spicules are invaginated into the spicular pouch. In most nematodes the spicules do not lie within the cloaca; they are extruded from the spicular pouch through the cloaco-spicular orifice, which is normally very near the anal opening. Exceptions to this occur in strongyles, trichuroids, and dioctophymatids.

2. Spicules

The spicules are not solid, flat, bladelike structures as is often supposed. Each spicule is crescentric or tubelike with a central cytoplasmic core (Fig. 5.8C). Recent studies have shown that dendritic nerve processes also extend the length of the spicules in some nemas. How widely throughout Nemata this innervation of the spicule occurs is unknown; but that they have a sensory awareness should not be surprising. Where described, the dendritic endings lie in a terminal pore, and therefore, the structure of the sense organ is in the nature of a sensilla coeloconica.

The basic number of spicules is two; however, many genera and species are known which either have one, or in a few instances no spicules. Among parasites, the distal tips of the spicules may be fused.

Spicules may be divided into three regions: the **head** (capitulum or manubrium), **shaft** (calomus), and **blade** (lamina) (Fig. 5.8A). Some-

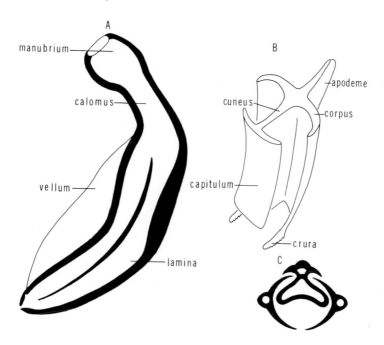

Figure 5.8. A. Spicule. B. Gubernaculum. C. Transverse section showing relationship of spicules to gubernaculum (original).

times the blade ventrally has membraneous winglike extensions called the **vellum** (Fig. 5.8A). A pair of muscles, the retractor and the protractor, are associated with each spicule. Both muscles are attached to the proximal end of the spicule. The retractors extend anteriad from the proximal depression in the manubrium to the body wall. The origin of the muscle on the body wall varies according to species and may be dorsolateral, ventrolateral, or subdorsal. The protractor muscles extend posteriad from the dorsolateral region of the manubrium and are inserted postanally on the body wall.

Terms for the major morphological shapes of spicules encountered among nematodes were proposed by Cobb. These terms are still used in taxonomic keys and are illustrated in Fig. 5.9.

3. Gubernaculum

When present the gubernaculum is presumed to function as a guiding apparatus for the spicules. It is a thickening of the dorsal wall of the spicular pouch in the region between the spicule head and the anal opening. In its simplest form it is merely a thickened plate called the **corpus**; however, it may become in some forms rather complex. When complex, or perhaps more accurately complete, it may have apodemes for muscle

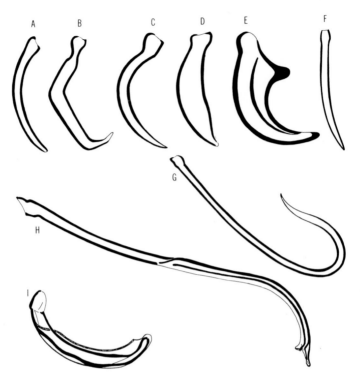

Figure 5.9. Spicular shapes. A. Arcuate. B. Hamate. C. Falcate. D. Fusiform. E. Cuniform. F. Setaceous. G. Sickle-shaped. H–I. Unequal spicules as seen among Spiruria (A–F redrawn from Cobb; G original; H–I redrawn from Maggenti).

attachments or have cuticular extensions between, around, or in front of the spicules (Fig. 5.8B, C). The main platelike portion of the gubernaculum, even in complex forms, is the corpus. Dorsally and posteriad there may be, in complete forms, a moderate to large **apophysis.** A ventral medial extension from the **corpus**, which extends between the spicules is called the **cuneus**. The corpus may also be strengthened laterally by two strongly sclerotized longitudinal pieces called the **crura**. A flanging of the cuneus along its length, like an I beam, is called the capitulum.

A gubernaculum composed of all the parts described above guides the spicules anteriorly, posteriorly, medially, and laterally. This type of gubernaculum can be seen in members of the genus *Cyatholaimus* (Chromadoria).

Musculature is not always associated with the gubernaculum; however, it is almost always present when the gubernaculum is more than a simple dorsal plate. Three types of muscle may be present: (1) retractor gubernaculi, (2) protractor gubernaculi, and (3) seductor gubernaculi. The retractor gubernaculi extend from the gubernaculum to the dorsal surface of the body. Protractor gubernaculi are paired muscles extending from the ventral body wall posterior to the anus anteriad to the gubernaculum. The seductor gubernaculi are also paired and extend from the lateral body wall to the gubernaculum.

4. Telamon

Confusion and misunderstanding have arisen about the telamon, which is a specialized structure found, so far as is known, only among strongyles. The **telamon** is formed from the cloacal pouch and not from the spicular pouch as are the spicules and gubernaculum. The telamon is a thickened immovable plate of the ventral cloacal wall. Its function in strongyles is to turn the spicules posteriorly when they are protruded. In strongyles the structure is necessitated because the spicular pouch orifice is not directly opposite the cloacal (anal) orifice as in other Nemata. What is most often confused with a true telamon is the capitulum of the gubernaculum, i.e., the anterior flange of the cuneus, which may appear in front of the spicules.

IV. Spermatogenesis

The first studies of spermatogenesis in animals were conducted by Van Beneden in the early 1880s. It was in these studies that meiosis, or the reduction division, was first seen and described; the study animal was a nematode, *Parascaris equorum*. For several years after nematodes served as the base material for cytological and genetic studies.

At the distal end of the testis, diploid spermatogonia are produced from the original primordial germ cell. These descendants undergo a series of mitotic divisions until a certain definite number is reached (Fig. 5.10). In this process spermatogonia (by mitosis) give rise to somewhat smaller cells called primary spermatocytes. The primary spermatocytes now proceed to undergo a special kind of nuclear division that results in the production of haploid gametes. The formation of gametes is accomplished by two consecutive reduction divisions in a process called meiosis.

A. Primary Spermatocytes: The First Meiotic Division

1. Prophase I

The reticulum of the primary spermatocyte nucleus resolves itself, in early prophase, into the diploid number. At this time the chromatin material is threadlike and may already be in the process of doubling. These slender threads now come to lie side by side, forming a monoploid set of conju-

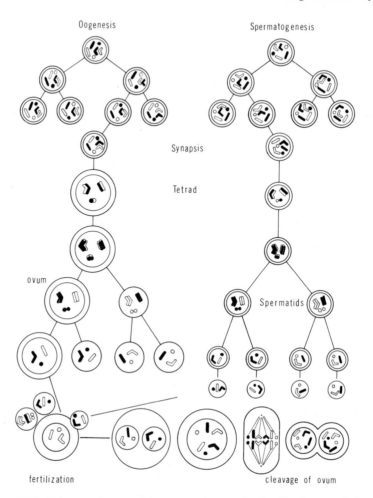

Figure 5.10. Scheme of oogenesis, spermatogenesis, fertilization, and cleavage (redrawn from Huettner).

gating pairs of homologous male and female derived chromosomes. This process of pairing is called synapsis.

   The paired threads (chromosomes) shorten and thicken and each thread is now clearly doubled, but held together by a centromere. Thus, four strands are now associated in a tetrad of four chromatids for each pair of original chromosomes. In each tetrad two chromatids separate from the other two so that each chromatid has a pairing partner. The paired chromatids, still in a tetrad, continue to shorten and thicken in a dense matrix. This stage is called diakinesis and is the stage when chromosomes can most easily be counted and characterized.

2. Metaphase I

The tetrads align at the cell's equatorial plate.

3. Anaphase I

The paired chromatids now separate from the other two of the tetrad and these dyads go to opposite poles of the cell.

4. Telophase I

The cell now divides forming two cells, each with its own complement of dyads. Each cell has $1n$ chromosome, but $2n$ chromatids, and each cell is called a secondary spermatocyte.

B. Secondary Spermatocytes: The Second Meiotic Division

Each cell resulting at the end of telophase I now proceeds through the second reduction division.

1. Prophase II

The dyads of the preceding telophase have their chromatids still in association.

2. Metaphase II

The dyads align on the equatorial plate.

3. Anaphase II

The chromatids of each dyad separate from each other and each becomes a complete chromosome.

4. Telophase II

The distinct chromasomes move to opposite poles of the cell. The resulting haploid cells are now called spermatids.

Each primary spermatocyte produces two secondary spermatocytes and each secondary spermatocyte produces two spermatids. A primary spermatocyte thus yields four spermatids. It is apparent from the small size of the spermatid nucleus that its chromatin content is only about one-half that of the primary spermatocyte.

C. Spermatozoa

The spermatids undergo maturation and morphogenesis culminating in the formation of spermatozoa. In nemas the spermatozoa are generally described as nonflagellated, frequently amoeboid cells. Refringent bodies are often described in nema sperm. These were originally thought to represent stored food products; however, because they react as if constituted of ribonucleoproteins, it is now postulated that they may be derived from Golgi bodies.

In shape nema sperm varies from near spheroid to elongated, tail forms with a swollen headlike region. Ascarid sperm are rather cuboidal and are distinguished by peripheral vesicles and a central nucleus surrounded by mitochrondria (Fig. 5.11). Whereas, *Aspicularis* and *Nippostrongylus*

Figure 5.11. Sperm types. A–D. *Aspicularis*. E. *Axonolaimus spinosus*. F. *Enoplus communis*. G. *Ascaris*. H. *Nippostrongylus*. I. *Rhabditis strongyloides* (A–D redrawn from Lee and Anya; E,F,I from Chitwood; G,H from Jamuar).

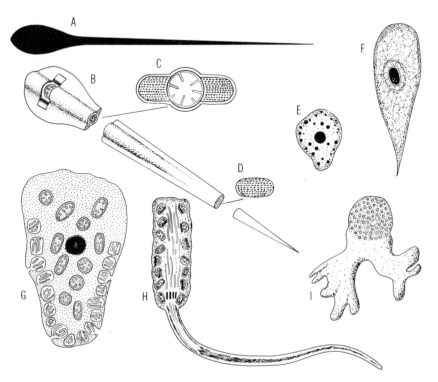

both are pseudoflagellate (Fig. 5.11A–D, H). When pseudoflagellate the dispersed nuclear material seems to be associated with the tail rather than the head of the sperm. There appears to be no relationship between the form or shape of the sperm and taxonomic categories. However, most secernentean plant parasites seem to have spherical to ovoid or ellipsoidal sperm, which are amoeboid where observed. Movement of sperm is not seen until they have been placed in the uterus. When in this position the sperm of *Ascaris* and some plant parasites are seen to move in an amoeboid fashion.

V. Oogenesis

The early stages of oogenesis are similar to those that occur in spermatogenesis. The production of oogonia proceeds from the original germ cell by mitosis and in turn these descendants continue to multiply by mitosis. As these oogonial cells, produced by mitosis, move down the ovary they increase in size (Fig. 5.10).

A. Meiotic Division

1. Metaphase I

After synapsis the tetrads align along the equatorial plate. At the metaphase plate in oogenesis chromosome separation (anaphase) may be either prereductional or postreductional. In prereductional separation, which is common among nemas, homologous pairs separate during the first meiotic division. Postreduction separation is the phenomena of having the homologous pairs separate in the second meiotic division.

2. Anaphase I

The aligned chromosomes (tetrads) undergo longitudinal splitting and the dyads move away from each other.

3. Telophase I

In oogenesis identical cells do not arise as in spermatogenesis. In the female this first reduction division produces an egg nucleus and the first polar body. The latter may again divide forming two polar bodies.

In the second meiotic division the egg nucleus divides again. This division produces the egg pronucleus and a second polar body, which is eventually extruded from the egg. The pronucleus and polar body are both haploid. The pronucleus is now in a condition to receive the sperm pronucleus. The sperm generally enters the egg before the formation of the first polar body.

When the sperm enters the oocyte, the oocyte nucleus moves toward the center of the cell and undergoes maturation division. After the second reduction division the sperm and egg pronuclei fuse. The eggshell is generally present at this time and two polar bodies may be seen outside the vitelline membrane.

4. Types of Reproduction

Among Nemata the most common form of reproduction is that described above, i.e., cross-fertilization or **amphimixis**. The second most common type of reproduction is **parthenogenetic**. Parthenogenesis among nemas both free-living or parasitic may be either meiotic, facultative meiotic, or mitotic.

Meiotic parthenogenesis proceeds after the original mitotic divisions, which produce the primary oocytes. In **meiotic parthenogenesis** synapsis takes place and the reduced chromosome number appears at first maturation prophase. The diploid condition or somatic chromosome number (tetraploids, etc.) is restored by chromosome duplication at anaphase I. This is the only reductional and restoration division; no second maturation division occurs and thus eggs are produced by meiotic parthenogenesis.

In **facultative meiotic parthenogenesis** the diploid chromosomal complement in the reduced oocytes is restored by fusion of the second polar nucleus with the egg pronucleus.

Mitotic parthenogenesis occurs when there is no pairing of homologous chromosomes during prophase. Therefore, the somatic chromosome number is maintained. As such, maturation consists of a single mitotic division resulting in the formation of the first polar body and a diploid egg pronucleus; therefore, fertilization is unnecessary and embryogenesis proceeds normally.

Pseudogamy or pseudofertilization is a modification of parthenogenetic reproduction. It may be the precursor step to the development of parthenogenesis. In pseudogamy the sperm enters the oocyte and activates further development with syngamy. The sperm nucleus degenerates after penetration and further oogenic development takes place by parthenogenesis. This form of reproductive activity is seen in some species of root-knot nematode (*Meloidogyne*).

5. Eggs

Eggshell deposition takes place after the sperm has entered the egg and usually before the first cleavage division. Even when cross-fertilization does not take place, the shell is evident when the egg is in the uterus. The eggs of nematodes are remarkably similar in size, and as a rule range from 50–100 μm in length and 20–50 μm in width. This range in egg size is true for *Placentanema gigantissima* (8 meters long) and *Paratylenchus* spp. (0.3 mm in length). A notable exception to this is *Deontostoma timmerchioi* (Enoplia:Leptosomatidae) from Antarctica. This nematode, measuring some 40 mm in length, has eggs that range from 0.87–1.1 mm by 0.24–0.35 mm. When held up to light, the eggs are easily visible to the naked eye.

There has been some confusion surrounding the nomenclature applied to the various layers of the eggshell. Most confusing is the question of the vitelline membrane. To some it is the inner lipid layer and to others it is the outermost triple-layered membrane produced by the oocyte itself.

Foor in the mid-1960s proposed that the external triple layer (30 mμ in *Meloidogyne* and 0.5 μm in *Ascaris*) be recognized as the vitelline membrane. This membrane seems to be closely related to the epicuticle of the general body. The chitinous layer then becomes a secondary deposition under the vitelline membrane. Thus, it would be similar to the serosa layer of some orthopteran insects. A difficulty that arises with Foor's proposal is the presence of a triple layered unit membrane below the egg shell in *Ascaris* and other nematodes. If a biological unit membrane occurs around the cytoplasm, then the outer triple layer cannot be the vitelline membrane. Until the issue is resolved, this text will follow traditional nomenclature.

The traditional concept of the egg recognizes three layers in the eggshell, and in some free-living and parasitic nemas there may be an additional external fourth layer. The three layers produced by the egg itself are the external triple layer reminiscent of the epicuticle of the body. Immediately below this is the chitinous layer. The chitin found in a nema egg is the only chitin present in nematodes. The next layer is variable and is lipid; some authors call this the vitelline membrane. Underlying all these layers is the unit membrane of the original oocyte. When polar bodies are cast, they are lodged between the vitelline membrane (lipid layer) and the unit membrane of the oocyte.

It is the fourth layer contributed by the uterus that, when present, lends distinctiveness to nematode eggs. Such features as sculpturing, spination, and polar filaments, including the elaborate byssi of mermithids, are allegedly protein secretions of the uterus. The entire subject of egg layers and uterine shells needs study and clarification. The chemical constituents of the eggshell vary in proportions and composition among Nemata, but most contain protein, chitin, carbohydrate, lipid, polyphenols, and assorted minerals.

VI. Embryology

Nemic embryology is truly a study of cell lineage. The development in nematodes is determinate, i.e., each blastomere can be identified in embryogenesis as the stem cell of a particular organ or part of an organ. The fate of each cell in embryogenesis is foreordained.

It is customary to indicate the cleavage cells as "P" or "S" cells. P cells indicate parental germinal cells and S cells are somatic stem cells. The fertilized egg prior to cleavage is designated P_0 (Fig. 5.12).

The first cleavage produces two cells, generally near equal in size. The anterior cell is designated S_1 and is destined to form the greater part of the ectodermal epithelium. The posterior cell is P_1; this cell is less differentiated, but its potential is to form the remainder of the embryo.

From the second cleavage on, each family of cleavage cells (Fig. 5.12) (S_1, S_2, S_3, etc.) has its own cleavage rhythm. The second cleavage is transverse in the S_1 cell, forming daughter cells A and B. Subsequently, P_1 also divides transversely and the resulting cells are S_2 and P_2. S_2 may also be designated as EMSt, because it is destined to give rise to the somatic musculature, part of the esophagus, and the entire intestine (mesenteron). The P_2 cell divides transversely, giving rise to S_3 and P_3. The third somatic stem cell S_3 is destined to form the ectodermal epithelium of the posterior

Figure 5.12. Generalized diagram showing fate of nemic blastomeres during embryology. Abbreviations: S, somatic stem cells; P, parental germinal cells; Ec, ectoderm; End, endoderm; Es, entoderm, EpG, epithelium for germinal cells; PG, germinal cells (after Chitwood).

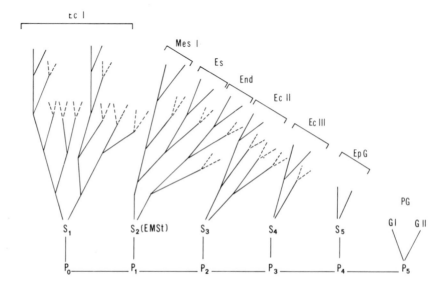

part of the body. This ectoderm is referred to as secondary ectoderm. The next parental cell division results in S_4 and P_4. The line of S_4 descendants form the proctodeum or rectum, and in some species, these cells form the mesoderm or ectoderm of the posterior ventral part of the body. This ectoderm is called tertiary and resulting mesoderm is secondary. The final parental cell cleavage produces cells S_5 and P_5. These are the cells contributing to the genital primordia; the S_5 cells (S_5I and S_5II) are epithelial and the P_5 descendants (GI and GII) are germinal.

Each of the various cleavages described do not proceed simultaneously; as such, there is no regular doubling of the cells. The individual lineages proceed at independent rates, and each lineage contributes to separate portions of the embryo. The S_1 lineage forms most of the body epithelium (hypodermis), which comes to be arranged in dorsal, lateral, and ventral chords. This group also contributes to the formation of the esophagus, the nervous system, and the excretory system when present. The S_2 lineage develops into the major portion of the mesoderm that forms the longitudinal muscles, transverse muscles, and some of the isolation tissues. Furthermore, this same group of cells forms the esophageal musculature and the intestine. The S_3 line forms the posterior body epithelium and contributes to the nervous system and musculature of the posterior body. The rectum and rectal glands are products of the S_4 lineage; some muscular tissue may also be formed. The two S_5 cells and the P_5 cells form the germinal primordia. A theoretical first stage larva would have 1024 cells at maturity.

The above is a history of cell lineage, but tells us little of the process of gastrulation. Gastrulation is the process whereby the endoderm and mesoderm enter into the blastula or placula and become surrounded by ectoderm. A blastula begins to form in some species in the 12–16 cell stage of the embryo; it is at this stage that a blastocoel (cavity) appears within the embryonic cellular mass. In other species a blastocoel is not formed; instead there develops two layered cell plates. This stage is followed by epiboly characteristic of gastrulation; therefore, this type of "blastula" is called a placula.

There are a variety of ways in which the ectoderm may proceed to envelope the endoderm-mesoderm: (1) the endoderm cells may retain their relative positions, **synectic**, or (2) they may not retain their relative positions, **apolytic** (Fig. 5.13). Ectoderm envelopment may be embolic or epibolic. **Emboly** is only possible if there is a blastocoel and the process of invagination, and may proceed synectically or apolytically. **Epiboly** occurs in the absence of a blastocoel and occurs when the ectoderm grows over the endoderm which may or may not retain its relative position. Epibolic-apolytic (change in relative position) gastrulation is unknown among nematodes. All the other listed forms do occur. *Parascaris* undergoes embolic apolytic gastrulation; *Rhabdias* and *Nematoxys* proceed through embolic synectic gastrulation; and *Camallanus* undergoes epibolic synectic gastrulation.

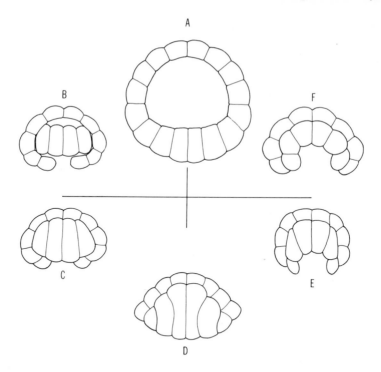

Figure 5.13. Methods of gastrulation. A. Coeloblastula. B. Epibolic-synectic gastrulation as in *Ascaris* and *Parascaris*. C. Epibolic-apolytic gastrulation unknown in Nemata. D. Placula or sterroblastula. E. Epibolic-synectic gastrulation as in *Camallanus*. F. Embolic-synectic gastrulation as in *Rhabdias*. Those above the horizontal line are embolic; those below epibolic; on the left of the vertical they are apolytic, and on the right, synectic (redrawn from Chitwood).

VII. Postembryonic Changes

Subsequent to becoming first stage larvae, nematodes are subject to a series of molts prior to attaining full adulthood. It appears that all nemas undergo four molts. In most Adenophorea all molts occur outside the egg, so that, it is the first stage larva that emerges from the egg after hatching **(eclosion)**. However, most Secernentea emerge as second stage larva since the first molt occurs within the confines of the egg.

The molting process has not been experimentally followed in many instances. Most observations have been made with plant parasites or closely related forms. Few observations are available for Adenophorea. From the observations that have been made, it appears that in those species possessing an epi- and exocuticle, such as most Tylenchida and Aphelenchida, then the exocuticle is dissolved during molting and only the epicuticle is shed (Fig. 3.25). In nemas possessing a more complete complement of cuticular layers, for example, those that have the meso- and endocuticles, then no dissolution is noted, and the entire cuticle is shed (Fig. 3.26). There no longer seems to be much doubt that the hypodermis secretes the new cuticle; however, the source of the molting fluid remains a subject of debate. Some believe that the excretory gland plays an important role in producing the molting fluid. It has been speculated that the hemizonids may play a role in molting, but these are commissures and not likely to function in a neurosecretory fashion.

It has been historically taught that **eutely** (cell constancy) from first larva to adult or adult to adult is characteristic of most nematodes. Of course, the concept would not apply to the reproductive system. It is becoming more and more apparent that the concept is no longer applicable and advocates of the concept must continually retreat to different organ systems. For instance, they must now exclude the reproductive system, hypodermis, intestine, and body musculature. There is little or no difference in the nervous system from first larva to adult with the exception of sensory receptors and genital papillae. The same constancy of structure from larva to adult is seen in the esophagus of nonparasitic forms. However, there are, especially among Adenophorea, great increases in both intestinal cells and muscle cells throughout the growth period. In Secernentea, such as *Ascaris*, there may be, in addition to the increase in muscle cells, a great increase in the individual length of a muscle cell. Even the hypodermis is subject to change and increases with the growth process. Cuticular changes are most often associated with the development of the secondary sexual characters, especially on the male, such as spicules, gubernaculum, and caudal alae (bursae), and the vulva and vagina of females. Other structural changes that occur with change to adulthood are the development of preanal ventral suckers on the males of some ascarids, the greatly enlarged amphids and phasmids on spirurid parasites of annelids, and the increase in ambulatory setae and tubes in Draconematida.

Chapter 6
Plant Parasitism

More than 200 years have passed since Needham in 1743 described the first plant parasitic nematode from wheat galls. Yet only within the twentieth century has agricultural nematology come into its own. However, the full extent of the damage nematodes inflict on plants and crops is still not fully appreciated. Vast areas of nematological research in forestry, range, pasture, native plants, and plant succession remain virtually untouched. No plant or plant part seems to be free from nematode attack. The toll nematodes take on crops in the United States is estimated at $4.5 billion which is only slightly behind that estimated for insects. In reality they may equal or even exceed insects because nematode damage is so insidious it is difficult to measure.

Plant parasitism occurs in both classes of Nemata. Among Adenophorea plant parasitism is recorded only in the order Dorylaimida where it is confined to two families (Trichodoridae and Longidoridae), representing some 200 nominal species. The form of parasitism in Dorylaimida is restricted to below ground ectoparasitism, and it is within this group that all known nematode vectors of plant viruses occur.

It can be safely stated that the greatest array of plant parasites and parasitic habits occurs in Secernentea among the members of the orders Aphelenchida and Tylenchida. In these orders, which also include insect parasites, over 2000 plant parasitic species are recorded, and they occupy every possible niche the plant offers. All parts of the plant, above and below ground, are subject to attack by nematodes, which may be either ectoparasitic or endoparasitic.

Plant parasitic nematodes generally fall into a size grouping similar to nonparasitic soil nematodes (free-living). As such they average about 1 mm. The smallest reported plant parasites are approximately 0.25 mm (Secernentea:Criconematidae) and the largest over 12 mm (Adenophorea: Longidoridae). Among Secernentea the largest known plant parasites are approximately 3 mm in length.

The majority of plant parasitic nemas have a direct life cycle without alternate host or vectors. **Facultative parasites**, that is, those that can subsist either as higher plant parasites or fungal feeders, are found among the species in Tylenchidae and Aphelenchoididae. Some species of plant parasites use insects as **phoretic hosts** (transport host), vectors, or in at least one instance, the insect is actually an alternate host. In the latter case one generation of the nematode is parasitic in the insect and the alternate generation is plant parasitic.

The life cycle differs between plant parasites in Adenophorea and Secernentea in respect to the hatched larva, or, as it is sometimes called, **infective larva**. In Adenophorea it is the first stage larva that emerges from the egg and in Secernentea it is the second stage larva (the first larval stage and molting having occurred within the egg). Therefore, a simple direct life cycle is characteristic of Adenophorea. From size ranges within a population, it has been assumed that *Trichodorus* (Adenophorea) emerges from the egg as a second stage larva; however, this has not been confirmed by direct observation. The size ranges recorded as larval stages could be growth changes between molts. On the other hand, minor complexities are common among Secernentea. In some instances the larvae and males do not feed (*Rotylenchulus*); in others, the larvae are migratory ectoparasites and the adult female is a sessile ectoparasite (*Tylenchulus*); in still other cases only the second stage larva and adult female feed, and the third and fourth stages are sessile and nonfeeding (*Meloidogyne*).

Normally males and females look alike except for their sex organs and secondary sexual characteristics such as the male spicules, bursa, etc. However, in some, a condition known as *sexual dimorphism* occurs in which the males and females differ on gross morphological features. For example, females of root-knot nematode (*Meloidogyne*), citrus nematode (*Tylenchulus*), cyst nematode (*Heterodera*) and reniform nematode (*Rotylenchulus*) are swollen into lemon shapes, pear shapes, globe shapes, or kidney shapes, whereas the males remain wormlike. Another form of sexual dimorphism occurs in burrowing nematode (*Radopholus*): both sexes are wormlike, but the male lacks any feeding apparatus. In still another example among Criconematidae the female is heavily annulated but the male is only lightly so and has no feeding apparatus. Other examples will be discussed under specific diseases.

Much of the damage caused to plants by plant parasitic nemas, in addition to mechanical damage caused by their feeding and migration, is attributed to secretions of the esophageal glands. It is alleged that esophageal gland secretions aid in dissolving intercellular cement, act to break down cell walls, or cause suppression of cell division, cell proliferation (**hyperplasia**), giant cell formation (**hypertrophy**), or death. It is also assumed that they act in extraoral digestion. Since numerous observers have seen the flow from glands enter the stylet and the plant, we can assume that any or all of the above functions are likely to occur. There is also morphological evidence to support the hypothesis that esophageal glands play an important part in plant parasitism. Coincident with the biological change from ectoparasitism to sessile endoparasitism is an enlargement of the esophageal glands to the extent that in some forms such as root-knot (*Meloidogyne*) and cyst nematodes (*Heterodera*) the glands greatly overlap the anterior intestine. To a lesser extent this development is repeated in every family of plant parasitic nematodes.

Common to plant parasitic nemas is a stomatal armature capable of piercing cells. In both classes (Adenophorea and Secernentea) this armature usually takes the form of a hollow stylet or spear. The exception is among Trichodoridae in Adenophorea, which utilizes a protrusible solid "tooth" to pierce cells. Food is then drawn into the stoma and esophagus. For a further discussion of stomatal armatures among plant parasites see Chapter 4, Section II,A.

In Adenophorea little can be said about evolutionary development of parasitism morphologically or biologically. All are ectoparasites of subterranean plant parts; all have rather elongated spears or a mural "tooth." No great development is noted in the esophageal glands and no special areas for valves are developed within the esophagus. However, evolutionary trends in morphology and biology are evident among Secernentea in gland development, spear development, and parasitic habits.

I. Adenophorean Plant Parasitism

The adenophorean plant parasites are a separate and independent line of development that does not follow the general scheme of the evolution of plant parasitism by nematodes. They are discussed here because all the known plant parasites in the subclass Enoplia are obligate ectoparasites of the belowground parts of plants. Plant parasites among Adenophorea are found only in two families, Longidoridae and Trichodoridae of the order Dorylaimida, and only three genera are currently important: *Trichodorus*, *Longidorus*, and *Xiphinema*; if one recognizes the splitting of *Trichodorus*, then the additional genera *Paratrichodorus* and *Monotrichodorus* must be considered. Other genera occur in these families but are not recognized as currently being of any great economic importance as plant parasites. Within Nemata it is only among these plant parasitic dorylaims that the ability to vector plant viruses is known (Table 6.1).

Table 6.1. Nematode Vectors of Viruses

Nematode	Virus	Crops infected
Longidorus attenuatus	Tomato black ring	Celery, globe artichoke,
Longidorus elongatus		lettuce, peach, potato,
		raspberry, strawberry,
		sugar beet, tomato

Table 6.1. Nematode Vectors of Viruses (*continued*)

Nematode	Virus	Crops infected
Longidorus elongatus	Raspberry ringspot	Blackberry, raspberry, red currant, strawberry
Longidorus macrosoma		Cherry
Longidorus martini	Mulberry ringspot	Mulberry
Xiphinema americanum	Tomato ringspot	Blackberry, cherry, grapevine, peach, raspberry, tobacco
	Tobacco ringspot	Bean, blueberry, gladiolus, grapevine, tobacco
Xiphinema diversicaudatum	Cherry leaf roll	Cherry, blackberry, elm, rhubarb
	Hop strain	Hop
	Strawberry latent ringspot	Black currant, cherry, celery, peach, plum, raspberry, rose, strawberry
	Raspberry ringspot	Cherry
	Arabis mosaic	Cucumber, grapevine
Xiphinema index	Grapevine fanleaf	Grapevine
Xiphinema coxi	Arabis mosaic	Cherry, cucumber, grapevine, raspberry, rhubarb
	Cherry leaf roll	Cherry, blackberry, elm, rhubarb
	Strawberry latent ringspot	Black currant, cherry, celery, peach, plum, raspberry, rose, strawberry
	Tobacco ringspot	Bean, blueberry, gladiolus, grapevine, tobacco
Xiphinema italiae	Grapevine fanleaf	Grapevine
Xiphinema vuittenezi	Cherry leaf roll	Cherry, blackberry, elm, rhubarb
Paratrichodorus anemones	Pea early browning	Pea, lucern
Paratrichodorus allius	Tobacco rattle	Tobacco
Paratrichodorus Christiei		
Paratrichodorus teres		
Paratrichodorus pachydermus		
Paratrichodorus porosus		
Trichodorus primitivus	Pea early browning	Pea, lucern
Trichodorus similis		
Trichodorus viruliferus		

A. Longidorus, Xiphinema

The genera *Longidorus* (Fig. 6.1) and *Xiphinema* (Fig. 6.2) are closely
related and their approach to plant parasitism is so similar that a single
discussion of both will suffice. Morphologically the two genera differ
primarily in the structure of the spear which is greatly elongated in both
genera. As previously noted in Chapter 4, Section II,A the spear is composed of two parts: the anterior portion originating from the cheilostome
(odontostyle) and the posterior extension from the esophastome (odontophore). It is in the structure of the odontophore that these two genera
differ. In *Xiphinema* the odontophore is in the form of a well-developed
cylinder with three elongate flanges (Fig. 6.2A); in each flange is a sensory
organ called the gustatory organ (Fig. 4.10C). In *Longidorus* the odontophore lacks well-developed flanges; a gustatory organ is also reported in
Longidorus even though the flanges are not developed (Fig. 6.1A).

Under the right conditions these nematodes can cause severe plant
damage. The root damage caused is dependent on the host and nature of
its root system. Their attack may result in a variety of symptoms. On
trees, especially pines and firs, their feeding (*X. bakeri* on Douglas fir)
induces "coarse root" or "rope root." This occurs when the growth of any

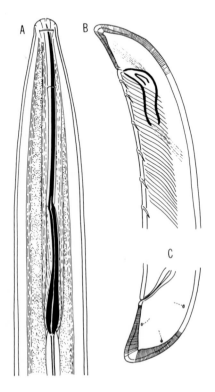

Figure 6.1. *Longidorus* sp. A. Female
anterior. B. Male tail. C. Female tail
(compilation).

Figure 6.2. *Xiphinema index.* A. Female anterior. B. Esophagus. C. Male tail. D. Female tail (redrawn from Thorne and Allen).

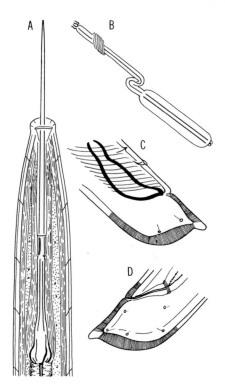

lateral root is inhibited as it breaks through the cortex or while it is just emerging. As a result the root system is limited largely to the main roots, which for the most part are devoid of any lateral branches or rootlets. On other hosts such as grapes, roses, or lettuce, they may induce "terminal galls" or "curly tip." Both are the result of a hyperplastic reaction of the root tip caused by the feeding of these nemas. Other nemas feeding similarly do not cause this reaction; therefore, it is assumed that the galling is induced by esophageal secretions of *Xiphinema* or *Longidorus*. It is interesting to note that these galls become attractive feeding sites for others of the same inducing species. Severe attack by any or all of these nematodes on young seedlings can result in death of the plant.

The damage caused by trichodorid types (Fig. 6.3) seems to be limited to the **stubby root** condition. The stubby root condition is the result of lateral root growth being stopped each time the rootlet attains a moderate length. This inhibited rootlet will itself attempt to produce a lateral root, which in turn, is inhibited from growth by the nematodes feeding. Eventually the entire root system is characteristically stubbed. These nemas can be especially serious on truck crops and often corn, fig, and date palms.

Regardless of the damage these genera (*Xiphinema, Longidorus,* and *Trichodorus*) can induce on their own, they are perhaps better known

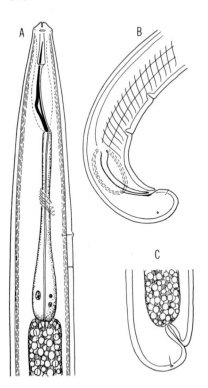

Figure 6.3. *Trichodorus proximus.* A. Female anterior. B. Male tail. C. Female tail (modified from Allen).

for their ability to vector plant viruses (Table 6.1). Hewett, Raski and Goheen reported the first plant virus transmitted by a nematode: Fan Leaf of Grape vectored by *X. index.* In no known instance is the virus persistent in the nematode. The virus is eliminated from the nematode at each molt. Therefore, to again be infective, each life stage must feed on a viruliferous plant. However, adults can remain viruliferous for 10 months or more. The virus is carried by the nematode primarily in the esophagus or structures derived from esophageal tissue (odontophore). In *Trichodorus* virus particles are found in the stomatal region (cheilostome), but whether they play a part in transmission is unknown.

Within the esophagus and esophastome specifically, the virus particles are arranged in an orderly monolayer, attached to the cuticular lining. The implication is that the forces or bonds that act on the particles associated with the cuticle of esophageal origin are nonexistent in cuticle of external origin (cheilostome). Surface charges on virus particles may be the means whereby they adhere to esophageal cuticle. That there are differential isoelectric points between external and internal cuticle has been demonstrated. Therefore, this preferential location in the esophagus can be explained, but remains unconfirmed, by the availability of charged receptor sites on the esophageal tissue and surface charges on virus particles.

B. Trichodorus

In *Trichodorus*, which lack a hollow axial spear, the "tooth" is used to puncture plant cells. The released juices are then ingested down the food canal that is ventral to the tooth (6.3A). During the process of feeding and exploration *Trichodorus* produces a tube of unknown origin and chemical constitution. This tube appears instantaneously in the cell whenever the tooth punctures a cell wall. The tube may act as a gasket or "pseudospear" between the cell and the oral opening. As such, it would facilitate nutrient flow in the absence of the hollow axial spear characteristic of other plant parasitic dorylaims. Additionally, it has been reported that *Trichodorus* is most efficient as a virus vector when probing and exploring rather than during the feeding process. It may be that this plug or "tube," always left at each probe, may be the mechanism of virus transfer.

The feeding of all these dorylaim parasites causes a devitalization of the root tip, and in most instances the affected tips show no discoloration, necrosis, or other evidence of injury. Even so, only a few types of root injury will affect plant growth more quickly than this devitalization of root tips.

II. Secernentean Plant Parasitism

In order to discuss evolutionary trends in the development of plant parasitism among Secernentea an examination of the order Diplogasterida is necessary. The diplogasterids are omnivorous nematodes, and among the various forms included in the group one finds predators, bacterial feeders, and fungus feeders, many of which are insect associates. In addition, this group represents the first appearance, within Nemata, of a valved muscular metacorpus and enlargement of the glands of the postcorpus coincident with muscular reduction. Thus, both morphologically and biologically the order lends itself to understanding the development of insect and plant parasitism among Tylenchida and Aphelenchida. Among derived diplogasterids not only is the esophagus grossly identical to that of tylenchs and aphelenchs, but the stomatal armature is styletlike and appears to be protrusible. An example of this is seen in *Tylopharynx*. As will be seen later, these same characteristics are important to the development of insect parasitism. That both insect and plant parasitism may have evolved from similar groups of nematodes is not really surprising when one considers that species radiation among nematodes, angiosperm plants, and insects was closely correlated.

The preadaptive features, mentioned above, set the stage for plant parasitism along two lines, both emanating from fungus feeding nematodes. One line of development can be derived from fungus feeders in either the soil environment or as associates in insect galleries. These forms, which have small slender spears and esophagi with nonoverlapping glands, are the most likely precursors of aboveground plant parasitism and insect parasitism. A second line, also derived from fungus feeders, developed large stout spears well adapted to feeding on subterranean plant parts but of such dimensions as to exclude them from any longer feeding on fungi. These two lines will be followed separately. First to be discussed are the subterranean plant parasites.

A. Nematodes as Parasites of the Subterranean Parts of Plants

The developmental sequence in the evolution of subterranean plant parasitism proceeds in two directions from facultative parasitism. One evolutionary sequence, exemplary of Criconematoidea, follows a line of development in which all life stages are ectoparasitic or emulators of endoparasitism (citrus nematode) and as such are continuously exposed to the soil environment. The evolutionary trend is toward ever-increasing contact, by adult females, with the plant; this line culminates in sessile ectoparasitism. The first step is from migratory ectoparasites such as *Paratylenchus* or *Criconemella* to semisessile ectoparasites as *Hemicycliophora* and *Cacopaurus*, which induce plant modifications that encourage a sedentary existence, to sedentary ectoparasites where the adult female has established a specialized relationship that inhibits movement or migration as in *Tylenchulus semipentrans* (citrus nematode).

Criconematoidea show morphological and physiological adaptations to the increasing sedentary life: (1) Those that migrate have short to moderately long spears, and no special plant response, such as cellular feeding sites are developed. (2) Attendant with increased obesity and restricted movement is a great lengthening of the spear. This allows the sedentary animal to probe deeply and widely for food cells and does not require a plant response. Few, if any, special feeding cells (plant response) are evident. (3) The truly sedentary forms such as the citrus nematode induce the plant to produce special feeding cells around the nemas head. In these forms the spear is again small and delicate because the special plant cells are not subject to necrosis, and deep probing by the nema is eliminated.

The other evolutionary sequence leads to endoparasitism among Tylenchoidea with larval exclusion from the soil environment coincident with total dependence on a root environment for survival except in periods of dispersion. A side shoot to this development is exemplified among Tylenchoidea by *Rotylenchulus reniformis*, which developed from migratory ectoparasitism to a sessile ectoparasite. As such, it parallels the development in Criconematoidea. In this biological sequence of plant parasitism the first step is a biochemical relationship between the plant and nematode. The chain of events is from ectoparasites as *Dolichodorus* to migrating endoparasites (*Pratylenchus*), which migrate within roots in advance of necrosis; to endoparasites, which either limit a necrotic plant reaction to a critical time of larval hatching (*Nacobbus*) or avoid necrosis altogether by establishing specialized relationships (Heteroderidae) such as galls and giant cells (special plant nurse cells).

Each step in these two evolutionary lines of subterranean ectoparasitism and endoparasitism by nematodes will now be discussed with examples.

1. Facultative Plant Parasites

Facultative plant parasites are those forms that can feed either on plant roots or fungi without interference to their life stages or development. The nemas commonly placed in this category possess all the morphological equipment necessary for plant parasitism, namely, a protrusible spear or stylet. Most often they are accused of being plant parasites by circumstantial evidence because they are collected around the roots of plants. It is assumed, because of their morphology, that they feed on fungi, root hairs, and other thin-walled plant cells. This seems a safe enough assumption since the spear aperture ($0.2 \mu m$) is too small to accept anything other than cell juices or the very smallest of bacteria. Though there are many alleged plant parasitic nematodes in this category, only a few are confirmed by direct observation or experimentation.

The subterranean facultative parasites are found in the orders Tylenchida and Aphelenchida. Among the Tylenchida all recognized facultative parasites are limited to the superfamily Tylenchoidea. Within the Aphelenchida they are apparently restricted to the family Aphelenchoididae. It is also from these taxonomic groups that the aboveground plant parasites evolve. Two species in Tylenchoidea and one in Aphelenchoididae will be presented as examples of subterranean facultative parasites.

a. Tylenchoidea

a1. Coslenchus costatus The most familiar nematode among the Tylenchoidea, specifically Tylenchidae, is probably *Coslenchus costatus* (Fig. 6.4). Taxonomically this species has experienced many name changes; in the literature it appears under the genera *Tylenchus*, *Anguillulina*, and *Aglenchus*. Both males and females are small wormlike nemas seldom exceeding 0.5 mm. They are distinctive because the external cuticle is marked by both longitudinal and transverse striae, giving the overall impression of a corncob (Fig. 6.4B). Their spears are very delicate and only about 12 μm long. The esophageal glands are enclosed in the postcorpus, which does not overlap the anterior intestine. Males are extremely rare; therefore, parthenogenetic reproduction appears to prevail.

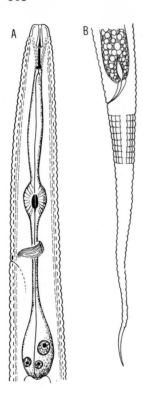

Figure 6.4. *Coslenchus costatus.* A. Female anterior. B. Female tail (modified from Meyl and CIH Plt. Par. Nem.).

There are not many references in the literature relating to this nematode as a plant parasite, most likely because its damage to plants is minimal. It has been observed feeding ectoparasitically on various grass roots without any resulting root lesion. On at least two plants it has been observed feeding endoparasitically. On both *Paeonia officinalis* and *Papaver somniferum* it causes small gall-like swellings as a result of feeding endoparasitically. Though unconfirmed this species allegedly is also a fungus feeder. In addition to the above, it has been found in nature on roots or in the rhizosphere of many plants including strawberries, poppies, soybeans, rice, lucerne, apple, fig, grape, Sitka spruce, several grasses, pineapple, and citrus.

In laboratory studies on roots of *Lolium perenne* its life cycle took 27–35 days at 18°–20°C. Eggs hatched in 5–7 days at 25°C This nema is very slow moving and according to Thorne can be recognized under low magnification because the long tail is held straight during movement.

a2. Cephalenchus emarginatus   The tylenchoid *Cephalenchus emarginatus* differs from *Coslenchus* by having only transverse annulations interrupted by six longitudinal incisures on the cuticle covering the lateral area (Fig. 6.5). In different populations males may or may not be abundant. Therefore, it is presumed that both parthenogenetic and amphimictic reproduction occurs.

Figure 6.5. *Cephalenchus emarginatus.* A. Female anterior. B. Female tail. C. Male tail (redrawn from CIH Plt. Par. Nem.).

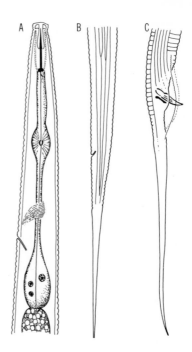

C. emarginatus is a facultative root ectoparasite that feeds on epidermal root cells and/or fungi. It reproduces exceedingly well on several species of conifers, but without causing visible root damage. However, when large populations develop on *Picea sitchensis* (tideland spruce) considerable root stunting occurs, but without obvious root necrosis. *C. emarginatus* also feeds on maples, figs, alfalfa, and grapes. In the laboratory it has been reared on *Penicillium expansum*.

The life cycle is short for this species, being completed in 5–6 days at 25°C.

b. Aphelenchoidoidea

b1. Aphelenchoides parietinus *Aphelenchoides parietinus* was first recognized over 100 years ago by Bastian. Since that time numerous forms have erroneously been placed in this species. Fortunately, clarification of its taxonomic status was presented by Franklin in 1950. As we now understand *A. parietinus*, it is a rather robust nematode whose length is generally less than 1 mm (Fig. 6.6). Males are unknown and reproduction is parthenogenetic. The stylet, as in most members of *Aphelenchoides*, has basal thickenings that can be scarcely called knobs (Fig. 6.6A).

A common environment in which to find *A. parietinus* is yellow lichen where it feeds on the fungal constituent. They can also be found in decaying plant matter where they are probably feeding on fungi. Though found

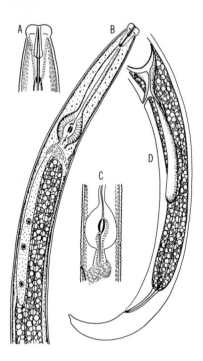

Figure 6.6. *Aphelenchoides parietinus.* A. Female head. B. Female anterior. C. Metacorpus. D. Female tail (modified from Thorne).

feeding on the cells of higher plants, they have little or no pathogenic effect. However, when numbers are high, they may cause sufficient damage to meristematic tissues of germinating plants, before they emerge from the soil, to cause "blind buds" or distorted foliage as the plant emerges. Normally, their feeding on roots kills only surface cortical cells and this limited necrosis produces yellowish roots.

Because these facultative ectoparasites, and sometimes endoparasites, cause little noticeable damage, they have received no real attention by plant nematologists. This is unfortunate because understanding these forms, which contemporarily represent ancestral forms of parasitism, could add a great deal to our knowledge of how obligate plant parasitism began. Also they may, because of more simple relationships with the plant, offer clearer information about host–parasite relationships.

2. Obligate Ectoparasites

The development of an obligatory dependence on higher plants emerged early in the evolution of nematodes as plant parasites. In obligate parasitism the energy utilized for the nematode's development comes from the plant at the direct expense of the plant. Unlike the facultative parasite, the obligate ectoparasite is often associated with severe root damage. Many of the species assigned to this category are capable of suppressing root growth, causing root death or inducing, by their feeding, root galls. Root galls caused by ectoparasitic nematodes result from hyperplasia, and whether they be caused by adenophoreans or secernenteans, they have the peculiar characteristic of being attractive feeding sites for nemas of the same species. Other symptoms associated with ectoparasitism are rots, surface necrosis, root lesions, and "coarse root."

a. Migratory Ectoparasites

Obligate subterranean ectoparasitism occurs in two superfamilies of Tylenchida:Tylenchoidea and Criconematoidea. Ectoparasitism is scattered among the forms in Tylenchoidea which also includes ecto- or endoparasites. However, among Criconematoidea only ectoparasitism is known. Within this group it is developed to a most sophisticated level, often being confused with endoparasitism. It differs from endoparasitism in that, even with such sedentary parasites as citrus nematode, all life stages are exposed to the soil environment, whereas in endoparasitism all life stages, except those of dispersion, are isolated from the soil by the protective root environment.

Among Criconematoidea the parasitic habit ranges through migratory ectoparasitism to semisedentary ectoparasitism to sedentary ectoparasitism. The genera that exemplify this sequence are *Criconema* (=*Ogma*), *Criconemella* (=*Criconemoides*, =*Macroposthonia*), *Hemicycliophora*, *Paratylenchus*, *Gracilacus*, *Cacopaurus*, and *Tylenchulus*. A consistent feature of Criconematoidea is the lack of a functional feeding apparatus among males.

The genera *Criconema* and *Criconemella* contain some of the most exotic forms of soil inhabiting plant parasitic nematodes. They are distinguished not only by body shape, but by their different forms of body musculature, which permits "earthworm" locomotion. Of the two genera, *Criconema* is the more ornate (Fig. 3.6B). The heavy body annulations may possess short retorse triangular plates, elongate triangular plates, or elaborately fringed spines. *Criconemella* lacks annular elaborations in the adult female, but does exhibit the heavy annulation which accounts for their common name "ring nematodes" (Fig. 6.7). The number of body annules is generally about 50, but may in some species exceed 150.

The somatic musculature of adult females and larvae, rather than being oriented to the longitudinal axis of the body, is oblique. Therefore, when viewed laterally, the muscle fibers form a chevronlike pattern (Fig. 4.5). This arrangement of the somatic muscle fibers permits the nematode to lengthen and shorten the body during locomotion. Their movement, however, is slow and sluggish. Males, when observed, retain a more wormlike shape, the body annulation is less pronounced, and their spear and esophagus degenerate (Fig. 6.7C, D).

Ring nematodes are generally stout and range in length from 0.2 mm to 1 mm. The spear of the females is generally large and stout, ranging in

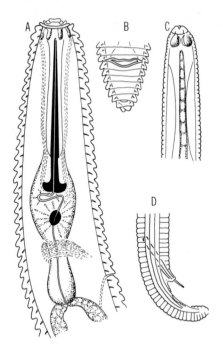

Figure 6.7. *Criconemella xenoplax.* A. Female anterior. B. Female tail, ventral view. C. Male anterior. D. Male tail (modified from Raski).

length from 30 to 122 μm and is exemplary of the Russian hypothesis that obligate ectoparasitism emerged as spears became too large to be useful in feeding on fungi. The shortness of the body and the length of the spear enforced architectural changes in the esophagus. The corpus and metacorpus appear to form a single enlargement. When the stylet is retracted, the spear knobs and tortuous lumen of the procorpus are immediately anterior to the metacorporal valve (Fig. 6.7A). This enlarged corpus is followed by a short isthmus and a small postcorporal bulb that does not overlap the anterior intestine. The form of this esophagus is so typical as to be called a "criconematid" esophagus. When the stylet is protruded the procorpus is extended, the tortuous lumen straightened, and at such times the esophagus is not too unlike other tylenchs.

a1. Criconema (=Ogma)   The genus *Criconema* is unsubstantiated as an important plant parasite. Most often it is recorded from undisturbed natural situations, such as forest or areas high in moisture. Their morphology would seem to exclude anything but an ectoparasitic mode of attack (Fig. 3.6B). As nematological research extends into native plants and forestry, their role in plant parasitism may be more fully appreciated.

a2. Criconemella (=Criconemoides, =Macroposthonia)   The genus *Criconemella* and its related genera or subgenera can be considered together. All are sluggish migratory ectoparasites. The long stylet allows them to reach cortical cells below the root epidermis (Fig. 6.7A). In some cases on older roots they can even reach the secondary phloem. The observation that has been made indicates that they stay at one feeding site for prolonged periods. Feeding has been observed all along the root, from root tip to the root hair zone and beyond. Sometimes they can embed their anterior end several cells into the root; they have also been seen completely within the root. Such rare occurrences should, however, not be confused with endoparasitism.

Though known to attack a variety of plants, they attain their greatest importance as parasites of woody perennials. Serious damage has been reported, however, on peanuts and carnations by *C. curvatum*. On herbaceous plants their feeding results in reduced and decayed root systems, which is reflected in stunting of the plant, reduced flower production, and alterations in mineral content in the leaves. Most notable are the reductions of nitrogen, calcium, potassium, and phosphorus. In leguminous plants such as vetch, they are capable of interfering with bacterial nodulation. In one instance, nodulation was reportedly reduced by 77%.

Most confirmed observations on the feeding habits and resultant damage have been made on woody perennials, specifically, deciduous fruits and nuts. The lesions formed on walnuts are not unlike the lesions caused by *Pratylenchus*. It has been speculated that the lesion caused by *Criconemella* results not as a necrosis associated with direct cell feeding but as a plant defense mechanism. The cause of the lesion is believed to result from increased phenol oxidase activity. Externally, larger infected roots are darkened and often contain longitudinal cracks that extend through the periderm to the secondary phloem. However, it is the death of fine feeder roots that is considered the most serious effect of direct feeding. It is because of excessive killing of the feeder roots that plants are stunted, mineral levels lowered, and the plants made more susceptible to water stress.

In addition to damage attributed to nematode feeding, of special importance in stone fruits, is the interaction of the nematode *C. xenoplax* (Fig. 6.7), and the causal agent of bacterial canker. One of the most serious disease problems of peaches in California and the southeastern United States is bacterial canker caused by the bacterium *Pseudomonas syringae*. Even though the canker is an aerial disease, the spread and severity of the disease is directly related to the distribution and population density of *C. xenoplax* in the soil. Serious bacterial canker seldom occurs in the absence of the nematode. The severity of canker is increased in trees stressed by having their roots severely damaged by the feeding of this nematode. Therefore, the current recommended control depends on controlling the nematode rather than the bacterium.

All the included species of *Criconemella* have a high reproductive potential over an extended period of time. The life cycle from egg to egg takes approximately 30 days. During her lifetime an adult female can lay 8–15 eggs per day. In one series of experiments on walnuts an initial population 20×10^3 nemas increased in 18 months to 1×10^6. Optimum ranges of temperature for development are reportedly around 25°C. Development seems to be favored in light sandy soils, with moderate to high moisture. If nothing else, their body structure and form of locomotion would be favored by a sandy soil.

According to the whims of taxonomy, other genera or subgenera that would be included in the same discussion are *Nothocriconema, Lobocriconema, Discorcriconemella, Xenocriconemella,* and *Criconemoides.* The genus *Criconemella* once included many of these genera, and *Macroposthonia* is now placed in genus inquirendum.

b. Gall-Forming Migratory Ectoparasites

Sexual dimorphism is characteristic of all Criconematoidea. In *Hemicycliophora* and *Hemicriconemoides*, the males, when present, lack a stylet and have a degenerated esophagus (Fig. 6.8C). Males are most often distinguished from each other by the shape of the spicule. *Hemicycliophora* males usually have sickle-shaped spicules (Fig. 5.9G) and an leptoderan caudal alae (bursa) (Fig. 6.8D), whereas in *Hemicriconemoides* the male

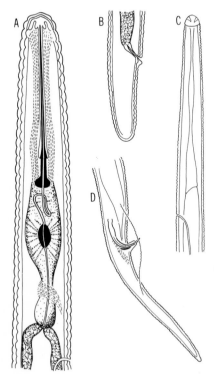

Figure 6.8. *Hemicyliophora arenaria.* A. Female anterior. B. Female tail. C. Male anterior. D. Male tail (redrawn from Raski).

spicule is rather straight and the bursa usually absent. However, there are intermediate forms in each genus.

The common name applied to members of these genera is "sheath nematodes." This refers to the additional external layering of the cuticle which gives the appearance that the females have two cuticles (Fig. 6.8A). In fact, this was the interpretation until 1970 when Johnson et al., in ultrastructure studies, elucidated the nature of this extra layering that overlaps an otherwise typical tylenchid cuticle. Males also possess the extra deposition, but it is thin and connected to the normal cuticle, and therefore, seldom recognized by light microscopy. In the female this excess deposition is connected only at the head and near the tail; thus a loose fitting sheath appearing like a retained cuticle of a previous stage is characteristic.

The ectoparasite *H. arenaria* is distinctive among Criconematoidea because of its capability to induce root galling. Not all members of the genus *Hemicycliophora* are gall formers. In fact, only two other gall formers are reported, *H. similis* and *H. nudata*. Since the other species in the genus, as well as the closely related genus *Hemicriconemoides*, affect plants in a manner not greatly different from any other migrating ectoparasite, attention will be directed to the one well-documented gall-inducing species *Hemicycliophora arenaria*.

b1. Hemicycliophora arenaria   The genus *Hemicycliophora*, known to science since the early 1900s (Fig. 6.8), was not recognized as having any economic importance until the late 1940s when Steiner reported it caused root stunting on slash pine (*Pinus caribaea*). Extensive root injury by *H. arenaria* on an agricultural crop was first observed by Van Gundy on rough lemon. Subsequently, it was discovered that the nematode is endemic on native desert plants, and with the reclamation of the southern California deserts, *H. arenaria* gained prominence as a parasite of agricultural crops.

Females feed ectoparasitically and lay their eggs singly in the soil. Each egg has a gelatinous coating that causes them to adhere to soil particles and roots. The optimum temperature range for reproduction is between 30° and 32.5°C, which coincides with their natural desert existence. Within this range they can complete a life cycle in 15–18 days. The highest recorded population increase occurred on tomatoes: in three months 150 females produced 1.5 million progeny, an exceptional rate of increase. The normal rates noted on acceptable hosts varied from 100- to 1000-fold. Though males do not feed, they are reproductively functional but are not alawys required for reproduction, which can occur parthenogenetically (Fig. 6.8C, D).

Initially feeding inhibits root cell elongation; this inhibition is soon followed by hyperplasia (cell number increase) that results in the production of the typical terminal gall. In the individual feeding site cell proliferation is rapid. Thus, the nemas are not only continuously killing cells by feeding but stimulating gall growth and the formation of new food supplies. The galls, as is characteristic of galls caused by ectoparasites, are attractive feeding sites for other individuals of the causal species.

When actively feeding, the spear is inserted for 60–70% of its length, or about 50 μm, into the root gall. Preferred feeding sites are near root tips or in regions of lateral root primordia. The latter regions are recognized even though there is no surface manifestation of a lateral root. During feeding, the nema is tightly attached to the root or gall by a polysaccharide plug. This plug must be broken from the root before the nema can retract the spear and move to a new feeding site. In the process of freeing themselves from the root, the body activity becomes so violent that they have been observed to break their spear.

The host range for this nema is quite diverse, and in addition to native desert plants includes rough lemon, sweet lemon, limes, tomatoes, beans, peppers, celery, squash, and grape (Tokay). On good hosts *Hemicycliophora* can retard plant growth as much as 30% with comparable reductions in yield.

The close association and attachment of this nematode to gall tissue places them almost in the category of being sessile. The additional ability to stimulate feed cells is also an important preadaptive feature for a sessile existence.

c. Obligate Sessile Ectoparasites

c1. Paratylenchinae   The Paratylenchinae demonstrate a smooth transition from subterranean migratory ectoparasitism to sessile ectoparasitism. The nominal genera involved are *Paratylenchus* (Fig. 6.9), *Gracilacus*, and *Cacopaurus* (Fig. 6.10).

Some authors question the division of Paratylenchinae into three genera because morphologically the three form a continuum of character development without clear lines of distinction. As noted above, biologically the three genera also merge in a smooth transition to sessile ectoparasitism. It might be best if upon revision the three were treated as subgenera; however, in nematology such taxonomic placements seldom last beyond the date of publication.

The separating characteristic is based primarily on the length of the stylet, which among the subfamily ranges from 16 μm to well over 100 μm. *Paratylenchus* is defined as having on the average spears less than 36 μm; *Gracilacus* spears are greater in length, ranging from 48–119 μm; *Cacopaurus* is distinguished by cuticular ornamentation and degree of obesity (Fig. 6.10C).

Males as well as the fourth stage preadult of *Paratylenchus* do not feed. In a few species the spear is evident in males but is so poorly developed that it probably has no function. The nonfeeding preadult is the most resistant stage and accounts for 90% of stages found in winter. Reportedly

Figure 6.9. *Paratylenchus hamatus.* A. Male tail. B. Female tail. C. Female anterior (modified from Thorne and Allen).

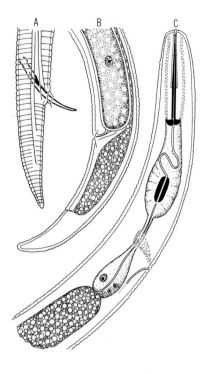

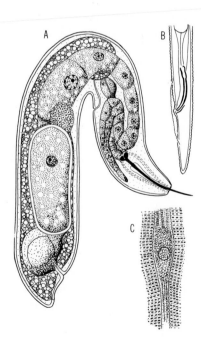

Figure 6.10. *Cacopaurus pestis.* A. Adult female. B. Male tail. C. Cuticular pattern on caudal region of female (modified from Thorne).

they can survive up to four years in the absence of a host. Molting to the adult stage allegedly only proceeds in the presence of stimulatory substances given off by suitable host roots.

Feeding by adults is limited to epidermal cells and the base of root hairs. Younger larval stages often can be seen to feed far out on the root hairs. Damage is minimal except when populations are high. Under favorable conditions, which vary according to species of nematode and plant, these nematodes are capable of increasing to tremendous numbers. In greenhouse experiments populations of 100–125 nematodes per gram of soil are not unusual. When populations attain these levels root damage is severe. Situations of this type are often attained in greenhouse cultures but occur only rarely in the field.

Even when vermiform, these nemas are sluggish in movement and females become even more sluggish after feeding for several days. Among all the genera obesity is common but is especially characteristic of *Gracilacus* and *Cacopaurus*. When this occurs the nema is committed to a sessile existence.

The feeding of *Cacopaurus* (Fig. 8.10) reportedly results in the swelling and rupturing of cells in the feeding site. It is believed that the nema feeds on the resulting exudate rather than on cell contents directly. This exudate eventually covers the female and her eggs are deposited into it. It is further alleged that hatched larvae also feed on this exudate. Recognizing the biology of other criconematoids does not lend confidence to these allegations. Rather it would seem that the "food exudate" is in reality no more than a "gelatinous matrix" secreted by the female into which eggs are layed. Such an interpretation is consistent with observations of the Tylenchulidae. In either event, the alleged exudate feeding should be reexamined and confirmed or rejected.

Paratylenchinae attack often with great severity a wide variety of herbaceous and woody plants. These include native plants, ornamentals, and agricultural crops far too numerous to list.

c2. Tylenchulidae   The most sophisticated development of exposed sessile ectoparasitism among Criconematoidea is exemplified by the citrus nematode, *Tylenchulus semipenetrans* (Fig. 6.11). Since its discovery on orange trees in California in 1912, this nematode has become recognized as the most important nematode parasite on citrus throughout the world. However, in many areas of the world it remains unrecognized or ignored. This can only be attributed to the stubbornness of citrus pathologists who fail to acknowledge that a nematode could so devastate a tree. As a result much time and effort is foolishly wasted looking for other causes, while the real culprit continues its destruction unimpeded. Unfortunately citrus nematode is not an isolated example of such naïveté.

This nematode has an unusual life cycle among plant parasites. Only females (larvae and adults) feed. The male larva matures in one week

Figure 6.11. *Tylenchulus semi-penetrans*; adult female (modified from Guiterraz).

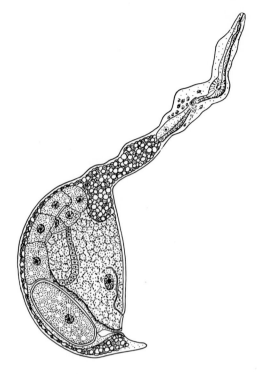

without feeding. Their life support is what is available to the second stage larva at hatching. The preadult female larvae are migratory ectoparasites through the fourth stage. Upon molting to the fifth or adult stage female, the biological habit changes from migratory ectoparasitism to a highly developed form of sessile ectoparasitism that emulates endoparasitism. The young female penetrates the root for the length of its esophagus or half again this depth, i.e., from 120 to 180 μm. On most roots this is sufficient to reach the pericycle. The remainder of the body, approximately one-half to two-thirds of the total length, remains exposed to soil environment. This posterior region of the body increases in diameter from approximately 16 to 180 μm, which in turn means roughly a 25–30 time increase in exposed body volume. Surprisingly, this occurs in the span of little more than one week.

The posterior placement of the excretory pore is characteristic of the genus *Tylenchulus*. In *T. semipenetrans* it is located near the vulva or at about 80% of the body length (Fig. 6.11). The excretory cell in these nemas is exceptionally large and may occupy 30% of the body volume (Fig. 6.11). This cell functions not only in a normal manner, but additionally secretes the matrix into which eggs are laid. It is for this reason that a "food exudate" produced by the plant rather than a matrix by the female is questioned in relation to *Cacopaurus*.

Among tylenchulids the development of the rectum and anus is unusual. In citrus nematode there is no evidence of either a rectum or anus (Fig. 6.11); in other species in the same or related genera an anal slit may be visible in the cuticle, but the rectum is undeveloped. A nonfunctioning anus is not as contradictory to normal body functions as it at first seems; however, certain compensations must be active. In Tylenchulidae the diet consists entirely of plant fluids and the size of the spear opening precludes the entrance of almost any solid material or food that would leave a nondigestible residue. Because the amount of nondigestible particles ingested must be very small, less than 0.3 μm, and the nematode's life span is very short, whatever particles are ingested can be stored in the intestinal mass without impairment of body functions. Soluble metabolic waste and other unneeded solutes are eliminated by the enormous excretory cell (Fig. 6.11). That the system is capable of exuding large volumes of material is demonstrated by the production of the gelatinous matrix.

The condition produced in citrus by the attack of *T. semipenetrans* is commonly referred to as a "slow decline" and infested roots are readily recognized by their dirty appearance. Even vigorous shaking in water will not free the root of the soil particles that adhere to the matrix material exuded over the females, eggs, and roots.

When infested trees are planted in noninfested fertile soils, symptoms may not be evident for 20 or more years. However, throughout the citrus growing regions of the world where virtually all land has become infested, symptoms appear in 5 to 15 years. Reduced terminal growth is the first indication of an infestation and is soon followed by reduced vigor, yellowing, and drying of leaves and twigs. These visual symptoms are coincident with lowered production of fruit and a decline in fruit quality. In advanced unchecked infestations, such as occur in many parts of the Middle East, dieback and defoliation becomes severe, fruit production drops to near nothing, and the trees soon die. Extreme situations as just described are not the norm because trees are usually removed when conditions of decline are severe. Young trees planted on infested land in old tree sites may fail to survive the first year.

The nematicidal controls developed by R. Baines of the University of California preserved the citrus industry throughout the world. However, because of the banning of the fumigant DBCP new techniques must be developed. Currently the industry must rely on preplant fumigation and hopefully clean or resistant rootstock such as trifoliate orange and troyer citrange. The availability of these measures are also largely due to the efforts of Baines. In many areas the nematode has developed biotypes that break down tree resistance. This is always a danger when dealing with species of nematodes that can reproduce parthenogenetically. Once a biotype is developed, it can reproduce itself without the ameliorating effects of cross-fertilization.

In his last years Baines was concentrating on developing methods for the efficient use of organic phosphates or carbamates against this nematode. As a result of his monumental efforts, there is continued hope for the industry throughout the world.

There are other representatives of Tylenchulidae that are of lesser importance, either because of limited distribution, or the economic importance of their hosts. In specific cases, however, any of these could be serious. The included genera are: *Sphaeronema*, *Trophonema* and *Meloidoderita*.

3. Obligate Ecto–Endo Parasites

a. Tylenchidae, Hoplolaimidae

Obligate ectoparasitism, at least at the generic level, is a very ephemeral phenomenon among Tylenchoidea. In most taxa, whether the species acts ectoparasitically or endoparasitically, is more dependent on host root characteristics than any inherent quality within the nematode. As earlier noted, even the facultative parasites among Tylenchidae can be endoparasitic on certain hosts. A few forms do seem to be exclusively ectoparasitic such as *Belonolaimus* (Fig. 6.12A–C) and *Dolichodorus*, and one genus, *Rotylenchulus*, is exceptional in that the ectoparasitic habit most closely resembles that attributed to Tylenchulidae of the Criconematoidea. Among the more important agricultural pests are the genera *Tylenchorhynchus* (Fig. 6.12D–F), *Merlinius*, *Hoplolaimus*, *Rotylenchus*, *Helicotylenchus*, *Scutellonema* (Fig. 6.12G–I), *Belonolaimus*, *Dolichodorus*, and *Rotylenchulus*. These and other obligately parasitic genera, currently of lesser importance, are distributed throughout the families Tylenchidae and Hoplolaimidae.

With the exception of *Rotylenchulus*, which will be discussed separately, the types of damage and life cycles are sufficiently similar that all can be discussed as a single topic.

The genera included in this discussion all have direct life cycles and all stages feed. Though not really confirmed, it would appear that reproduction may be either amphimictic (cross-fertilization) or parthenogenetic. More often reproduction is amphimictic and takes place in the soil environment. However, when the nemas are functioning as endoparasites, reproduction takes place within the root. The latter is an important phylogenetic preadaptation that is essential to the development of obligate endoparasitism. In these instances of facultative endoparasitism all life stages must be able to sustain themselves within the root environment. In general, whether living ecto- or endoparasitically, the life cycle takes approximately one month.

Feeding occurs almost anywhere along the root system except the root cap. These nemas have been observed to feed on root hairs, between root hairs, in the region of elongation, and even on mature roots and

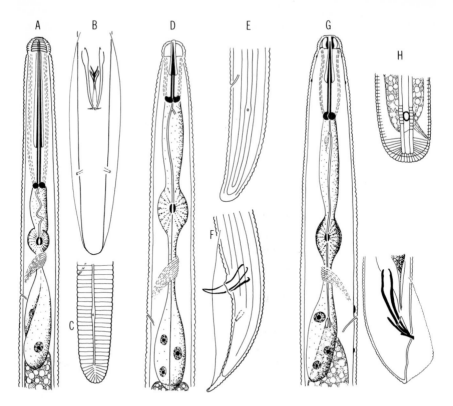

Figure 6.12. A–C. *Belonolaimus gracilis.* A. Female anterior. B. Male tail, ventral view. C. Female tail (modified from Steiner). D–F, *Tylenchorhynchus cylindricus.* D. Female anterior. E. Female tail. F. Male tail (modified from Allen). G–I. *Scutellonema validium.* G. Female anterior. H. Female tail. I. Male tail (modified from Sher).

tubers; only rarely are they observed feeding on root tips. While feeding, and this often depends on the host or nature of the root, they may be found outside puncturing roots with the spear only, partially embedded, or totally embedded. In regions of roots where several nemas are concurrently feeding, the root may become slightly swollen, spongy, and discolored; eventually the cortex of such roots is sloughed off. Commonly their feeding results in numerous small, brown, necrotic root lesions.

When acting endoparasitically, their penetration of the root injures both cortical and endodermal tissues. Their feeding causes injury primarily in the phloem and parenchyma. Injury due to feeding is not confined to the cells immediately being fed on. Numerous reports refer to injury in advance of, and some distance from, the feeding site. One of the common distant effects is xylem tylosis, in which xylem parenchyma cells are induced to enlarge and force their cytoplasmic extensions to protrude through the xylem pits into the xylem vessel. Tylosis can eventually cause the blockage of xylem vessels.

External examination of the root system generally shows malformation of the primary root and a reduction or absence of secondary and tertiary roots. With *Belonolaimus* (Fig. 6.12A–C), damage may be confined to the upper soil layers, i.e., the nema population is highest in the top 30 cm. Under these conditions, plants form a dense root growth near the soil surface and roots are unable to penetrate beyond depths of more than 10 cm. Plants so attacked can be easily lifted from the soil, and their root system appears as if sheared off at 8 to 10 cm.

When attacking yams, *Scutellonema* (Fig. 6.12G–I) induces a condition known as "dry rot." The initial damage is confined to shallow subdermal lesions, which at first are yellow, but soon become brown or black. These shallow lesions caused by the feeding of the nematode are soon invaded by secondary organisms such as mites, fungi, or bacteria. The resulting complex often destroys the entire "tuber."

Members of the *Tylenchorhynchus* (Fig. 6.12D–F) group attack a wide variety of plants among which are tobacco, cotton, corn, and sugarcane. Generally damage is associated with a stunting of the root system, which as with other ecto–endoparasites, is expressed aboveground by yellowing of the foliage, stunted tops, and sometimes associated with defoliation and wilt. Under severe conditions such as high populations, young sugarcane sets can be killed.

b. Rotylenchulus reniformis

Unusual among this group is the genus *Rotylenchulus* (Fig. 6.13), for it most closely resembles *Tylenchulus* (Criconematoidea) in its life cycle. Economically the most important species is *R. reniformis*. It is difficult to place them in a classification based upon type of parasitism. Some authors classify them as semiendoparasitic, but because of the close similarity to citrus nematode and their larval development in the soil, I prefer to class them as sedentary ectoparasites. All larval stages and males are found in the soil, and as far as is known, they do not feed. When held in water in the laboratory the larvae rapidly pass through three molts to become adult females and males. The young female is vermiform and her reproductive system is immature. Shortly after the molt to adulthood these females seek and penetrate host roots. Generally, only the anterior of the body becomes embedded, though in rare instances they have been seen to completely embed themselves within root tissue. Within a week the females swell to the typical kidney shape and the reproductive system matures (Fig. 6.13). At this time males are strongly attracted to the females and several males can be found embedded in the gelatinous matrix exuded by the female.

Only the females are parasitic; the male spear and esophagus are poorly

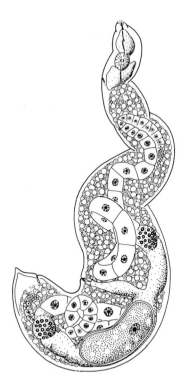

Figure 6.13. *Rotylenchulus reniformis*; adult female (modified from Linford and Oliviera).

developed for feeding. In *R. reniformis* reproduction is amphimictic, but in other species as *R. parvus* parthenogenesis predominates. The complete life cycle takes from 24 to 29 days. Eggs hatch in about eight days; larval development to adulthood is an additional eight to nine days, and females begin laying eggs about nine days after root penetration.

Females feed primarily in the pericycle. In the region of the feeding site and some distance from it, the cells of the pericycle are hypertrophied and uninucleate. These cells are about twice the size of normal cells, somewhat elongated, but rather regular in shape. The hypertrophy is evident some fifteen cells on either side of the nematode's head region. In addition, the cell walls of the endodermis becomes thickened and lignified.

As a result of their feeding, the roots of the host plant become discolored and malfunction. Young plants attacked prior to tap root formation may be killed. The feeding sites are a portal of entry for both *Fusarium* and *Verticillium*, which then act synergistically with the nematode causing greater damage than the organisms would cause singly. Though seldom realized, these nemas have a tremendous biotic potential because of the short life cycle and egg productivity. Egg production, which can begin 7–10 days after root entrance can, on a good host, average between 75–120 eggs per day; one half of these are potential females. In a period of 3.5 months a female, theoretically, could give rise to 13 million offspring. If only one-tenth of the females were to survive each generation and reproduce, the offspring resulting from one female over the period of one year would be 270 billion.

The host range for *R. reniformis* is extensive and includes cotton, soybean, tea, pineapple, and castor beans. *R. parvus*, another important economic species attacks sugarcane, corn, cowpeas, and kidney beans.

4. Obligate Endoparasitism

Obligate endoparasitism among Tylenchoidea may be either migratory or sedentary. In Pratylenchidae both phenomena occur; however, as far as is known among Heteroderidae, only sedentary endoparasitism occurs.

a. Pratylenchidae

The most important genera in this family, either because of ecology or economics, are *Pratylenchus, Hirschmaniella, Radopholus*, and *Nacobbus*. These nemas illustrate the transition from migratory endoparasitism to sedentary endoparasitism, and their effect varies from lesion formation to galling. Galling by members of this family may be associated with either the migratory or sedentary way of life.

a1. Pratylenchus   The genus *Pratylenchus* (Fig. 6.14) encompasses more than 60 nominal species, and worldwide they are likely to be the second most important group of plant parasitic nemas, being exceeded as parasites only by *Meloidogyne* spp. in Heteroderidae. Commonly members of this genus are referred to as "lesion" or "meadow" nematodes. The former is to be preferred because, though common, it at least speaks to symptomatology. It is a waste of time to apply common names to parasitic nematodes that have meaning only in the mind of the proposer. The most accurate and usable common names generally gain usage when they evolve from symptoms or crop damage as indicated by growers. Therefore, such names as "pin" nematode, "lance" nematode, "stylet" nematode, "dagger" nematode, and "sting" nematode serve no useful purpose and should be discarded.

In *Pratylenchus* both males and females are vermiform and differ only in sexual characters. They are easily recognized by the sclerotized lip region, overlapping glands, and generally darkened intestinal contents. Females have only a single anteriorly directed ovary and a postuterine sac of variable length.

The species *P. penetrans* (Fig. 6.14) is probably the most widespread and important. It occurs on some 400 hosts, both woody and herbaceous.

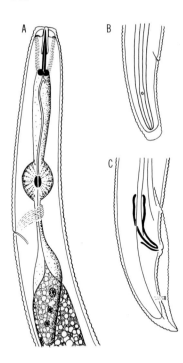

Figure 6.14. *Pratylenchus penetrans.* A. Female anterior. B. Female tail. C. Male tail (modified from Sher and Allen).

This does not mean that the other species are unimportant, but they have either limited host ranges or limited distribution. The better known species and their prime hosts are *P. vulnus* (deciduous fruits and nuts), *P. brachyurus* (cotton, peanut), *P. coffeae* (banana, coffee), *P. scribneri* (potatoes), and *P. zeae* (corn).

The mode of attack is similar for all species. Feeding is within the cortex of the root generally with the total body embedded. However, feeding does take place during penetration which may take from four to six hours. Both penetration and migration involve intra- and intercellular movement. Feeding results in cell death and breakdown, and externally the roots become covered with lesions that may be measured in millimeters, or on large roots in centimeters. Often on small to moderate roots, the lesions girdle the root and cause their death. The lesion is most pronounced in the endodermal layers, though lesion formation occurs in other cells as well. The extent of damage depends on the phenolic content of the cells, or their ability to synthesize phenols after injury. The oxidation of these phenols causes the characteristic browning of the lesion. Lesion formation thus varies with the host; peaches have high concentrations of amygdalin in their roots and hydrolysis of this compound releases HCN which can cause extensive root damage in a matter of hours. Top symptoms include chlorosis, stunting, defoliation, and twig dieback, as well as reduced yields.

All stages and both sexes are important in producing root damage. The nematodes are not found within the root lesion, but in advance of its formation. Though there may be some change with season or host, most of the population resides within the root. Because of this, soil samples may be very misleading and root samples should always be examined.

a2. Radopholus and Hirschmaniella   *Radopholus* and *Hirschmaniella* are very close and in gross respects resemble *Pratylenchus*, however, females in both genera possess two opposing ovaries. The two genera differ from each other primarily by the phenomenon of sexual dimorphism in *Radopholus*. In this genus the male is degenerate and does not feed (Fig. 6.15B); also the esophageal glands overlap the intestine dorsally (Fig. 6.15A). Among *Hirschmaniella* both males and females are fully and functionally developed, and the esophageal glands overlap the intestine ventrally. Like other Pratylenchidae the head skeleton is most strongly developed as the cheilostomal cylinder and basal plate which peripherally has posteriorly directed extensions.

a2a. Radopholus similis   *Radopholus similis* (Fig. 6.15) occurs as an important root parasite in the tropics and subtropics where it is the recognized causal agent of blackhead of banana, "yellow disease" of pepper (*Piper nigrum*), and spreading decline of citrus. *Radopholus similis* is

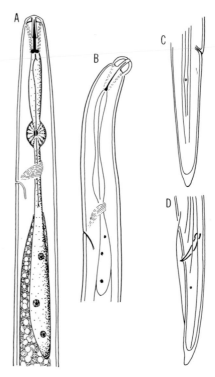

Figure 6.15. *Radopholus similis*. A. Female anterior. B. Male anterior. C. Female tail. D. Male tail (modified from Cobb).

subject to the production of biotypes. The biotype that attacks bananas reportedly does not attack citrus. However, the citrus biotype is known to attack banana. Other hosts include greenhouse ornamentals.

It has not been clearly demonstrated that the biotypes on foliar ornamentals can attack citrus. However, when dealing with a species with a propensity to develop biotypes one can never feel safe with biotype host range lists or the development of resistant hosts. The plasticity of the species will foil the best of endeavors.

Blackhead of banana is important throughout most banana growing regions of the world and was likely to have been disseminated with banana corms. The typical injury induced by the nematode is root and rhizome (corm) necrosis. As a consequence of the death of the primary anchoring roots the plants tend, when supporting a maturing bunch, to topple over in high winds. In addition with the impairment of root absorption efficiency, water stress is increased. The end result is reduced bunch weights, size, and leaf growth. In areas of heavy infestation, the life of a plant may be reduced to as little as one year.

The histopathology of the roots indicates cortex destruction and cavities resulting from nematode feeding. As necrosis sets in, the nemas move either in advance of the consequent lesions, or leave the roots to seek new feeding sites. The internal cellular destruction is also conducive to the development of secondary organisms. Of particular economic importance is the disease complex resulting from the invasion of *Fusarium oxysporum* f. *cubensis*, the causal agent of "Panama disease." When combined with *R. similis* the damage expected from either singly is doubled and infection rates may increase from 40 to 80%.

The diseases caused by *Radopholus* can be devastating and this is unusual among nematodes as plant parasites. One of the more serious disasters has been noted with pepper vines in Indonesia. The common name first given to the disease in the 1930s was "Black Pepper Yellows." This is a spreading disease, also characteristic of the genus, which initially is noted by localized vine chlorosis that yearly increases until 90% of a given crop may be lost. In the span of one year a vine becomes completely chlorotic and at this point growth ceases, dieback becomes evident, and death of the vine soon follows. Examination of the roots reveals that all feeder roots are destroyed and there is extensive necrosis of older roots.

The history of this disease on the island of Bangkra serves as an excellent example of the destruction and economic loss this nematode can cause. In the 1930s when yellowing was first observed, the island supported 22 million vines. In a period of 20 years pepper gardens were so severely hit by the nema, that only 2 million vines remained. Where formerly gardens lasted several decades, they now persist for only three to five years.

The disease that has received wide publicity in the United States is "spreading decline of citrus," even though the disease is limited to Florida citrus and occurs only on well-drained deep sandy soils along the central ridge of Florida. In these soils the nematode is rare in the upper 60 cm, but abundant below this level. It is here that most damage is perpetrated. *Radopholus* is a poor competitor and therefore finds more suitable conditions at these deeper soil depths where there is little or no competition from other soil microorganisms.

Part of the notoriety of this disease is due to drastic measures undertaken and publicity given to early attempts in Florida to control the disease, i.e., pull and treat. Under this regime trees were burnt within the infested grove, roots were removed, and burned as was practical. This was followed by heavy soil fumigation and the land was left bare for a minimum of two years. Why this nematode has not caused problems in other citrus growing areas has never been satisfactorily answered. However, thanks to its economic importance in Florida we have available considerable information on the biology, pathogenicity, and ecology of *Radopholus*. How the nema attacks citrus seems a good model for both *Radopholus* and *Hirschmaniella*.

Root attack initiates near root tips; especially attractive is the region of cell elongation and root hair production. When penetration is delayed, surface damage in the root is evident as shallow craterlike depressions, but when root invasion is rapid, there is little indication on the root where entry was made except for a microscopic hole slightly larger than the diameter of the nematode. In about one week these areas of invasion become discolored and subsequently crack over the region of penetration. In about two weeks lesion formation is manifested. When the nemas remain in the same part of the root, their tunnels and cavities form a network throughout the cortex near the phloem and cambium ring which is the preferred feeding site, though all types of parenchyma tissue may be attacked.

In the surrounding tissues of the lesion, starches are reduced and the cells become loosened, even though the area may be free of nematode activity. Deposition of wound gum imparts a tan to amber color to parasitized roots.

The most striking response of roots to the feeding of *Radopholus* is the formation of tumors in the pericycle. This has significance because it illustrates among Pratylenchidae a sequence of preadaptations to inducing plant cells to react to nema secretions through hyperplasia and hypertrophy. As will be evident, this ability among Pratylenchidae is clearly exemplified by and necessary to the completion of the life cycle by members of the genus *Nacobbus*, commonly called the "false root-knot nematode." *Radopholus* can initiate these tumors in the pericycle and primary apical meristem. This pericycle reaction occurs when the nematode passes through the endoderm and locates in the root stele. The hypertrophied cells, though increased in size, show no increase in cell content density or enrichment. The hyperplastic cells are uninucleate thin walled and have a finely granulated cytoplasm. These tumors are apparently attractive feeding sites as is evidenced by the nemas' propensity to penetrate and burrow through the tumors. If root tips are invaded they may become swollen. Normally the root tumors are not visible externally, but are confined within the crushed cortex. This type of response is not limited to citrus; tumorous nodules have also been noted on coffee roots attacked by *Radopholus*.

Nematicides are difficult to use against this nematode because of its depth of attack in the soil, as well as its protection within the deeper root tissues. However, since the roots in the upper 60 cm of soil are virtually free from this nema, relief in Florida is attained by irrigations in the period between January and May when rainfall is low. Irrigation provides moisture to the upper root system, which compensates for the loss of the deeper roots. Utilizing this regime of irrigation without soil tillage avoids the normal wilt and fruit drop experienced in the decline syndrome.

a2b. Hirschmaniella   *Hirschmaniella*, though similar to *Radopholus*, is not of such great economic importance. Because so many of the known species are associated with rice, they are often referred to as "the rice root nematodes." Though occurring in large numbers they appear to cause little suffering to the plants. However, reduced tillering and head set are reported, but apparently are of little concern. Originally *Hirschmaniella* spp. were implicated in "Mentek" disease of rice; however, the presence of the nema is coincidental and the real causal agent is a virus. Root symptomatology is not unlike that seen with *Pratylenchus* or *Radopholus*.

a3. Nacobbus   In its larval stages, *Nacobbus* looks far more like a member of *Pratylenchus* than either *Radopholus* or *Hirschmaniella*. In the adult stages sexual dimorphism is dramatic: males remain fully functional and vermiform (Fig. 6.16C, D), but the females become sedentary and saccate (Fig. 6.16A).

The life cycle is unusual in that both migratory endoparasitism and sedentary endoparasitism are always essential to the completion of the cycle. Since sugar beet is the best known economic host of nematodes in the genus, discussion will be directed to *N. batatiformis*.

Overwintering is primarily accomplished by eggs that hatch soon after the growing season commences. The hatched larvae enter plant roots near the region of the root tip; it is not uncommon to have four or more larvae concentrated together. Larval feeding is not unlike that of other Pratylenchidae, in that their feeding and migratory habit result in cell death and eventual necrosis. As necrosis progresses, the nemas migrate in advance of the encroaching lesion. These activities continue until the final molt of the young female to adulthood. Young females are vermiform and are attracted to larger roots than younger larvae.

Upon penetration females make their way to the central cylinder. Once in position, changes occur within the female and the root. Females become swollen or saccate. The anterior region, anterior to the metacorpus bulb,

Figure 6.16. *Nacobbus dorsalis.* A. Adult female. B. Female head. C. Male head. D. Male tail (modified from Thorne).

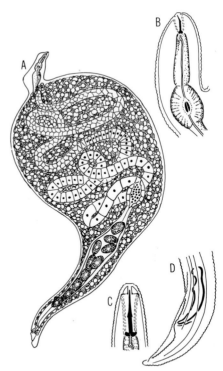

remains narrow but with some evidence of swelling. The main portion of the body becomes saccate, but the posterior fourth of the body again narrows to the almost terminal vulva. Female root penetration and feeding induce root cells to hyperplasia and hypertrophy. This hypertrophy occurs in the stelar region of the root and hyperplasia is initiated in cortical tissue; thus a gall with greatly enlarged cells at the feeding site is formed. The giant cells may number in the hundreds and the area they occupy becomes spindle-shaped with orientation following the root axis. These enlarged "nurse cells" are often multinucleate probably as a result of nuclear division without concomitant cell wall formation.

The elongated posterior region of the female extends toward the periphery of the gall. An opening is formed through which the eggs are discharged into a gelatinous matrix produced by the female. Males may be found in the matrix, and this is interpreted as indicative of fertilization occuring after the initiation of gall formation, but has not been confirmed. More importantly when the females cease to feed and die, gall breakdown and necrosis begin. However, because eggs are layed in a matrix outside, the hatched larvae are not subjected to necrotic tissue. Therefore, the life cycle can be broken into two phases: first, the larval phase characterized by migration in advance of necrosis; second, the sedentary adult female phase that avoids necrosis by biochemical adaptation. The latter results in a specialized relationship limiting a necrotic plant reaction to a critical period of egg laying or hatching that forces larval migration to new roots.

b. Heteroderidae

The family Heteroderidae can be divided into two groups of great economic importance: the root-knot nematodes (*Meloidogyne*) and the cyst nematodes (*Heterodera*, *Globodera*, and other minor genera). All members of the family as adult females are saccate obligate sedentary endoparasites with only one mobile or dispersing stage, the second stage or infective larva. Males, where present, are vermiform and mobile, but only in the second stage, and as adults. The intermediate male larval stages are saccate and sedentary as they are in the female.

b1. Meloidogyne   Root-knot nematodes (Fig. 6.17) were first introduced to the world of science in 1855 by M. J. Berkeley in England when he studied galls on the roots of cucumber. The first studies in the United States were conducted independently by J. C. Neal and G. F. Atkinson in 1889. Neal's investigation revealed that among growers the disease condition had been recognized in Florida even before 1805. From the time of

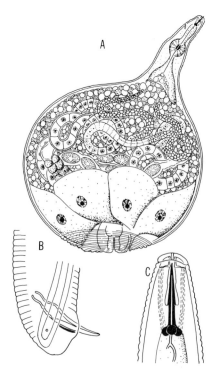

Figure 6.17. *Meloidogyne* sp. A. Adult female. B. Male tail. C. Male head (after Maggenti, Annu. Rev. Microbiol. 35).

their discovery until 1949, all species were treated as either *Heterodera marioni* or *H. radicicola*. In 1949 B. G. Chitwood differentiated five species and one subspecies as differing from *Heterodera* and he placed these together in the genus *Meloidogyne*, first proposed by Goldi in 1887. Chitwood introduced the unique concept of species differentiation based on the annular pattern of the perineal region of the female. This remains the basic tool used in the identification of the species of *Meloidogyne* (Fig. 6.18). Other characteristics used are the position of the excretory pore, spear shape, tail length of second stage larvae, chromosome number, males, and host plant keys.

A concept that continues to plague nematode taxonomy is the use of a host plant key as a procedure for identifying unknown *Meloidogyne* species. The key is a standard dicotymous approach that relies on host specificity. The host plant key has some value at the gross level, but should not be taken seriously when species identification is required. The unreliability of the host plant key lies in the danger associated with species that can reproduce parthenogenetically, that is, their propensity to produce biotypes. The phenomena is not limited to parthenogenetic species, but their mode of reproduction is conducive to the easy establishment of biotypes. Nematologists working with root-knot soon become aware of this ever-present danger, or they proceed blithely describing new species. It is for this reason that there are more than forty nominal species, and why this genus presents the greatest problems to species identification among all Nemata.

Leaders among the important economic species of wide distribution are *M. incognita*, *M. javanica*, *M. arenaria*, *M. hapla*, and *M. exigua*. It will suffice, however, to limit the discussion to a generic level since the principal differences between species is not mode of attack but rather hosts and morphology. The list of plants known to be susceptible has become so

Figure 6.18. Perineal patterns of *Meloidogyne*. A. *M. javanica*. B. *M. hapla*. C. *M. incognita* (original).

A   B   C

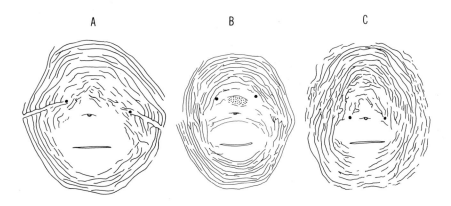

large that it includes nearly all cultivated varieties and numerous native plants.

The infective stage is the second stage larva that hatches from the egg. Once finding or being attracted to a root, the larva begins penetration near the root tip in the nondifferentiated region. The larva moves inter- and intracellularly to the root stele. From this time on begins the formation of the notorious galls. The gall is formed by both hypertrophy and hyperplasia. Most notable within the gall are the giant plant cells or "nurse cells" formed about the head end of the nematode. These develop from undifferentiated cells, generally of the pericycle, in response to the nematode feeding.

The invading second stage larva, upon settling near the stele, increases in size over a period of two to three weeks. At this time the sausage shaped larva is referred to as being "spike-tailed"; it may be either male or female. Inside the second stage cuticle two rapid molts now occur in the absence of feeding. Emerging from the final molt is the adult female or male, as the case may be.

Females continue to feed and grow rapidly for the next seven to ten days. Egg laying begins generally about seven days after the final molt. It is alleged that females cease to feed soon after egg laying begins. This reportedly corresponds to a decrease in giant cell size once egg laying is initiated. Morphology of the adult female lends some confidence in these observations. The intestine of the female, as is common to many plant parasitic nematodes, lacks a defined lumen. The ovaries are intertwined throughout this intestinal mass. However, contrary to common belief the ovaries are not haphazardly coiled and flexed in the female body. The two ovaries are carefully laid in the body so that the youngest developing oocytes are at the anterior most part of the female body (Fig. 6.17A). Each gonad consists of an ovary with a short germinal zone followed by a region of growth with oocytes in tandem. During egg maturation the intestinal contents are used up in ovarial nourishment from the posterior of the body to anterior. Thus, until the very end of the females life the developing oocytes are bathed and nourished by intestinal tissue contents.

The anal opening, in adult females, does not function in defecation; rather the six anal or rectal glands become greatly enlarged and function to secrete the gelatinous matrix into which eggs are deposited. In dry conditions, when water evaporates from the matrix, the gelatinous mass shrinks and hardens and the egg mass changes from colorless to reddish-orange. The hardened matrix may serve to prevent water loss from the eggs. Though commonly deposited to the exterior of the gall these egg masses, in woody roots or multiple galls, may be retained inside the root or gall tissue. This is also true of root knot in potato tubers. Under these conditions larvae may not, upon hatching, leave the plant environment but rather migrate first within the plant to seek new feeding sites.

Migrating or penetrating larvae apparently feed because those that enter and exit roots can only do so about three times. After three entrances and exits certain morphological changes occur, and if the larva is not settled it will die. These irrevocable changes are probably stimulated by the feeding process engaged in during penetration.

Little is known about the activity of the adult male (Fig. 6.17B, C). Approximately one week after the final molt to adulthood the vermiform male forces its way out of the gall. In many species of root-knot the males are not required for reproduction; parthenogenesis, either meiotic or mitotic, is a common form of reproduction. The number of males produced in any given species, population, or biotype is never constant, and is influenced by host, growing conditions, and time of year. When the plant is unsuitable or conditions are unfavorable, such as overcrowding, the population of males increases. The majority of these are not normal males, but the result of sex reversal (arrhenoidy). The sex reversed males are easily recognized because they possess two testes instead of the single testis normal to Secernentea.

More studies should be directed to the biology, formation, and function of males. It is not even known whether or not they feed. Reportedly they do not, and a partial explanation is given that one of the glands of the esophagus is reduced or not developed. It is also difficult to offer an explanation for sex reversal under unfavorable conditions if the males in these populations are nonfunctional. Usually among invertebrates such a phenomenon is explained as increasing the gene pool mixture in order to assure survival. It may be that this phenomenon had significance in evolution, but is now just a nonfunctioning carryover.

The observation that males do not function in reproduction with any genetic exchange may be in error. Studies with *Meloidogyne* are generally conducted with populations developed from a single egg mass. This is done in order to assure that the investigator is working with a single species. Herein may lie the error, for being capable of parthenogenetic reproduction the offspring are veritable clones. Thus, a situation parallel to plants may occur, that is, clonal males are sterile within their own clone (population). However, with other nongenetically identical populations they are fully functional. Therefore, it might be more fruitful and enlightening to study functional reproduction among *Meloidogyne* with separate populations rather than single egg mass cultures that may well be doomed to self-sterility. Such a situation might also explain the increased production of males under adverse conditions. Being a mobile dispersal stage they may then seek and find females of unlike genetic composition, thus renewing the gene flow. In any event, these speculations and observations should be investigated and greater attention given to the potential biological differences between males with one or two testes.

The gall itself is due to hypertrophy and hyperplasia of the root cortex and partly to the development of giant cells on which the nematode feeds. Though in some hosts there is little increase in the diameter of the invaded root, giant cells are always formed. In plants that respond strongly to invasion by the development of large galls, the cortex forms the major contributing tissue of the gall by cell proliferation. Under unusual conditions galls can be found on the aerial parts of plants. This is especially true of leafy ornamentals when the leaves touch the soil surface or when there is sufficient moisture to allow larvae to move onto the plant.

The formation of the giant cells within the gall has been the subject of much controversy. Two schools of thought persist: one adheres to giant cell formation by cell wall breakdown and coalescence (lysigenoma); the other school is based upon the development being coenocytic, i.e., nuclear multiplication (karyokinesis) without concomitant cell wall formation (cytokinesis). The school of cell wall breakdown is based on tradition and the emotions so invoked. No evidence has been presented at the nuclear or electron microscope level that supports the continued belief in giant cell formation by cell wall breakdown. On the other hand, nuclear division without cell wall formation has been carefully documented at both the electron microscope level and light microscope; these observations have been confirmed with more than 600 developing giant cells. In addition to observing the nuclear division without cytokinesis it has been established that the number of nuclei in giant cells follows a geometric pattern: $2n$, $4n$, $8n$, $16n$, $32n$, etc; none have been observed with intermediate or numbers following arithmetic progression.

The cell wall breakdown proponents offer the following as an explanation based on observations in *Vicia faba* (broadbean) whose $2n$ number of chromosomes is 12: 192 chromosomes ($32n$) in a single giant cell could be attributed to the fusion of nine adjacent cells, seven of which contained tetraploid nuclei and two of which contained diploid nuclei. Similarly, 48 chromosomes ($8n$) per giant cell could have arisen by fusion of two diploid cells and one tetraploid cell. Such a complex explanation being proposed without any evidence may seem preposterous. Tradition can be a powerful enemy of science. The so-called "cell wall" pieces within the giant cell (used by cell wall breakdown proponents to support their thesis) are manifestations of secondary wall deposition. These secondary wall depositions initially assume rodlike form. Subsequently these increase in length and become branched, eventually developing into anastomosing trabeculae. This is confirmed by electron microscopic studies. The pieces of wall in paraffin sections result from cutting through the rods or branching of the secondary wall formations.

Recently indirect evidence has been presented that also supports the hypothesis of nuclear multiplication without cell wall formation. In tests with Orzalin ($3,5$-dinitro-N^4,N^4-dipropylsulfanilamide), which inhibits nuclear division, it was noted that neither galls or giant cells were formed after root-knot larvae attacked treated cotton and tomato plants. If cell wall dissolution induced by secretions from the nema forms the giant cell, the inhibition of nuclear division should not have stopped giant cell formation. On the other hand, if giant cells form because of nuclear division without cell wall formation, the action of Orzalin is explainable. Similar observations have been made with other plant growth regulators affecting mitosis. Therefore, if cell wall breakdown does occur, it is insignificant to the formation of giant cells.

The damage observed on infected plants is a result of the galling of the root system and consequently disruption of the vascular system. The most obvious top symptoms are stunting and a relatively greater root to shoot ratio. When water is limited, infested plants manifest increased susceptibility to stress. Initially wilting can be overcome with irrigation, but as the season progresses and transpiration increases, infested plants often cannot recover, whereas uninfested plants either show no stress or readily recover with irrigation. Within the gall there is increased synthesis of proteins concomitant with the disturbance of the normal transport of substances between root and shoot.

Infestations of root-knot, as well as infestations of other plant parasitic nematodes, often predisposes plants to infections by other microorganisms such as fungi and bacteria. This increased susceptibility to other disease organisms often extends to resistant plants, this is well documented with increased *Fusarium* wilt in resistant tobacco cultivars. Other studies on tobacco indicate that unless the plant is predisposed by root-knot there is no reaction to *Alternaria tenuis*. Other interactions are reported with *Phytophthora*, *Verticillium*, *Rhizoctonia*, *Pythium*, *Curvularia*, *Botrytis*, *Aspergillus*, *Penicillium*, and *Trichoderma*. Similar responses are noted for several bacterial diseases including those induced by *Pseudomonas*, *Agrobacterium*, and *Corynebacterium*.

b2. Heteroderinae   *Heterodera, Globodera*: Cyst Nematodes. These two economically important genera and related genera of lesser importance are very similar in their plant attack tactics. These two genera are separated by the shape of the cyst: *Heterodera* normally produces lemon-shaped cysts (Fig. 6.19) and *Globodera* generally produces spherical cysts (Fig. 6.20). The cyst is the tanned cuticle of the mature female body in which eggs are retained. The resistant cyst is generally the most readily available life stage recovered and much of *Heterodera* or *Globodera* species identification is based on cyst characters.

The life cycle of all species of cyst nematodes are substantially the same and very similar to that of root-knot nematodes. The only mobile stages are the second stage larvae and the adult male. The second stage infective larvae enter almost any area of the root of susceptible plants; however, the region just posterior to the root tip seems to be preferred. After a short distance of migration they settle in the cortex with their anterior end near the central stele. The orientation most generally assumed has the head near the vascular cylinder and facing away from the root tip. Upon settling down larval feeding stimulates the plant to produce special feeding

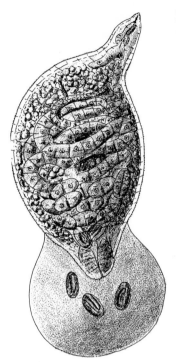

Figure 6.19. *Heterodera schachtii*; adult female (modified from Raski).

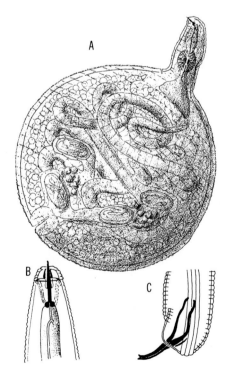

Figure 6.20. *Globodera rostochiensis.* A. Adult female. B. Larval head. C. Male tail (original).

cells like the giant cells of root-knot nematodes. Cyst nematodes only rarely induce galling; however, they always stimulate giant cells.

The life cycle, after completion of the second stage, differs between males and females. The development of the male through the third and fourth stage larva, like that of root-knot nematodes, proceeds inside the cuticle of the second stage larva without feeding. At the completion of the fourth molt the vermiform adult male emerges. Females do not undergo this type of metamorphosis, but continue to grow and become increasingly saccate, apparently feeding at each stage. The second stage molts to the third stage after approximately six days, then proceeds to enlarge and molts to the fourth stage in another six days. They only remain as fourth stage larvae about three days prior to molting to mature adults. By the thirteenth day the reproductive system is filled with eggs. As in *Meloidogyne*, the paired ovaries are not haphazardly laid inside the body, but are carefully organized to take full advantage of the nutrition supplied by the intestinal tissues.

Unlike root-knot nematodes, the mature female is seldom surrounded by plant tissue. At the time of initial invasion the larvae are completely embedded in root tissue but as the female larva develops and begins to assume the enlarged saccate form, plant cells are crushed and forced aside. Therefore, at maturity the swollen portion of the female body is on the outside of the root and only the neck and head remain embedded in the central root cylinder surrounded by giant cells. When all the eggs are produced the female dies, and due to the action of polyphenol oxidase on polyphenols in the cuticle, the cyst is formed. During the process females turn from white to yellow or light brown to dark brown or golden. Some cysts are covered by an additional deposit called the subcrystalline layer. Once formed, the cyst can persist in the soil for several years. A few eggs may be deposited outside the body; these may even be deposited in a gelatinous matrix. Eggs can persist in the cyst for several years and are stimulated to hatch in the presence of host plants. This retention of eggs inside the body of the dead female is correctly called *endotokia matricida*. Some eggs do continuously hatch. This percent per year hatch is the foundation of crop rotation. In the absence of a host the decline within the cyst may be 30–50% per year.

Some of the more important species are the sugar beet nematode, *H. schactii* (Fig. 6.19); cereal cyst nematode, *H. avenae*; pea cyst nematode, *H. goetingiana*; soybean cyst nematode, *H. glycines*; and golden nematode of potato, *G. rostochiensis* (Fig. 6.20). These vary from wide host ranges to narrow host ranges. All the above are serious pests and subject to a variety of control measures. The most common approach to the control of cyst nematodes is crop rotation or resistant varieties. Soil fumigation is used, but often crop value does not support such an approach.

H. schachtii: *Sugar Beet Nematode.* This nema occurs virtually wherever beets or crucifers are grown (Fig. 6.19). The origin of this nematode is unknown, but it appeared in western Europe soon after the introduction of sugar beets. This nema must be credited for the widespread interest in plant parasitic nemas, for without the devastation wrought on the sugar beet industry in Europe in the late 1800s, interest in nemas as plant parasites would have remained at a low level for years, if not decades.

Beets were introduced in Europe in the early years of the nineteenth century. For almost thirty years the industry flourished and expanded without any indication of problems. From 24 sugar processing factories in Germany in 1836 the industry grew to 184 factories in 1860, and by 1870 there were 304 factories processing sugar beets. In 1876 24 factories in Germany closed down, due to the losses induced by this nematode.

It was common practice to grow beets in the vicinity of the processing plant. This was the common practice in the United States as well. Tare soil from the beets delivered to the factory was returned to the beet fields. Waste water from the factories was pumped out over the surrounding fields for irrigation. These procedures rapidly caused an increase in sugar beet pests and aided widespread distribution, especially of sugar beet nematode.

The causal agent of sugar beet decline was first seen by Herman Schacht in the 1850s in fields in which it was believed the soil had become exhausted due to the repeated culture of sugar beets, a condition known as Rubenmudigheit or "beet weariness."

Cyst nematodes are one of the few parasitic nemas that can be seen with the naked eye. When infested beets are dug, the white female bodies, exposed on the outside of the root, are easily distinguished.

In 1871 Schmidt named the pest *Heterodera schachtii*, but it was nearly 20 years before the world was convinced of the true nature of beet weariness. Unfortunately, for the sugar beet industry but fortunately for the science of nematology in the United States, the history of Europe was repeated in the western United States. In the late 1920s an intensive program of crop rotation under supervision was initiated in the United States against this nematode. This program lasted for more than 20 years and contributed significantly to the public awareness of nematodes in the United States and resulted in public funds being directed to the budding science of nematology.

Globodera rostochiensis. While investigating sugar beet nematode in Germany, Kühn in 1881 found cyst nematode on potatoes. However, it was 40 years before it was described and 60 years before it was accepted as different from *H. schachtii*. Today we know this nematode as *Globodera rostochiensis*, the golden nematode of potato (Fig. 6.20). If plant nematology owes its origin to the sugar beet nematode, it owes its financial support and growth to the golden nematode. In the late 1940s to early 1960s it is likely that more money was directed to research on this nematode than all other plant parasites combined. It is an extremely devastating problem in potatoes, but fortunately because of strong quarantines it remains somewhat isolated in the United States. Strong restrictions also prevail in Europe. Of particular interest is the strong stand taken against this nematode in the Netherlands. On land where potatoes are grown for consumption, if the nema is found, the land can no longer be used to grow potatoes. Even though the nema is not found, potatoes must be grown under strict supervisorial rotations of four to six years. When potatoes are grown for alcohol they may be grown twice in four years on infested land providing one of the crops is a resistant variety and the soil is fumigated once in four years.

One of the most intensive areas of research has been the exploration for "hatching factors." The idea being that the most susceptible stage is the infective larva, in as much as other stages are protected by the root, the cyst, or the eggshell. The female upon bursting through the root is also subject to control by nematicides. The currently available organophosphates and organocarbamates are not really effective. Therefore, if a substance could be found to induce hatching from the cyst, then preplanting fumigation would be more effective.

Hatch stimulatory activity appears to take three forms with respect to general modes of action: metabolic, consisting of substances serving as substrates, cofactors, or coenzymes of conventional physiological systems; pseudometabolic, consisting of substances reacting competitively with physiological reaction systems; and ametabolic, consisting of substances with nonphysiological modes of action.

The stimulus behind this research stems from W. Baunacke's observation in 1922 that the leachings from host plant roots stimulated larval emergence. There is evidence that stimulatory substances in natural leachings can be consumed or utilized as substrate during the process of hatching. Inhibitors may also be present in cysts and in natural leachings together with activation from host and nonhost plants. The whole phenomenon is subject to complexities because of the diversity of plants whose leachings can stimulate larval emergence and of the numerous synthetic compounds capable of inducing hatching. In addition, there are numerous physical and environmental parameters important in the hatching process and in any given cyst lot's readiness and capability to hatch. Though still an active area of research, there is no conclusive answer to this complex phenomenon.

Even though the cyst nematodes are predominantly amphimictic they are subject to the production of biotypes, and therefore, caution must always be shown when introducing or developing resistant plant varieties.

B. Nematodes as Parasites of the Aerial Parts of Plants (Shoots)

A discussion of nematodes as parasites of the aboveground parts of plants, including tubers and bulbs, cannot completely exclude insect parasitism. In order to present this hypothetical evolution it is necessary to return our attention to the slender speared fungus feeders in the order Tylenchida and

Aphelenchida. Both orders contain taxa that withstand desiccation, and are parasites of the aerial parts of plants, insect associates, or parasites of insects.

These two orders are in some classifications treated as suborders in the order Tylenchida. However, it is becoming increasingly apparent that these orders exemplify convergent evolution. Therefore, if phylogeny is to be reflected in the classification, independent ordinal ranking is necessitated. Evidence indicates that the greatest development of Tylenchida as plant parasites of the aboveground parts of plants is coincident with the radiation of the Angiospermae (di- and monocotyledonous), while Aphelenchida more likely first parasitized Filicinae (ferns) and possibly gymnosperms.

As plant parasites they attack bulbs, stems, leaves, and seed heads. Indications are that within both orders aboveground plant parasitism and insect parasitism, in part, evolved from facultative fungus feeding insect associates. Other closely related species evolved this form of plant parasitism by being able to attack plant shoots before or shortly after shoots emerged from the soil. The ability of taxa in both orders to withstand desiccation is an important preadaptation to becoming successful aerial plant parasites or insect parasites.

1. Tylenchoid Parasites of Plant Shoots

The economically important aerial plant parasites in Tylenchida occur in the family Tylenchidae. Two genera are most significant: *Ditylenchus* and *Anguina*. Both genera can attack all aboveground parts (shoots) of plants and both are capable of inducing plant galls.

a. Anguina

The most important species of *Anguina* is *A. tritici* which holds the distinction of being the first plant parasitic nematode known to man (Fig. 6.21). It was discovered by T. Needham in 1743 from wheat seed galls. His letter was read before the Royal Society of London on 22 December 1743. The following is an excerpt from that letter:

> Upon opening lately the small black grains of smutty Wheat, which they here distinguish from blighted Corn, the latter affording nothing but a black Dust, into which the whole Substance of the Ear is converted; I perceived a soft white fibrous Substance, a small Portion of which I placed upon my Object-plate: It seemed to consist wholly of longitudinal Fibers bundled together; and you will be surprised, perhaps, that I should say, without any the least Sign of Life or Motion. I dropped a Globule of Water upon it, in order to try if the Parts, when separated, might be viewed more conveniently; when, to my great Surprize, these imaginary Fibers, as it were, instantly separated from each other, took Life, moved

Figure 6.21. *Anguina tritici.* A. Female anterior. B. Male tail. C. Young female tail (modified from Thorne).

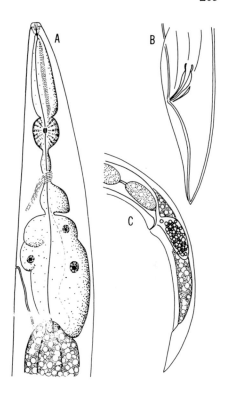

irregularly, not with a progressive, but twisting Motion; and continued so to do for the Space of Nine or Ten Hours, when I threw them away. I am satisfied that they are a species of Aquatic Animals, and may be denominated Worms, Eels or Serpents, which they very much resemble. This, if considered, will appear to be something very singular: But I have since repeated the Experiment several times, with the same success, and gratified others with the Sight of it.

The nematode remained, however, unconfirmed as the causal agent of the wheat disease known as "purples," "cockles," or "peppercorns," for more than 30 years when Rofferdi demonstrated the relationship in 1775. It was almost 60 years before it was named *Vibrio tritici* Steinbuch, 1799. This name was later changed to *Anguina tritici* (Steinbuch, 1799) Filipjev, 1936. Strange as it may seem, 114 years passed from the time of discovery until the life cycle was worked out by Davaine in 1857.

Each gall may contain from 500 to over 30,000 second stage larvae in an anhydrobiotic state. In nature the galls fall to the ground from the seedhead along with normal seed. Under agricultural situations the gall is sown with the good seed at planting. When the proper conditions of moisture and temperature exist for growing, the gall softens and the revitalized larvae escape into the soil to seek new plants.

It is not uncommon for the larvae to leave the gall in autumn and seek plants for overwintering. If plants are unavailable, the larvae overwinter in the soil where they can persist for at least seven months, or in the gall where they can persist for several years. Those larvae that remain in the gall under normal conditions make their exodus in spring. The nematode can invade young seedlings as soon as, but not before, the coleoptile loosens about the stem. Seedlings remain susceptible only as long as the terminal growing point is near the base of the plant. For this reason infection is enhanced by cool, moist weather; in hot weather the plant grows rapidly through the susceptible period, after which it cannot become infected.

Initially the larvae feed ectoparasitically even though they are protected by surrounding leaves and leaf sheaths. It is generally not until the formation of flower primordia that larvae actually penetrate into plant tissue. Occasionally, leaf galls occur and within the gall the larvae develop to maturity the same as in flower galls. Wrinkling and twisting of the leaves may result from feeding at the leaf base and is not necessarily a symptom of direct leaf invasion.

Soon after the formation of the flower primordia several larvae penetrate the tissue destined to become a seed. The penetration and feeding of larvae on this tissue stimulate the production of galls instead of normal seeds. Once the primordia is invaded the nemas rapidly develop to mature males and females. Soon after copulation the females begin egg laying. Each female produces several hundred eggs that hatch into second stage larvae. These are the larvae that survive in the gall for these nemas produce but one generation a year. However, these larvae can survive in the anhydrobiotic state for more than 25 years. Because the gall is produced from undifferentiated flower primordia more than one gall may replace a single kernel. Therefore, the number of galls in a given grain sample does not of necessity represent the number of kernels parasitized.

The effect on the plant is seen as stunting, wrinkled, rolled, or twisted leaves. The most obvious sign of the disease is the replacement of the grain by galls. If heavily attacked, seedlings can be killed. Though most commonly associated with wheat, this nematode also attacks rye, emmer (*Triticum dicoccum*), and spelt (*T. spelta*). Yields may be reduced as much as 65% and in the market, whole grain for sale may be as high as 3% galls.

This nematode is still common in many parts of the world, but is particularly serious in the Middle East, Eastern Europe, India, and Brazil. The disease is no longer known in the United States and has been virtually eliminated in Great Britain, Western Europe, Australia, China, and U.S.S.R.

The most effective control is mechanical separation of the seed and gall. Formerly, galls were separated by flotation in brine. However, in many areas of the world growers maintain their own seed and reinfestation is a way of life. In these countries flotation would still be a valuable tool for control. However, salt remains scarce or precious in many underdeveloped countries. Still another simple form of control is a hot water treatment of the seed; after 30 minutes at 50°C or 10 minutes at 54°C nemas within galls are killed.

There are several other species of *Anguina*, but most are of limited importance. A closely related genus *Cynipanguina*, which occurs on pasture and range grasses along the west coast of the United States, is of interest because the unmated adult male and female persist in the anhydrobiotic state and not the second stage larvae. Another interesting genus is *Subanguina*, which produces root galls on several grasses.

b. Ditylenchus

The most important and widely distributed species in the genus is *Ditylenchus dipsaci* (Fig. 6.22), and second in importance is *D. destructor*. The activity of these two is quite different and to some extent a clue to the differences is incumbent in their common names. This represents another of the few instances where common names can serve some slight service. *D. dipsaci* is known as "the stem and bulb nematode" and *D. destructor* as "the potato rot nematode." Both of these nematodes and most, if not all *Ditylenchus* can sustain themselves on fungi; however, the facility with which this can be accomplished is variable with species or races. For

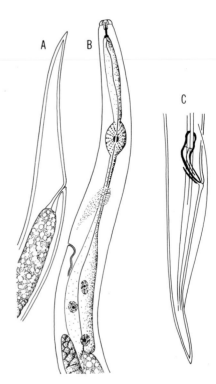

Figure 6.22. *Ditylenchus dipsaci.* A. Female tail. B. Female anterior. C. Male tail (modified from Thorne).

example, *D. destructor* can be reared in the laboratory on a wide variety of fungus species, but *D. dipsaci* can only be raised on a few species and then there with only a moderate increase in population. Two "successful" fungi for rearing *D. dipsaci* are *Verticillium theobromae* and *Cladosporium* spp.

b1. Ditylenchus dipsaci

The stem and bulb nematode, *D. dipsaci* (Fig. 6.22), is subject to the production of biotypes often referred to as "races." These races—some 11 have been named—vary from near stable to highly plastic. Dependence on race stability in a crop rotation scheme can be extremely dangerous. Few scientists today persist in the belief that the so-called "races" are truly identifiable or consistently reliable. This is not to say that awareness of the phenomena is not useful; on the contrary, recognition of this plasticity and its nature is extremely important.

Since its discovery by J. Kühn in 1857 in seedheads of Fuller's teasel (*Dipsacus fullonum*), *D. dipsaci* has been reported to attack more than 450 species of plants. Though found in soil, *D. dipsaci* is more often recovered from plant tissue where it passes generation after generation as an internal parasite of bulbs, stems, leaves, and occasionally as a seed associate.

Invasion of plant tissue generally occurs while the seedling is still below the surface. Feeding is largely confined to the parenchymatous tissues in stems, leaves, and bulbs. Because the preferred anhydrobiotic stage is the fourth stage larva, this stage is referred to as the infective stage. Injury is due not only to cell death, but to the breakdown of the interlamellae and interference with the normal growth habit. Infested stems become swollen and stunted and invaded leaves are often malformed. In cereals, such as oats and rye, heavy infestations cause excessive tillering and a swelling of the basal portion of the plant, which results in a condition known as "tulip root." Garlic is primarily attacked in the region of the basal plate, and the anhydrobiotic state is found in the scales surrounding the cloves. Where the garlic infestation is serious, the bulb and cloves become separated from the basal plate and roots, which results in the loss of the crop. Death is also characteristic in heavily infested onions.

Inflorescence invasion commonly occurs in beans, clovers, alfalfa, onion, and teasel and this then becomes an important source for widespread distribution.

The most commonly used control is hot water dips. When properly administered these can be highly effective. For example, the hot water dip developed by B. Lear for the control of *D. dipsaci* in garlic has virtually eradicated the disease in California. Prior to the discovery of this treatment, growers could lose up to 95% of their plantings to this nema.

The hot water treatment referred to is quite simple, but as with all such treatments, time and temperature must be exact. In order to wet the material and to have nemas in a proper condition, loose garlic cloves should be soaked for 30 minutes in 1% formalin–0.1% detergent solution at 37.7°C (100°F). The stock should be then immersed in a 48.8°C (120°F) for 20 minutes. Cloves should then be immersed in a cold water dip for a few minutes, followed by air drying at 37.7°C (100°F) for about two hours.

On narcissus bulbs a spectacular sign of the disease often develops in storage, where reproduction can continue. In heavily infested bulbs, the fourth stage (preadults) migrate to the region of the basal plate where they emerge and go into the anhydrobiotic state. The continuous mass emergence of the larvae soon becomes visible to the naked eye as grayish-white fuzz which is called "nema wool" or "curds." The other common sign on growing narcissus is the "spikkle" which is a small leaf gall Spikkles are light colored and easily felt when the leaf is run between the forefinger and thumb. When the bulb is cut transversally, the damage caused by the nemas is seen as a series of brown concentric rings.

b2. Ditylenchus destructor

b2. Ditylenchus destructor Few nematodes have a greater host range than *D. destructor*; however, it has gained its greatest prominence as a parasite of potato tubers. It is probably safe to say that this nema has not received its due attention as a parasite. More often than not it goes unrecognized because symptoms in root crops are often mistaken for other microorganism rots. Because of its facultative nature, which allows it to survive on a wide variety of fungi, it is often discredited as an important parasite of higher plants. However, laboratory studies show that this nema can destroy plant tissues and leave potato tubers mere mummified shells in the absence of fungi. In nature invasion of plant tissues by this nematode is more often than not followed by secondary invasions by other organisms, principally saprophytes.

Taxonomically this nematode is one of the most important warning signs to aspiring systematists. This nematode clearly illustrates the dangers of using length and plastic morphological features as the basis of a new species. In this nematode, characters normally considered conservative or profound are subject to variation according to host. Variation according to host as illustrated by *D. destructor* should shake the confidence of those who believe host ranges or "specificity" are legitimate reasons to describe new species of plant parasitic nematodes.

Incredible morphological changes occur when this nematode transfers from one host to another. These changes are not, as many would propose, due to mixed species populations. The variations that occur have been too carefully documented for such a trite explanation. The expected or common condition for *D. destructor* is to have the esophageal glands noticeably overlap the anterior intestine dorsally, and this is the case when the nema is reared on potato. However, when the nema is transferred from potato to sugar beet or bulbous iris, the glandular extension usually develops ventrally. In addition to esophageal changes, these nemas are generally two-thirds the size of their potato progenitors, and the number of developing oocytes in the ovary is greatly reduced. These changes are so dramatic, according to Thorne, that if the nema were found in the field, without prior knowledge, they would undoubtedly be described as a new species. Depending on the host, mature females may be as short as 0.6 mm, or as long as 1.9 mm.

This type of change in size and morphology is not restricted to *D. destructor*. G. Martin worked in Rhodesia with a population of root-knot nematodes, reportedly *Meloidogyne javanica*, that when grown on potatoes had the vulva position shift from terminal to near equatorial as is normal in *Meloidodera*. When progeny of these nematodes were returned to tomatoes, the vulva assumed the normal terminal position. By varying growing conditions for sugar beets D. R. Viglierchio was able to increase the diameter of *H. schactii* cyst almost threefold over field populations.

These observations deserve serious thought and should make us question how many of the over 2000 nominal species of plant parasitic nemas owe their existence to host-induced variability.

The varied types of injury associated with *D. destructor* will be described with a few of the more important hosts.

In the early stages of potato tuber attack, symptoms are not visible by external examination. These early symptoms can only be seen by peeling the tuber and looking for small, white chalky spots marking the location of the nematode. Because entrance into the tuber occurs in the region of the eye and lenticles, even the initial lesion formed by secondary microorganisms, invading where the nema has fed, cannot be seen unless the tuber is peeled. As the disease progresses the skin of the tuber shrinks, dries, and cracks.

Many continue to insist that the damage is not due to the nematode but to invading fungi. They insist that the nematode is not feeding on the potato but on the fungi. However, the nema has been shown to damage the tuber on its own. Even when fungi do secondarily invade the tuber, the bulk of the nematode population is well in advance of the fungi.

The symptoms on bulbous iris can be seen externally, but final confirmation requires a microscope. Examination of the bulb often shows a shredding at the base of the outer fibrous husk that protects the inner fleshy scales. Removal of the husk reveals yellowish, gray, or black longitudinal streaks that begin in the region of the basal plate. These streaks are most numerous on the outer scale. In advanced infestations these may coalesce, and in late stages, the entire scale becomes involved. This rotting may then extend through the entire bulb. The basal plate itself may be honeycombed and grayish in color. If it grows at all the plant is stunted and yellowish, and flower production is inhibited.

This nema is a very serious pest to field-planted bulbous iris grown for cut flower production. In California planting stock (bulbs) for the cut flower industry are largely grown out of state and are acceptable for shipment into the state if the infestation, by visual inspection, is less than 2%. However, we have found stock accepted from the state of Washington in actuality to be 22–27% infested when subjected to laboratory methods.

Under these conditions, when field planted, economic flower production can be expected for only one or two years. Unfortunately little can be done to protect the California industry.

Hot water treatments will control the nematode, but the treatment should be done just after digging from the bulb producers field when bulbs are dormant. Treatment at this time avoids damage to flower primordia, and therefore, cannot be done when the bulbs arrive in California. Treatment should be the responsibility of the bulb producer. It is the responsibility of nematologists everywhere to educate and encourage the producers whether of bulbs, trees, or seeds to produce clean planting sources. If we do not accept this responsibility then the backseat accorded to nematodes as disease agents is our fault and not to be blamed on the apathetic attitude of growers and other disciplines. The simple hot water treatment of iris is a good example of where the need for chemical fumigation of the soil or chemical applications to bulbs could be avoided.

D. destructor can be important in sugar beets. The nema enters the beet at the crown and the first evidence of attack is the appearance of discolored lesions under the cortex. These areas continue to enlarge and penetrate deeply into the beet; eventually due to secondary invasion the entire beet is destroyed.

This nematode is distributed throughout the world, and more often than not, its damage goes unrecognized. Few growers bother to have isolated rots identified.

b3. Ditylenchus angustus   Another interesting *Ditylenchus* is *D. augustus*, the causal agent of "Ufra disease" of rice in the Middle East, India, and several other countries through Southeast Asia. Soon after rice seedlings are transplanted, the nematode can be found in the terminal buds. As the plant grows they ascend with it feeding ectoparasitically on the newly formed tissues. At the time of flowering and heading, they can be found feeding ectoparasitically on all tender parts of the plants. The greatest damage is the feeding on flower parts resulting in sterile twisted grains. The nematode overwinters in an anhydrobiotic state within the dry glumes and stubble. The recommended control is the burning and ploughing of stubble, or fallowing during the hot season.

2. Aphelenchoid Parasites of Plant Shoots: Bud and Foliar Nematodes

In the order Aphelenchida the most important genera of plant parasites are *Aphelenchoides* with 197 species, *Rhadinaphelenchus*, a monotypic species, and the recently recognized *Bursaphelenchus*. Discussion of *Rhadinaphelenchus* and *Bursaphelenchus* will be postponed to Section II, B, 3.

Among the numerous species of *Aphelenchoides* four stand out as important parasites of the aerial parts or shoots of plants: *A. fragariae* (Fig.

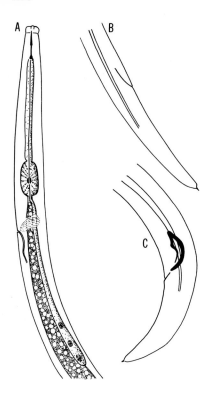

Figure 6.23. *Aphelenchoides fragariae.* A. Female anterior. B. Female tail. C. Male tail (modified from Allen).

6.23), *A. ritzema-bosi*, *A. besseyi*, and *A. subtenuis*. They will not be discussed separately, but this does not mean they are unimportant. The mode of attack and symptoms are similar, and differences are more characteristic of the host plant nature, rather than the species of nematode involved.

Most *Aphelenchoides* are facultative parasites and are able to sustain themselves, in the absence of a host plant, on fungi. However, few do well in the soil environment; therefore, survival is most often accomplished in a dormant or anhydrobiotic state in plant tissue.

The damage caused by these nemas may result from their feeding either ecto- or endoparasitically, and any given species can change its mode of parasitism according to plant or location on the plant. *A. ritzema-bosi* feeds ectoparasitically in the unopened buds of strawberry, but endoparasitically in the leaves of crysanthemum. When parasitizing black currant, *A. ritzema-bosi* is ectoparasitic in the buds but endoparasitic in the leaves.

The symptomatology of these *Aphelenchoides* can be lumped together. A common symptom is dead or devitalized buds. When the terminal growing point of an emerging seedling is attacked, a "blind plant" may result. This is not uncommon with even lesser pathogens as *A. parietinus* attacking cottonwood seedlings. In one instance of such an attack, state regulatory quarantine action was initiated before it was realized that the area in question was not a new finding of the dreaded "Broom Rape" parasite (Orobanchaceae) of tomato, but a blinded cottonwood seedling. Buds may become so heavily invaded that "blind stems" result on a portion of the plant. In still other instances the flower buds may be injured or killed and eventually fall from the plant. In the event that a bud or the growing point is not killed by the feeding nemas, the resulting stems, foliage, or flowers are twisted, crinkled, and distorted. In strawberries this can be reflected in "cat-faced" fruit. In those instances where *Aphelenchoides* feeds endoparasitically on the parenchymatous tissues of leaves or fronds the symptoms developed appear as spots or lesions. These leaf areas are commonly in the form of geometric patterns because the necrosis is bounded by the leaf veins. In these dry brown areas of foliage the nema can survive for several years in an anhydrobiotic state.

A disease in rice known as "white tip" is induced by *A. besseyi*. The activity of the nema is similar to that on other plants. The nema feeds ectoparasitically and the leaf tip of susceptible varieties of rice turn white or yellow for a distance of about 3.5 cm from the tip. These areas eventually turn brown or black as necrosis sets in. Some nemas make their way to the panicles which because of the nemas feeding produce small malformed kernels. Under the hull of these kernels the nematode goes into an anhydrobiotic state. They can remain viable in this location and state for at least 23 months.

Both *A. fragariae* and *A. besseyi* have been reported from the leaves and stems of various *Fiscus* spp. (rubber tree).

3. Aerial Plant Parasitism and Insects

We do not know the actual evolutionary pathways employed by nematodes in the achievement of insect parasitism or aboveground plant parasitism, but we can interpret from contemporary forms some of the more likely paths followed. We cannot say which came first, insect parasitism or the parasitism of the aboveground parts of plants. Perhaps it is safer to assume that both avenues were operative and probably concomitant. That is, in some instances insect parasitism among Diplogasteria developed directly from fungus feeding nematodes; in other cases insect parasitism developed from plant parasites. Plant parasitism, on the other hand may, in part, have developed from fungus feeding insect associates that were carried to aboveground plant parts, or aerial plant parasitism may have developed from the soil environment directly. Among modern forms there are examples that support all of these. Of these several possible lines of development, it appears from contemporary taxa and their habits that the fungus feeding Aphelenchida may have relied principally on insects for their introduction to the aboveground parts of plants (*Bursaphelenchus* and

Rhadinaphelenchus). Tylenchida, on the other hand, show more diversity, so that they are more likely to have developed their habits either directly from the soil environment (*Ditylenchus, Anguina*) or through the aid of insects (*Fergusobia*).

There are numerous examples of nematodes maintaining a phoretic (transport) relationship with insects. Often times the simple associations become rather elaborate; this is especially true with beetles. It is well established that such associations are common among bark beetles and wood boring beetles. Sometimes the nemas are encapsulated under the insect's elytra. This type of beetle, especially in the families Buprestidae, Meloidae, and Cerambycidae are not uncommon pollen feeders and therefore pollenizers of monocotyledonous plants like Liliaceae. In turn, the monocots (Liliaceae and grasses) are subject to parasitism by taxa in Tylenchidae (*Ditylenchus, Anguina*) and sometimes *Aphelenchida*. We know that members of both groups can be facultative parasites and must, over eons, have been deposited billions of times on plants by pollenizing insects. It is not inconceivable that some would have found conditions suitable for life sustenance.

Three contemporary species illustrate the hypothesis of a close interaction among aerial plant parasitic nemas, insect parasitism, and facultative insect parasites. The first, *Bursaphelenchus xylophilus*, illustrates an external phoretic relationship between an insect and a plant parasitic nematode; the second, *Rhadinaphelenchus cocophilus*, is an internal phoretic relation between an insect and a plant parasitic nematode; the third, *Fergusobia curriei*, involves parasitism of both insect and plant by the nematode.

a. Bursaphelenchus xylophilus (=lignicolus)

This species is plant parasitic on conifers and is carried phoretically from tree to tree by beetles (Fig. 6.24). In Japan where the seriousness of the nemas was first recognized, the Longhorn beetle, *Monochamus alternatus* (Cerambycidae), is the main transport host. The nematode is carried from tree to tree as fourth stage dauerlarvae either under the beetles elytra or in their tracheae. Large numbers of nemas are carried in this manner. A single beetle transports an average of 15,000 nematode larvae, and the maximum recorded is 175,000 larvae per beetle. The disease induced by these nemas is extremely serious; two-year-old trees can be killed in the span of a single summer. Experimentally trees can be killed by this nematode in 40 days.

The nematode can be recovered from all parts of the tree (stems, branches, and roots) where they feed on the parenchyma and epidermal cells of the resin canals.

During July and August the beetle lays its eggs under the bark of black pine (*Pinus thunbergii*) and red pine (*P. densiflora*) and several others

Figure 6.24. *Bursaphelenchus xylophilus.* A. Female anterior. B. Male tail tip with alae, ventral view. C. Male tail. D. Larval female tail. E. Adult female tail (modified from Mamiya and Kiyohara).

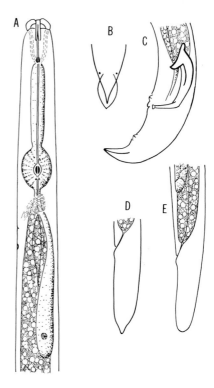

of the *P. sylvestris* group. The insect larvae, upon hatching, mine under the bark and eventually bore into the wood where they pupate and overwinter.

While the developing beetle is pupating, a change is seen in the nematodes abounding in the beetle gallery. The first change is seen among third stage larvae of the nematode. In preparation for dispersal the larvae become filled with large "oil" globules that obscure the internal structures. These larvae make their way to the beetle's pupal case where they molt to the fourth stage dispersal larvae. These are easily distinguished from normal fourth stage larvae by the absence of any feeding apparatus and by their tail shape, which instead of being bluntly rounded, has a sharply pointed digitate extension protruding from the rounded tail terminus (Fig. 6.24D, E). These larvae accumulate on the pupal case and align against each other perpendicularly, i.e., they all stand on their tails en masse. In this position they can be seen to wave like a wheat field in the wind.

During emergence of the adult beetle from the pupal case, each fourth stage larva begins to wave more vigorously and soon bends back on itself. Where it touches itself it forms an adhesion loop through surface water tension. With a strong muscular contraction the tension is overcome and the larvae is literally hurled from the pupal case onto the adult beetle as it is emerging. Larvae can hurl themselves in this fashion 20 times their length, or more than 1 cm. Once on the beetle they make their way to the tracheal system.

The adult insects that carry the dauerlarvae of the nematode emerge in June and early July. These adult beetles must undergo what is termed maturation feeding prior to the deposition of eggs. This maturation feeding takes place on the soft tissues and living branches of healthy plants and continues for about 30 days. It is during this time that the nemas quit the insect and begin their attack and development in the healthy trees. The fully ripe or mature adult beetles then move to suppressed or dead trees for egg laying. The suppressed trees are generally those infested with the nematode during the previous seasons maturation feeding. Thus, the trees weakened by the nematodes inoculated during beetle maturation feeding become the target of beetle oviposition the next year.

b. Rhadinaphelenchus cocophilus

Rhadinaphelenchus cocophilus (Fig. 6.25) is the causal agent of a disease known as "red-ring" of coconut and oil palms. The life cycle includes an internal phoretic relationship between an insect and a plant parasitic nematode. The males and females of *R. cocophilus* are exceedingly slender

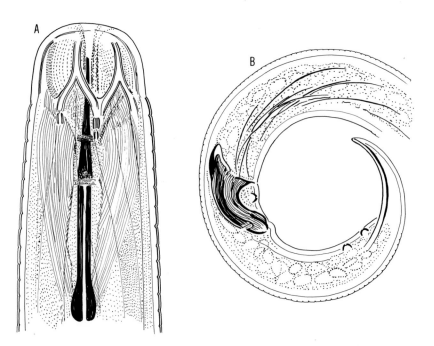

Figure 6.25. *Rhadinaphelenchus cocophilus*. A. Female head. B. Male tail (redrawn from Cobb).

nematodes. Their average length is about 1 mm, but they are 1/90 that in width. Even though males are present, reproduction is primarily partheno-genetic. The insect vector of the nematode is the palm weevil, *Rhyncho-phorus palmarum*; this is a very large weevil, the adult length averaging 34 mm, although infested weevil vectors are usually less than 30 mm.

The third stage nema larva is the infective stage for the inducement of the disease. They are the most common stage found in the red-ring region of the diseased palm. The red-ring lies about 2–5 cm below the trunk surface and is about 2–5 cm thick. The red-ring region extends as a cylinder from ground level to 5 or 8 feet up the trunk. This is the classic symptom of trees 3–10 years old. In older trees the ring may be quite diffuse and discoloration may extend to the center of the trunk. The greatest abundance of adult nemas and eggs occur in advance of the red-ring formation. As the red-ring progresses upward the feeding larvae stop their development in the third stage.

Larvae of the vector palm weevil feed on the diseased trunk tissue preferentially. During the course of their growth they consume large quantities of third stage nematodes along with plant tissue. In the period of development they consume over 500 gm of plant tissue and each 10 gm of plant tissue may house 50,000 larvae of the nematode. Therefore, an individual larval insect can consume 2.5×10^6 larvae during its develop-ment. The nematode persists in the hemocoel of palm weevil throughout its metamorphosis to the adult insect. The nematode shows no evidence of any change while in the insect hemocoel. In the adult palm weevil the nematode congregates in the vicinity of the ovipositor and is injected into the tissue of the tree when the insect deposits its eggs. The weevil oviposits in the soft parts of the crown, in the base of the tree, or in wounds or cracks found in the stem.

These two animals are well adapted to each other. Once within plant tissue the nematode completes its cycle in a few days; proceeding then from egg to egg every 9–10 days. The red-ring is evident 21–28 days after the deposition of the infective nematode by the weevil. The insect larval development is from 40 to 70 days; in other words, there are several generations of nematodes produced in this time, each contributing to the production of necrotic tissue sought after by the developing beetle larva.

In addition to the red-ring in the trunk, the petioles of the leaves also show a reddish discoloration in their center. As the disease progresses the lower leaves show a yellowing that progresses from the pinnae to the base. Eventually the leaves droop, and the crown develops the appearance of bud rot. The heart of the crown becomes liquefied and death of the tree soon follows.

At one time the burning of infested trees was suggested as a possible means of control. However, because of moisture, large portions of the trunks were not consumed in the fires and these portions began to ferment.

The fermenting trunks, with their still intact nema populations became extremely attractive to the weevil, and instead of a control measure, the burning enhanced the dispersion of the disease.

c. Fergusobia curriei

Fergusobia curriei (Fig. 6.26) illustrates alteration of generations between a plant parasitic generation and an insect parasitic generation: one parthenogenetic and one gametogenetic. These generations alternately occur in the eucalyptus gall fly, *Fergusonina tillyardi,* and *Eucalyptus camaldulensis* in Australia. Mating of the nematodes occurs in the eucalyptus gall. These impregnated females are the infective stage that penetrates into the gall fly maggot just prior to pupation. Male nematodes do not enter into the insect. During the pupation period of the gall fly the impregnated female nemas grow rapidly and they are found in association with the fat bodies of the pupating insect. Soon after the adult gall fly emerges the female nemas become gravid, and eggs are deposited in the hemocoel of the fly. After hatching the nema larvae make their way to the oviduct of the female fly. As the fly deposits eggs in eucalyptus tissue the larval nematodes are also deposited. These larval nematodes all develop into parthenogenetic

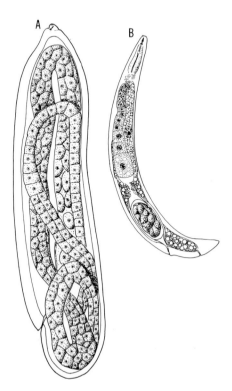

Figure 6.26. *Fergusobia currei.* A. Parthenogenetic female. B. Insect parasitic female (modified from Fisher and Nickle).

females. Eucalyptus gall formation begins before the insects hatch, but while the nematode is feeding on the plant, as such, it is not known what role the nematode may have, if any, in the formation of the gall versus the insect. Within the gall the nematodes proceed to molt three times to adulthood. The initial eggs produced by these parthenogenetic females develop into males possibly by arrhenotoky; later eggs produce females. At the time this is occurring the insect maggot has developed to the third stage and is ready to pupate. Thus, the cycle begins again when the maggot is penetrated by the impregnated females.

The production of males by the first layed eggs is extremely interesting and should be carefully investigated. This may be a case of male production as seen in other invertebrates, but unreported for nemas. The first eggs may be haploid and thus males (arrhenotoky). These males may then fertilize the females who now would produce diploid eggs, the source of new females.

Even though in many instances, perhaps even most instances, aerial plant parasitism and insect parasitism developed directly and independently from fungus feeding nematodes that became facultative parasites, the intimacy with plant parasitism cannot be overlooked as one operative mechanism. Likewise the role insects may have played in introducing nemas to the tops of plants, and vice versa, cannot be ignored. These are viable options and by recognizing the many pathways in evolutionary development of insect and plant parasitism we will be better able to comprehend the biology of parasitism than if we confine ourselves to a single authoritarian concept unsupported by contemporary forms.

Chapter 7
Invertebrate Parasitism and Other Associations

I. Introduction

The most publicized form of invertebrate associations with nemas are those of insects. This is partially due to the recognized potential of nematodes as biological control agents of pestiferous insects and additionally to the role insects play in vectoring nematode parasites of humans and other vertebrates. No one knows the total extent of nematode–insect interactions. They have been recorded to occur in 19 orders of insects and the estimate of insect species interactions exceed 3000; most of these encounters do not involve parasitism. These figures are also misleading because they are more representative of the numbers of interested nematologists rather than the true extent of nematode interactions. In the number of encounters, according to insect orders, Coleoptera lead, and are followed by Diptera, Orthoptera and Lepidoptera, with fewer encounters among Hymenoptera and the other orders. Among other invertebrates the largest numbers of encounters are recorded for Annelida and Mollusca.

Parasitic nematodes and associates of invertebrates, including insects, are found in both Adenophorea and Secernentea. Insect and invertebrate parasitism among Adenophorea is reported from two orders: Trichocephalida and Mermithida. The former are vertebrate parasites that often utilize invertebrates as intermediate hosts, while Mermithida encompasses some of the most important and best-known obligate parasites of insects. Mermithids also parasitize slugs and snails. It is generally accepted that both of these orders of nematodes evolved from a dorylaim-like ancestor whose contemporary counterparts exhibit a wide variety of feeding habits including predation, algal feeding, and higher plant parasitism.

Within Secernentea the associations are more widely spread and range from phoresis to obligate parasitism and occur in all three subclasses: Rhabditia, Spiruria, and Diplogasteria. Taxonomic relationships are not as influential in the determination of the type of association the nematode has with invertebrates as the feeding habit. The type of association encountered among nematodes and insects or other invertebrates stems from two ancestral lines: one evolved from bacterial feeders and the other from fungal feeders. Recognition of this has great significance to our understanding of the difference between an insect association and obligate insect parasitism. True obligate insect parasitism in Secernentea, excluding higher vertebrate parasites that use insects as vectors, originated among the fungal feeding nematodes.

A. Association vs. Parasitism

The difference between an association and parasitism, as used here, revolves around nematode life stage development. In an **association** there is no development of the nematode in or on the invertebrate or insect. If any development occurs, it is only after the death of the insect or invertebrate involved. In **parasitism** the nematode undergoes partial or total development at the expense of the living host and this may or may not involve measurable pathology.

Unequivocal obligate insect parasitism among Secernenteans has its ancestral origin among fungal feeding nematodes in the subclass Diplogasteria, where it occurs principally among Tylenchida and to a lesser extent in Aphelenchida.

Obligate insect parasitism, where the insect is the sole source of the energy and nutrition necessary for the completion of the nematodes life cycle, is unconfirmed among nematodes of a bacterial feeding origin. There is one alleged case of such a development with *Oryctonema genitalis*. There is as an offshoot from bacterial feeders, a very special form of obligate parasitism that is seen among the nematode parasites of vertebrates which utilize insects as vectors or intermediate hosts. The nematode does undergo some development within the insect. It will become apparent, under vertebrate parasitism, that in many instances this relationship is likely to have developed after the nematode was established as a vertebrate parasite. Even though these parasites may have developed from bacterial feeders, their association with insects occurred secondarily.

Among invertebrates, other than insects, there are many instances of obligate parasitism by nematodes with a bacterial feeding origin. These are especially evident among annelids and molluscs. It is also common for closely related nematodes to utilize these same invertebrates as intermediate hosts for vertebrate parasitism. **Facultative parasitism** where the nematode may alternate between successive free-living generations and successive parasitic generations is also known among bacterial and fungal feeding groups.

It can be concluded from the foregoing that in those examples where an insect is "parasitized" with a nematode whose ancestry is among bacterial feeders, the thrust of the line of evolution is toward vertebrate parasitism, not insect parasitism, and that obligate insect parasitism evolved from fungus feeding nematodes.

B. Associations Other than Parasitism

Two fundamental types of association occur between nematodes and invertebrates or insects. One is **phoresis** where the nematode is transported from one external substrate to another. The other form of association is obligatory to the life cycle of the nematode. In this situation the nema eventually utilizes the insect or invertebrate as a substrate for nourishment. Most of the reported associations are simply instances of phoresis where the role of the invertebrate or insect is merely to carry the nematode from place to place. A phoretic association may be either nonessential or essential and can be external, internal, or a combination of both.

Nonessential phoresis at its simplest could be interpreted as accidental transport. It could be that this phenomena is the first step in the development toward a parasitic habit. Encounters of accidental transport are seldom recorded because the observations are random and scattered. A fly, for example, visiting moist dung or other decaying substrates, where bacterial or fungus feeding nematodes are active, may acquire nematodes on their tarsal hairs or body. When the fly visits new substrates, the nematodes are free to leave the fly and continue their life. Such an association of short duration does not require any specialized resistant stage in the nematode's life cycle, nor is the phoresis necessarily essential to the continued existence of the nematode.

Essential phoresis, on the other hand, is necessary if the nematode is to continue its life cycle and preserve the population or species. Generally, essential phoresis involves a special stage being incorporated into the nematode life cycle. This need not be an anhydrobiotic state; other mechanisms for preservation include a sojourn in the insect's trachae or hemocoel. Most often it is the third stage larval nematode that is transported regardless of the mechanism of survival. In other invertebrates the nema may be carried in the nephridial system (annelids) or genital system.

An example of essential external phoresis is the association of the free-living rhabditid *Pelodera coarctata* and the dung beetle *Aphodius fimentarius*. The nema is a common associate of dung where it finds a rich supply of bacteria. As the substrate ages and begins to dry, third stage larvae of the nematode attach themselves to the exterior of adult dung beetles. Once attached, they quickly transform to a resistant state. In this condition they are called "dauerlarvae." Their attachment and arrangement on the body of the insect is not haphazard, rather they form dense geometric clusters on various portions of the insect's body. When the beetle visits a fresh dung pat, the nemas revive and quit the beetle in order to resume and complete the life cycle.

In Chapter 6, Section II,B,3, other examples of essential phoresis were presented. *Bursaphelenchus* exemplified external phoresis and *Rhadinaphelenchus* exemplified essential internal phoresis. There are many examples of such relationships, especially between tree inhabiting beetles and fungus feeding nematodes. In these instances the nematode may be transported under the elytra, in intersegmental folds, on the genitalia, or in the gut.

In other essential associations the insect or other invertebrate is victimized indirectly by the sojourning nematode. The association may be of an extended duration whereby the victim runs a normal or near normal life expectancy or of short duration where the nema vectors a disease agent to the victim which serves subsequently as a food source for the nematode.

These relationships are essential ("obligatory") because the nematode cannot exist in nature for long in the absence of a victim, be it an insect or some other invertebrate. It is among this type of association, especially where the nema vectors a lethal disease to the insect, that some of the most important biological control agents are to be found. Probably the best known forms are species of Steinernematidae, particularly those in the genus *Neoaplectana*.

The examples that follow are selected to represent the various associations described above, with the exception of phoresis. Most are potentially important biological control agents.

II. Facultative Parasitism

The ability to alternate between free-living (mycetophagous) generations and parasitic generations should not be confused as being simplistic or reflective of the primitive state. It may be true that this ability is ancestral; however, contemporary nematodes have often achieved a highly derived development to the parasitic phase of the life cycle. For this reason the terms "primitive" and "advanced" are to be avoided. For example, the five digit condition is a "primitive" (ancestral) condition among terrestrial vertebrates and as such the single digit (hoof) of the horse would have to be considered more "advanced" than the hand of primates with an opposable thumb. In this example the opposable thumb and the hoof are both derived; advancement is not inherent to the concept or to the needs of the animals involved.

A. Deladenus siricidicola

Among nematodes illustrating the phenomena of facultative parasitism *Deladenus siricidicola* (Fig. 7.1), which has alternate generations, parasitic in woodwasps of the genus *Sirex* and mycetophagous on the fungus *Amylo-*

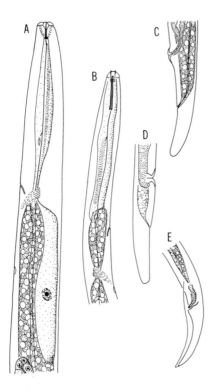

Figure 7.1. *Deladenus siricidicola.* A. Mycetophagous female, anterior. B. Infective female. C. Posterior region of mycetophagous female. D. Posterior region of infective female. E. Male tail (modified from Bedding).

stereum areolatum, has some fascinating features that elucidate how sophisticated facultative parasitism can become.

In the mycetophagous cycle, which occurs in dying or dead pine trees, the adult male and female nematodes are typically wormlike. The female (Fig. 7.1A, C) is close to 2 mm in length and the male is somewhat shorter, averaging nearly 1.5 mm (Fig. 7.1E). Mating among these mycetophagous forms proceeds within five days following the final molt to maturity and both the males and females mate more than once. In this free-living phase the males produce large amoeboid spermatozoa. Soon after copulation is completed the females begin egg laying. A single female may deposit over 1000 eggs in the span of her lifetime. Older females, who no longer are capable of oviposition, usually retain some unhatched eggs within their body. These eggs do not hatch until the female dies.

The population actively feeds, propagates, and grows wherever fungus is growing within the tree: tracheids between the wood and bark, in the galleries of larval wasp, and in resin canals. However, this migratory activity only proceeds when the internal tree moisture is less than 50%. As the fungal colony, especially that in the vicinity of the siricid wasp larvae ages and deteriorates, the nematode is induced to metamorphose to the parasitic generation.

Parasitic males are distinguishable from the mycetophagous males by the size and nature of the sperm. The sperm in the mycetophagous phase measures 10–12 μm and is described as amoeboid, whereas in the parasitic phase the sperm is only 1 μm in diameter. Even more striking is the difference in the total length of the females: parasitic females are only two-thirds the size of free-living females. The stimulus for these changes is unknown, but is allegedly triggered by a deteriorating fungal substrate.

Invasion of the woodwasp larva is accomplished only by the fertilized infective female. Copulation takes place in the exterior environment, and each infective female upon entering the woodwasp has been impregnated with some 15,000 spermatozoa. Infective females (Fig. 7.1B, D) can penetrate anywhere along the body, but generally invasion is concentrated laterally and posteriorly along the edges of the larval wasp's segments. Points of invasion are detected as melanized spots on the host surface.

Upon entry into the insect hemocoel some remarkable changes take place. Most notable is the postadult "molt" of the infective female. Initially the female undergoes a threefold increase in breadth accompanied by a slight decrease in length. Soon after she attains this size and shape the cuticle ruptures and the external surface of the hypodermis develops microvilli. This development of microvilli and the absence of an external cuticle is most unusual among Nemata. The obvious prime function is to facilitate direct absorption of nutrients through the general body surface. Coincident to this major advantage is the freedom of the nematode to increase in size without the restrictions imposed by a nonexpanding cuticle. (Other nemas such as the giant intestinal worm, *Ascaris*, compensates for body increase by continuously forming cuticle. *Ascaris* molts to adulthood when 2 mm long, but fully grown the length often exceeds 350 mm.)

The females having "molted" grow rapidly and full size is highly variable, ranging from 3 to 25 mm. Regardless of size no great change is noted in the reproductive system until the host enters pupation, which may be as long as two years after the initial invasion by the nematode. This indicates that ovarian development in the nematode may be under the hormonal or hemolymph control of the insect. During insect pupation the nematode's gonad becomes tremendously enlarged and each female is capable of producing 500–2000 eggs. Initially the larvae that hatch within the female nema escape by way of the vulva, but near the end of the insect's pupation, the female is packed with larvae and these young nemas make their way to the insect hemocoel by emerging through the body surface of the mother nematode. Once in the hemocoel the larvae make their way to the insect's reproductive system. During the adult female's development in the host little damage to the insect, beyond cuticle scarring and some fat body reduction, is noted. However, depending on the species of siricid involved, larval penetration of the insect's reproductive system near the end of pupation can cause ovarial suppression. In addition, the insect's eggs generally become sterile because almost all the insect's eggs are invaded by the nematode. Only a few nematodes remain free in the insect's ovaries and oviducts. In other species of siricids, egg invasion does not occur and the nemas remain in the ovaries and oviducts. The phenomena of entering or not entering the host's eggs also varies with different strains or populations of the same species of siricid. In any event, the nematode is deposited into the target tree at the time the insect attempts to deposit her own eggs. In male insects the nematode's development is a dead end.

In addition to being an interesting life cycle, the fascinating morphological changes have great significance in maintaining current Neotylenchidae as a valid family. Since the cycles can continue rather indefinitely, one or the other could be suppressed, in which case family relegation would be different. In this instance the free-living cycle would be assigned to Neotylenchidae and the parasitic cycle to a family of sphaerularoids. Both Bedding and Ruhm have pointed out this concept. Bedding showed that by prolonged culturing that favors the free-living cycle, the parasitic cycle could truly be suppressed. Perhaps more in-depth biological studies will prove that these are not isolated instances, but are, in fact, the rule among the so-called Neotylenchidae. In a communication to Bedding, Ruhm indicated that he had evidence that *Stictylus* (Neotylenchidae) was the free-living phase of *Sphaerulariopsis* (Allantonematidae).

Facultative parasitism among invertebrates other than insects does occur, but the development and pathology of the parasite are not as well documented as for *Deladenus* and *Sirex*.

B. Alloionema appendiculatum

This nematode *Alloionema appendiculatum* (Fig. 7.2) has been known for over 100 years, having been first described by Schneider in 1859 from the body of slugs, especially *Arion ater*. Like *Deladenus* several consecutive free-living generations may be interspersed between the parasitic generations. Unlike *Deladenus* several parasitic generations may occur in a single host. *Alloionema* is a bacterial feeder in the free-living cycle and, as discussed earlier, is thus more closely related to those forms that parasitize vertebrates. Evidence has been presented that would place them in Strongylida. This is consistent with strongyl parasitism, which among some forms, utilizes mollusks as intermediate hosts, i.e., *Angiostrongylus cantonensis*.

Males and females occur in both generations and no great morphological differences are seen. However, adults of the parasitic generation are much larger than those of the free-living cycles. Reportedly, free-living larvae also have caudal "appendages." The parasitic males and females quit the slug as immatures. Exit is generally through the foot. Once in the exterior environment they proceed to adulthood, mate, and produce progeny. Most often, unless conditions are unsuitable, the progeny develop to free-living

Figure 7.2. *Alloionema appendicula-*
tum. A. Female anterior. B. Male
posterior (modified from Chitwood
and McIntosh).

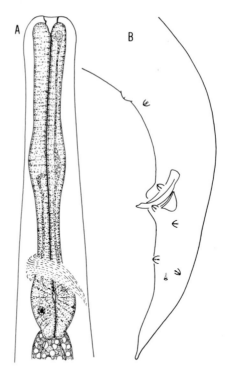

individuals. In the free-living state the cycle from egg to mature adult is
very rapid and has been observed to be completed in 3½ days.

When conditions become unfavorable in their environment, the free-
living generation declines and the larval stages become "encysted." In
order to survive and develop these "encysted" larvae must be taken in by
a slug. In the absence of a host, these larvae cannot survive more than
four months. Information is needed on how the "encysted" larvae enter
the slug.

III. Obligate Body Cavity and Tissue Parasites

A. Adenophorea

Obligate invertebrate parasitism among Adenophorea occurs in insects,
myriapods, and annelids. In addition, invertebrates may serve as intermediate hosts for higher vertebrate parasites; however, development of the
nema is limited generally to hatching or a single molt in the intermediate
host.

Of the Adenophorean parasites, Mermithida has received the greatest notoriety because of the numerous efforts to utilize them in the control of mosquitoes and blackflies that vector human diseases. There are some 60 nominal genera and 400 nominal species described in Mermithidae. Many exhibit some unusual features in their life cycle and a few of the better known forms will be discussed here.

1. Mermithidae

a. Mermis nigrescens

Mermis nigrescens seems to be limited to two families of Orthoptera: Locustidae, the grasshoppers; and Tettigoniidae, the long horned grasshoppers including katydids. The life cycle is extended over a two year period. Even though much of this time is spent in the soil, all nourishment for life sustenance was gained while in the insect.

Grasshoppers and katydids become infected by swallowing the nematode eggs that have been deposited by the female on plant foliage, especially grasses. This is a most unusual situation among Nemata. The adult females (Fig. 7.3A, B) reach lengths exceeding 10 cm; they are very agile and readily climb plants during spring or summer rains. The eggs are especially adapted to their exposed position on vegetation. Extending from the sides of the egg are long, filamentous branches called *byssi* (Fig. 7.3C).

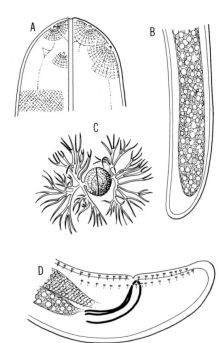

Figure 7.3. *Mermis nigresens*. A. Female head. B. Postparasitic juvenile tail. C. Egg with byssi. D. Male tail (modified from Nickle).

These multibranched stalks become entangled on the plant and thus hold the egg in place. The heavy dark covering of the egg is divided in half by an equatorial groove. This heavy outer covering is dark, almost black, and protects the enclosed embryo from harmful ultraviolet radiation as well as drying. Eggs can survive on foliage throughout the summer. When eggs are maintained within a moisture chamber, they remain viable for up to one year.

Upon being ingested by the grasshopper the heavy coating breaks apart at the equatorial groove. It is believed that this results from the digestive activity of the host. The shell proper is now exposed and it too splits along the equator. The hatched larva soon makes its way through the insect's gut and into the hemocoel. Penetration through the intestine and into the body cavity is facilitated by the stomatal spear of the larva. The sex ratio of *M. nigrescens* is influenced by the number of invading larvae that become established in the insect. If less than five invade they usually develop into females. However, if the insect is invaded by large numbers of larvae, then sex reversal occurs and all develop into males (arrhenoidy).

Once established the larvae undergo a remarkably rapid increase in size that is concomitant with an extensive development of the intestinal system. The invading female larva increases from 0.3 mm to 10 cm in the span of two months; male larvae develop to half this size in four to six weeks (Fig. 7.3D). At this point in the development, which is still pre-adult, the nematodes force their way out of the insect which is in turn killed by this action. On becoming free from the insect, which usually occurs from late summer to early autumn, the nematodes burrow into the soil. At this stage the nematode, still not fully mature, burrows into the soil to depths ranging from 15 to 45 cm. This is an individualistic endeavor and only rarely are males and females found together. The burrowing may, in exceptional situations, extend as far as 60 cm below the soil surface.

The following spring the nemas undergo their final molt to adulthood and at this time copulation may or may not take place; mating is not necessary for the production of viable eggs. Egg production and egg maturation proceeds throughout the spring, summer, and early autumn. By July females exhibit a brownish color due to the accumulation of eggs and by September they are black except at the anterior and posterior extremities. A fully gravid female can contain in excess of 14,000 eggs. No eggs are layed at this time; it is not until the following spring that oviposition begins, usually in May, and continues throughout the summer, but by August egg laying nearly ceases. Females ascend vegetation during daylight rains. Eggs are not deposited in mass, but are released singly as the female moves over the plant. Should the rains stop, the females coil up and drop to the ground where they immediately burrow below the surface to await another shower. Under certain conditions females may again overwinter and resume egg laying the following spring.

Egg laying is controlled by light stimuli; egg laying is not known to occur at night. When placed in the dark, oviposition stops, but promptly begins again upon exposure to light. The anterior extremity of adult females is pigmented reddish-brown and this is alleged to be a light sensitive organ. Circumstantial evidence is also presented by the fact that males, which do not have anterior pigmentation, are never seen on the surface of the soil. But it should also be remembered they have no comparable function to egg laying.

The eggs having been layed sets the scene for the cycle to begin anew.

The effect upon the insect is to suppress gonadal growth, especially the ovaries, and finally the instigation of the insect's death by the emergence of the preadult nematodes.

b. Agamermis decaudata

The host range of *A. decaudata* is reportedly wider than that of *Mermis nigrescens*. In addition to parasitizing Locustidae and Tettigoniidae they attack Gryllidae (crickets) and this or a closely related species has been recovered from leafhoppers and beetles.

When the preadults emerge in the late summer or fall the host is killed. Emergence occurs directly through the body wall between the segments. Immediately upon falling to the soil surface the preadults burrow to a depth of 5 to 15 cm. During the first winter the preadult males and females are isolated from each other in soil cavities. In the spring both males and females undergo their final molt to adulthood. At this time the mature males (Fig. 7.4D) actively seek out the females (Fig. 7.4B, C). Each female soil cavity soon houses several males and all become intimately entwined into a "knot." Copulation, which is essential for the production of viable eggs, now occurs. By midsummer females proceed to lay eggs within the cavity. Egg laying continues until the onset of winter. Eggs are not layed in a mass and soon they accumulate on the walls of the cavities and cover the outside of the entombed adults. Each female lays between 2500 and 6500 eggs.

The eggs remain dormant throughout the winter. With the warming of the soil in spring, hatching begins and continues to increase in rate well into the middle of summer. The hatched larvae are unusual in that ⅘ of the posterior body is devoted to food storage (Fig. 7.4A). The region of storage is clearly marked by a nodule. Anterior to the nodule are all the vital organs of the body; the "nodular" portion of the body contains a row of large cylindrical cells, which may be modified intestinal cells; no anus has been observed. These active, infective larvae make their way out of the cavity and up to the soil surface where they proceed to climb nearby vegetation during periods of high moisture such as dew or rain. Coincident with the high hatch rate of midsummer is the emergence of the host. The

Figure 7.4. *Agamermis decaudata.* A. Preparasitic juvenile showing node. B. Female head. C. Young male tail. D. Male tail. (A, redrawn from Christie; B–D redrawn from Nickle).

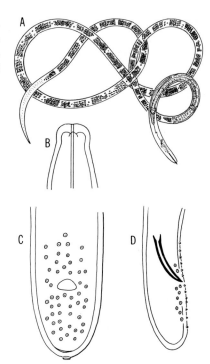

nematode actively seeks out and penetrates the newly hatched insect nymphs. Penetration into the hemocoel occurs directly through the body wall. The most common sites of penetration are under the edges of the pronotum and between the abdominal segments; however, they can penetrate anywhere the integument is thin and membranous. Penetration is effected by the stomatal stylet and is likely to be aided by esophageal secretions.

Once a portal of entry has been found the larva proceeds to enter the insect's body. An unusual thing happens to the larval nematode at this point: the ⅘ of the nema's body posterior to the node separates from the anterior body and is left on the outside of the insect. This discarded portion served the purpose of nourishing the larva in the external environment while it awaited the arrival of an appropriate host. Once inside the body cavity of the insect the same phenomenal growth noted in *Mermis* takes place. The preadult males remain in the host insect from 1 to 1½ months; females remain a little longer, persisting in the host for from 2 to 3 months. At the time of emergence fully developed preadult females may be 46.5 cm in length and the preadult males 12 cm. Upon emergence and falling to the soil the cycle begins anew.

In addition to causing the death of the host when emerging, severe damage is noted in the ovaries. The damage is greater than noted for

Mermis, because invasion by *Agamermis* is usually into nymphs, whereas *Mermis* can be ingested by any age of the host and the older the host the lesser the damage. Externally no damage is noted, however, the abdomen may be distended and the behavior of the insect is notably affected. Infested individuals are often sluggish and incapable of sustained flight.

c. Romanomermis culicivorax

Romanomermis is probably the most publicized nematode in the whole scheme of biological control of mosquitoes with nemas. In the literature prior to the mid-1970s this particular nema is referred to in the genus *Reesimermis*; this error was corrected by Ross and Smith in 1976. Students seeking more information should check the literature on *Reesimermis nielseni*.

This mermithid is included here because of its importance in biological control and because it represents an aquatic life style that can be readily manipulated by man.

The preadult female, upon emerging from the mosquito, molts to the full egg laying adult (Fig. 7.5A, B, D) in 10 to 30 days and each female

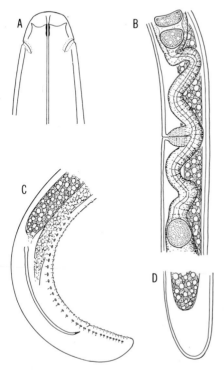

Figure 7.5. *Romanomermis culicivorax.* A. Female head. B. Female vulval region. C. Male tail. D. Female tail (A redrawn from Nickle; B–D modified from Nickle).

lays from 1300 to 4500 eggs. Males (Fig. 7.5C) mature earlier than the females. The nemas are considerably smaller than the other mermithids that have been discussed. The mature female averages 17 mm and the male, somewhat smaller, averages 10 mm in length. Eggs are layed some 10 days after mating and, in water, hatch in about seven days. Both males and females may remain active for at least six months.

The hatched larvae are large as compared to other mermithids. These infective larvae attain a length of 1.0 to 1.8 mm. They are strong swimmers that actively seek out their victim. Penetration is directly through the cuticle and can be completed in seven minutes. If the nematode does not find a mosquito larva to invade, they die in three to four days. Once in the mosquito, larval nematode development in the thoracic region is rapid. The parasitic stage lasts only five to ten days and is completed in the third or fourth mosquito instar. The nematode is not carried through the mosquito's pupation and, therefore, has never been found in adult mosquitoes. The mosquito instar does not survive the emergence of the preadult nematode.

As the pond water recedes and the mosquito population declines, nema eggs remain unhatched in the soil. Because the nema eggs can survive in this manner, the biological control scheme persists. Once introduced into an area, the nematode then becomes a permanent resident. Also, because the egg stage is the resting state, the nema can be laboratory reared and the eggs stored in sand. This inoculated sand can then be sent to areas for seeding ponds where mosquitoes occur. The recommended seeding rate is about 1000 infective eggs per square meter of water surface. The host range is very large and includes some 58 species of mosquitoes.

The nema is also under investigation for the control of the blackfly (*Simulium damosum*) which vectors Onchocerciasis. The nematode does not develop in the simulid, but kills the fly larva by penetration. The obvious drawback to this as a biological control agent is the lack of persistence in the environment. Continuous reseeding is required and often exceeds practicality. Currently of more interest is a mermithid parasite, *Mesomermis flumenalis*, of blackflies.

2. Tetradonematidae

Members of this family differ from Mermithidae in that development to adulthood and mating occurs within the host. The known hosts of tetradonematids are principally among Diptera, both aquatic and terrestrial; one unusual species is reported from beetles. The major morphological feature separating this family from Mermithidae is the reduced number of stichocytes on the esophagus; generally the number of these glands, in Tetradonematidae, is limited to three or four.

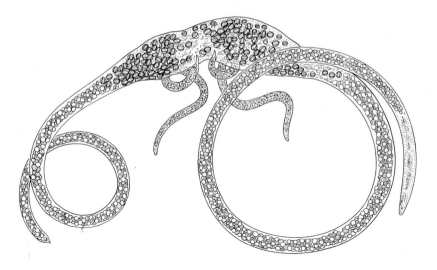

Figure 7.6. *Tetradonema plicans*; adult female with two males (modified from Hungerford).

a. Tetradonema plicans

The life cycle of *Tetradonema plicans* (Fig. 7.6), which is rather typical of the family, parasitizes fungus gnats and root gnats in the families Mycetophilidae and Sciaridae, respectively.

Eggs normally are deposited in the external environment, but in heavily parasitized hosts, eggs may hatch in the insect body. In the external environment eggs hatch within 24 hours, but how the infective stage enters a new host is controversial. It is unknown whether the infective larvae or unhatched eggs are ingested by the host and access to the hemocel is through the gut, or whether the larvae can penetrate the host's cuticle directly; perhaps all occur. Once inside, however, development is rapid, usually adulthood is achieved in eight days. During the four weeks of parasitism following mating, females are capable of producing in excess of 10,000 eggs. Fully gravid females ladened with eggs emerge from the host through the cuticle. The host fly is killed at this time. Upon deposition of eggs the cycle begins anew.

b. Heterogonema ovomasculis

The species *Heterogonema ovomasculis* (Fig. 7.7) is apparently unique among Nemata and the only unequivocal verified example of hermaphroditism in a naturally occurring nematode. It is also unusual for tetradonematids to parasitize a host other than among Diptera. *H. ovomasculis* is found, in nature, as a parasite of beetles in the family Nitidulidae, com-

Figure 7.7. *Heterogonema ovo-masculis*; adult hermaphroditic male (redrawn from Van Waere-beke and Remillet).

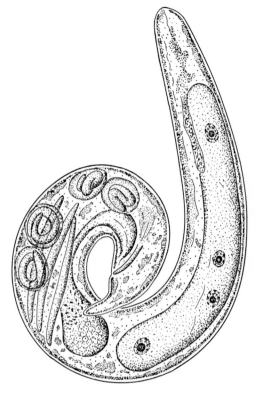

monly known as the sap-feeding or dried fruit beetles and in Cucujidae, saw-toothed grain beetles.

This nematode has two parasitic generations; in the first only females are present. These females, which are parthenogenetic, lack a vulva and young are released when the female body bursts. In the second generation both males and females are present. Unique, however, are the males of the second generation. The single larval gonad of the male divides into two parts: the anterior region becomes a functional testis and the posterior part gives rise to a functional ovary that produces a small number of larvae. The second generation females have a vulva, but larvae and eggs customarily burst free through the body wall of the mother in the same manner as the larvae of the first generation.

Eggs ingested by a susceptible host hatch in the gut, and the larvae penetrate into the host's hemocoel. Each larva develops into a partheno-genetic female that lacks a vulva. The eggs of this female hatch within her body, and the ensuing larvae burst free through the body wall of their mother. These larvae develop into the males and females of the second generation. After mating and becoming fully gravid, these females pene-trate from the hemocoel back into the host's gut. In the insect gut the female nematodes burst, thus freeing the eggs, which pass out to the environment through the insect's anus.

Males do not die after mating, but proceed to produce eggs in their own bodies (Fig. 7.7). The future and purpose of these eggs and larvae is unknown.

The result of nematode infection is sterilization and death of the host.

B. Secernentea

Obligate internal tissue and body cavity parasites in this class are limited to the orders Tylenchida and Aphelenchida. The majority of known species are found among the Sphaerulariina (Tylenchida) in four families: Sphaerulariidae, Allantonematidae, Fergusobiidae, and Iotonchiidae.

1. Tylenchida

a. Sphaerularia bombi

This is one of the oldest known insect parasitic nematodes and since first seen by Reaumur in 1742 has been subject to many investigations. One of the fascinating features of this parasite of bumblebees is its ability to completely evert the reproductive system, which then proceeds to enlarge independently of the original female body (Fig. 7.8).

Up to this point discussion has been directed to nematode parasites of pestiferous insects. This is not the case with *S. bombi* which parasitizes and sterilizes bumblebees, one of the most important noncultured pollinizers of native and cultivated plants throughout the world. Because of their long tongues, bumblebees are able to pollinate clovers, alfalfa, and many other deep blossoms that short-tongued bees cannot pollinate.

The queen bee of several species of *Bombus* is parasitized by this remarkable nematode, which reportedly also attacks some vespid wasps. The vespids most likely are attacked when they utilize old bumblebee burrows to overwinter.

When the parasitized queen bumblebees emerge from their overwintering burrows, the nematode is a mature female and may already be producing eggs. Even at this early date the queens exhibit abnormal behavior patterns. Flight patterns are altered and sustained flight is not observed. Rather short flights are alternated with crawling, and pseudoburrowing is interjected into their behavior.

By June the parasite, or rather the prolapsed reproductive system, is matured and third stage larvae are evident in the insect's hemocoel, midgut, and hindgut. Again the insect behavior is abnormal. The parasitized queens fly low over the hibernating grounds often trying to dig burrows,

Figure 7.8. *Sphaerularia bombi*; female attached to enlarged prolapsed uterus (original, drawn from a photograph by Poinar).

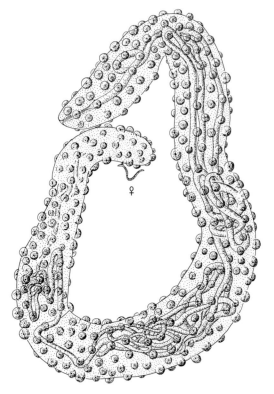

some of which reach a depth of 2 cm. Each time the bumblebee alights and attempts to dig, hundreds of third stage larvae are discharged or exit through the anal opening. These larvae molt during July and into August through the fourth larval stage to adults. During this same period infected bees die from the infestation. The adult nematodes mate and the impregnated females become the infective stage.

In autumn and fall when healthy queens are seeking hibernation sites, they are attacked by the infective female nemas. Natural attrition of the nema population must be extensive, since normally only one female is found per bumblebee queen. Records do show that captured queens may harbor over 70 nematodes; however, this is apparently rare.

It is evident that not only is there a change in behavioral pattern of flight but also the cycle of hibernation, that is, infective bees attempt to hibernate or dig burrows some 3–4 months before healthy bees. The change of activity allows the nema time to develop to the infective female and thus pollutes the hibernation grounds sufficiently in advance of normal hibernation procedures.

The development within the insect is a remarkable feature. The nematode upon prolapsing the uterus becomes an appendage to its own reproductive system. Often the prolapsed uterus may be only six times the length of the original female, but more often when only one or a few females enter the hemocoel, the prolapsed reproduction system may be 30 times the original length of the infective female and 300 times the volume.

For the parasitized bee the end result is death. If the bee dies before or during the time that nemas are quitting the body, the larval nemas may go into an anhydrobiotic state. It is not known how long they can survive in this state or whether it is an important mechanism in the total picture of the parasite's life cycle. The damage resulting from this parasite is most dramatic in the destruction or inhibited development of the insect's reproductive system; additionally fat bodies are reduced. No other class in bumblebee society has been shown to be attacked except the queens.

Most other members of the suborder Sphaerulariina have similar life cycles, that is, the infective stage is generally an impregnated female. It is also common to have some degree of uterine prolapse, though not quite to the degree exhibited by *Sphaerularia bombi*.

The family Allantonematidae is closely related to Sphaerulariidae, but in the absence of uterine prolapse, the reproductive system takes over most of the body so that gravid females are saclike or sausagelike and given over totally to the support of their reproductive systems. When they take on these forms they are often difficult to recognize as nematodes because all familiar structures of the anterior end become aberrant.

A notable occurrence that can happen in the life cycle of some Allantonematidae is an alternation between a gametogenic generation and a parthenogenetic generation, such as occurs in *Heterotylenchus autumnalis*, a parasite of cattle face flys.

b. Heterotylenchus autumnalis

Heterotylenchus autumnalis (Fig. 7.9) is being investigated, rather extensively as a potential biological control agent for the cattle face fly *Musca autumnalis*. Cattle face fly is an introduced species to the United States which normally occurs in Europe. The parasitism of the face fly by *H. autumnalis* holds additional interest in that the nematode was probably introduced into the United States along with its host in 1953.

Unfortunately, the term free-living has been applied to those transient phases of males and females that occur outside the host body in cow dung. This is a misnomer that implies feeding and development in the external environment which does not occur. After leaving the adult face fly, when she attempts to oviposit on dung, the liberated male and female nematodes mate. The infective impregnated female of this gametogenic generation then seeks out face fly larvae developing in the dung pat. Entry of the infective female is directly through the cuticle of the host maggot.

After attaining the hemocoel, this gametogenic female matures and be-

Figure 7.9. *Heterotylenchus autumnalis.* A. Adult female. B. Male tail (modified from Nickle).

gins egg deposition. The larvae that hatch from these eggs all develop into parthenogenetic females that are smaller than the mature parent infective female. The infective females range in length from 2.8 to 7.6 mm, while the parthenogenetic females are only 1.2 to 1.7 mm in length. The parthenogenetic female lays numerous eggs which, upon hatching in the host's hemocoel, develop into fourth stage females or sometimes adult males. These nemas make their way to the insect's reproductive system where they form packets of males and unmated juvenile females. Their position in the ovaries is that normally occupied by eggs. These nemas are the so-called free-living adults and preadults deposited during mock oviposition in manure; thus the life cycle is renewed.

2. Aphelenchida

Most of the associations with insects among aphelenchs are phoretic; in some, facultative parasitism has developed, but only in a few forms is obligate parasitism recorded. Three genera appear to be obligate parasites which retain their ability to feed on fungi during certain stages, but remain incapable of completing a life cycle in the absence of an insect host: *Peraphelenchus*, *Entaphelenchus*, and *Praecocilenchus*. It is phylogenetically of interest that these obligate parasites are limited to insect hosts among Coleoptera, namely, Silphidae, Staphylinidae, and Curculionidae.

a. Peraphelenchus

Among the parasitic species of *Peraphelenchus* only the mature impregnated female is found or can be found in the external environment (Fig. 7.10C). All other stages and all development occurs within the developing staphylinid beetle. Infective females penetrate the beetle larva directly and once inside they grow to the swollen parasitic females. The produced eggs hatch and develop into males and females that mate within the host body. The mated females penetrate into the insect's alimentary canal and exit through the anus. These females then actively seek another beetle larva to resume the cycle.

b. Entaphelenchus

Entaphelenchus differs in that it is not the adult that quits the insect, but third stage larvae. In the external environment the larvae undergo two molts to attain adulthood (Fig. 7.10B). It is not known whether there is

Figure 7.10. A. *Praecocilenchus*; female head. B. *Entaphelenchus*; female head. C. *Peraphelenchus*; female head. D. *Praecocilenchus*; male spicule. E. *Praecocilenchus*; tail of mature parasitic female (A–D redrawn from Nickle; E redrawn from Poinar).

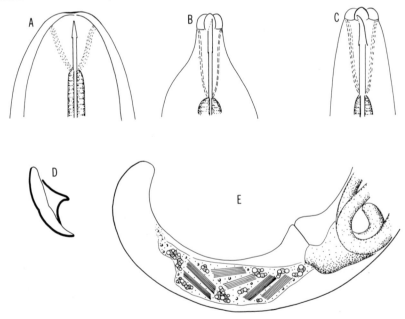

any feeding in this change to adults; however, the real point is that such feeding does not need to interfere with the obligate nature of the parasitism. A complete cycle apart from the insect has not been observed, indicating that these nemas are not facultative parasites. After mating, which occurs in the external environment, the fertilized females invade the pupal stage of the beetle. Beetle larvae have never been reported to be parasitized.

c. Praecocilenchus

Slightly different again to the foregoing genera is the cycle exhibited by *Praecocilenchus rhaphidophorus*, the only known species in this genus (Fig. 7.10A, D, E). The principal host is the adult palm weevil. Within the host hemocoel the impregnated infective female undergoes a significant growth and expansion. The eggs produced are not released into the host, rather they are retained inside the female nema where they hatch. Even more interesting is the fact that they also develop to adulthood within the mother's uterus. Mating presumably occurs also within the mother nematode. The fertilized females then make their way out of the mother's body and proceed to the external environment either by way of the host's reproductive system or alimentary canal.

In general the effects of these aphelench parasites on the host are minimal. They can, in large numbers, interfere with egg production or ovarial development. Death is not an inevitable consequence of aphelench parasitism.

C. Annelid and Mollusk Parasites

Little is known about annelid and mollusk parasites, most unusual nematodes. Generally they are placed in the superfamily Drilonematoidea. This superfamily has been historically placed in Rhabditina of Rhabditia (Fig. 7.11). Their morphology and habit, however, makes them more reminiscent of Spiruria and I recently proposed this move. Their habits and morphology are not too unlike those of other spirurid parasites of birds that utilize annelids or slugs as intermediate hosts. In some species the vulva is anterior to the base of the esophagus, a character not uncommon to other Spiruria. Somewhat unusual in nemic biology is the commonplace occurrence among these parasites to be in permanent copula with the much larger female. For unknown reasons the amphids and phasmids are often greatly enlarged, and the stoma is often armed with enlarged evertible hooks. Nothing is known of their life cycle or the damage they inflict. Larval stages have not been observed, or if they have, they were not recognized. Research into these nemas is much needed and may have great significance to our understanding of the subclass Spiruria, which is the only totally parasitic higher category in Nemata.

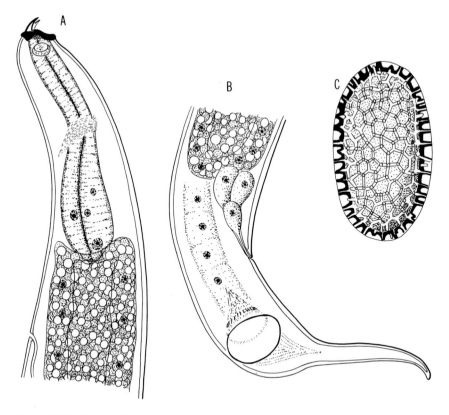

Figure 7.11. *Sucamphida robustum.* A. Female anterior. B. Female tail. C. Egg (modified from Timm).

D. Parasites of Leeches

The family Myenchidae is a small and little known group of nematodes, which in addition to leeches, have been recovered from amphibians. It is uncertain whether or not amphibians enter the cycle inadvertently or are themselves definitive hosts for these nemas (Fig. 7.12).

Within the adult leech, the adult nemas are located in the connective tissues and the larva are generally located in muscle. Adults have also been recovered from leech cocoons. It is presumed that the infective adult female starts the cycle within the cocoon. Two avenues for entrance into the cocoon are offered. One method would be via leaving the leech's reproductive system and entering the cocoon while this structure is being formed, and still encompasses the body of the adult leech. The other explanation is perhaps more plausible because of the location of the fertile female nematode in the leech. Since the adults are found in the subepidermal connective tissues, it has been proposed that they exit directly through the leech's body wall while the cocoon still encompasses the body of the

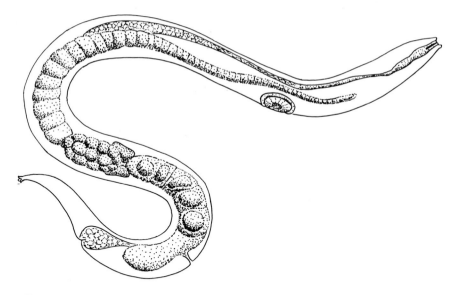

Figure 7.12. *Myenchus bothryophorus* (modified from Schulberg and Schröeder).

adult leech. Nematode eggs apparently are layed within the cocoon and the larvae hatching from these eggs infect the young leeches.

IV. Essential Nonparasitic Associations Including Vectors of Insect Diseases

These are the most difficult associations to understand, principally because they involve bacterial feeders. It is possible they might represent a diverse origin of invertebrate parasitism. However, they may equally illustrate a sideline development that is a dead end to direct parasitism. This is deduced because the true tissue parasites of insects are found in Diplogasteria (Tylenchida), which developed from fungus feeders, not bacterial feeders. The other tissue parasites are in Adenophorea or Spiruria; the feeding habits of their progenitors are unknown.

A. Nonpathogenic

1. Rhabditis maupasi

Rhabditis maupasi (Fig. 7.13) is an intimate associate of earthworms and has been described a number of times from *Lumbricus terrestris*, the common earthworm. Within the earthworm the nemas are found in the nephridia or in the coelom. In the nephridia they are located in the bladder region

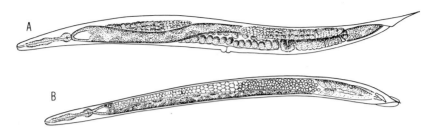

Figure 7.13. *Rhabditis maupasi.* A. Adult female. B. Adult male (modified from Johnson).

near the nephridiopore. Reportedly nemas may sometimes be found in every tube, and each tube may house from two to 12 or more nemas. Sometimes *R. maupasi* can also be recovered from the seminal vesicles. In these locations the nematodes are active and not in some resistant state; always they are active larvae. They remain as larvae so long as the earthworm lives. Upon the worm's demise the nematodes continue their development to adulthood within the body cavity and may, before the total destruction of the earthworm's carcass, undergo several generations. The nemas are most likely feeding on the bacteria that multiply in the decaying tissues and are not themselves feeding directly on the tissues of the earthworm. Once the worm is destroyed the nematodes enter the soil and await the passage of another earthworm. Survival in the soil is in the juvenile stages and these are the same stages found in the nephridia. There is neither development of the nematode nor harm done to the earthworm while the nemas sojourn in the nephridia.

Apparently the nematodes enter the worm through any natural opening; however, those that end up in the coelom are walled off by amoebocytes or blood cells. These encysted nemas are worked backward, by the natural movements of the earthworm, to the posterior extremity. Here they accumulate with other foreign bodies from the coelom. It has been observed that when sufficient foreign and encysted material accumulates, the earthworm is induced to form a stricture that severs the last segment from its body. Upon decomposition this detached portion may then release the encysted nematodes, as well as other coelomic parasites back into the soil environment.

B. Pathogenic

1. Neoaplectana

All the known species of *Neoaplectana* are vectors of insect diseases (Fig. 7.14). The disease is always bacterial. *Neoaplectana* is a member of the family Steinernematidae; the only other family, also in Rhabditia, that

Figure 7.14. *Neoaplectana glaseri.*
A. Female anterior. B. Female
tail. C. Male tail (modified from
Steiner).

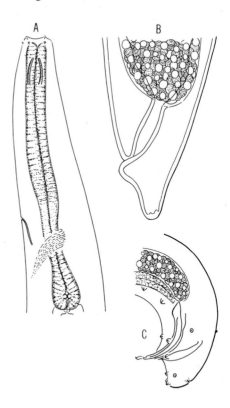

exhibits this type of symbiotic relationship with bacteria is Heterorhabditidae. The relationship might better be classed as one of mutualism, because both the nematode and bacteria benefit each other and are unable to sustain themselves, under natural conditions, in the absence of a victim insect.

The bacterium, which is assigned to *Xenorhabdus* (= *Achromobacter*) *nematophilus*, is carried in an expanded pouch in the anterior intestine of infective juvenile nematodes. The infective stage enters the host via the mouth or other natural openings; however, recently it has alleged that direct cuticular penetration is possible. Upon gaining entrance into the insect's hemocoel the bacterial pellet is ejected from the nematodes alimentary canal into the insect's hemolymph. Within the pellet the bacteria are nonmotile, but upon exposure to the insect's hemolymph they become motile. Multiplication of the bacteria is rapid and the insect dies from septicemia, generally, in less than 48 hours, though some hosts may linger for days. No development of the nematode occurs until the insect, whose every body cavity is invaded by bacteria, is dead or near death. Subsequent development of the nematode to adulthood is rapid, generally being accomplished in three to five days at ambient temperatures between 22° and 25°C.

Several generations may occur within the host insect and infective larvae begin emerging in about one week. These larvae are very active and can migrate 3–6 meters in a few months. Under favorable conditions infective larvae can survive in the soil up to three years. In the 1930s several attempts were made to use *Neoplectana glaseri* as a biological control agent of the Japanese beetle, *Popillia japonica*. However, in culture the nemas often lost their infectivity after seven or eight transfers on artificial media. If passed through host insects infectivity remained. It was not known at the time that elements of parasitism on the part of the nematode existed independent of the bacterium. The nema cannot continue to develop and grow in the absence of steroids supplied by the insect.

Interest still remains in the use of these nemas as biological control agents because of their wide host range and ability to disseminate and survive in the soil.

V. Obligate Associations of the Alimentary Tract of Invertebrates

The great majority of nematodes that have obligate associations of the invertebrate alimentary tract belong to the superfamily Oxyuroidea and are limited to two families: Thelastomatidae and Rhigonematidae. One species of Diplogasteridae occurs in the gut of the black field cricket. These families are found in associations with insects, myriapods, and diplopods. The nature of this symbiotic relationship is unknown, but pathogenicity only occurs where there are sufficient numbers of nemas to rupture the gut.

The nature of the host gut is apparently important to successful colonization. They prefer terrestrial and a few aquatic arthropods with well-developed digestive systems. Insects, diplopods, and myriapods with such alimentary canals afford almost complete digestion of the ingested food, a slow passage of the food through the gut, a relatively long stasis of the fecal pellets in the rectum, and a rich bacterial flora. Because of these requirements, these nemas are seldom found in caterpillars of Lepidoptera, or orthopterans such as Locustidae where food passage is rapid and only partially digested.

The life cycle is simple; all development generally takes place within the host gut and only the eggs deposited in the gut are passed with the host feces to the external environment. No egg development is recorded in the exterior environment for the Oxyuroidea; the diplogasteroid reportedly hatches and the larvae are ingested by a new host.

Chapter 8
Vertebrate Parasitism

I. Introduction

All vertebrate animals are subject to annoyance or disease from the activities of parasitic nematodes. No organ or tissue is free from attack and neither invasion of the air nor waters of the earth freed higher animals from the relentless evolution of nematodes as parasites. Current data record nematode parasites of vertebrate animals as more numerous in species than all other helminthic parasites combined.

The World Health Organization rates the four most important contributors to human suffering as malaria, malnutrition, hookworm, and schistosomiasis. Though malaria did decline from some 350 million cases in the 1940s to 75 million cases in the early 1970s, it is now again on the rise so that more than 150 million cases are expected in the early 1980s. There is no question but malnutrition and hookworm are also increasing. Some of the reasons for the higher rate of incidence in most parasitic diseases are increased world population, which in turn influences sanitation, education, environmental controls, and monitoring programs. Craig and Faust clearly point to the human factors contributing to increases in parasitic diseases; not least among these are administrative bureaucracy, human behavior, ignorance (lay and official), increased international travel, and political unrest. Therefore, there exists in the science of parasitology no room for complacency.

No parasitic disease should be viewed independently. Any disease is going to enhance the depredations of malnutrition; likewise malnutrition enhances the depredation of any disease parasitic or otherwise. Nematodes certainly must be considered among the most important contributors to malnutrition, not only because of their direct parasitic habits of sucking blood or denuding intestinal areas of microvilli, but also because of the great toll they take on the production of agricultural commodities as plant parasites.

The interaction between malnutrition and helminthic infections, which includes nematodes, cestodes, trematodes, and some other lesser groups, cannot be ignored. Changes or deficiency in diets are important factors governing susceptibility to certain helminths, spontaneous evacuation of worm burdens, resistance to infection, or efficacy of treatment. Malnutrition can interfere with antibody production, or decrease inflammatory reactions, thus lowering resistance. Dietary deficiencies combined with helminthiasis may cause loss of appetite (anorexia), vomiting, diarrhea, impaired liver function, neutralization of digestive enzymes, impaired absorption, and hemorrhaging.

A. Language of Parasitology

As in any form of human communication an agreement on terminology is essential to understanding. The most difficult concept is comprehending what constitutes a parasite. The distinctions or categories of animal interactions is a continuous spectrum of relationships linked by intermediates. The separations, therefore, are arbitrary and without exact boundaries. With nematodes the situation is further complicated because some nematodes illustrate highly sophisticated adaptations to parasitism, others appear to have only inaugurated the parasitic habit and some of these are not irreversibly committed, while still others exemplify only preadaptations to parasitism.

Some would include mutualism and commensalism with parasitism. It seems preferable to classify all permanent interspecific associations as **symbiosis**. Symbiosis then is an encompassing term for commensalism, mutualism, and parasitism. Of these three, **commensalism** is the least obligatory since both parties to the association can often survive independently. Among nematodes many such relationships occur between nematodes and insects. A relationship of this kind with nematodes falls into the category of transport or phoretic commensalism. An example of this is the relationship between the nematode *Pelodera coarctata* and the dung beetle *Aphodius fimentarius*. The beetle transports the nematode from dung pat to dung pat where the nema feeds on bacteria and develops independently. Other examples are those monhysterids and chromadorids that spend their entire life in the gill chambers of crustacea, an ideal habitat for living on filtered food particles. They cause no harm to the crustacea and presumably could live independently on bottom detritus.

Mutualism is a more intimate association that results in benefit to both participants. An example among nemas occurs with *Bursaphelenchus xylophilus* and cerambycid beetle. *Bursaphelenchus* needs the beetle for dispersion from infested to uninfested black pine. For egg laying and development the beetle needs weakened or diseased trees, which the nema provides (see Chapter 6).

Parasitism is limited in this text to that phase of symbiosis whereby the energy utilized for nematode (**parasite**) development comes totally or in part from another animal (**host**) at the direct expense of that animal. The parasitic habit may be either facultative or obligatory. The **facultative parasite** has the ability to live either a parasitic or free-living existence; development and completion of its life cycle is not dependent on a host. **Obligate parasites**, on the other hand, cannot exist without a host during all or some portion of their life.

All known nematode parasites of vertebrates are considered endoparasitic; as such the relationship of the nematode to its host is designated as an infection. **Endoparasitism** is sometimes further divided into two categories: **superficial endoparasites**, which live in the gut lumen or wall, and **somatic endoparasites**, which attack the deeper tissues of the body.

The host is the nonparasitic member (**victim**) of a parasitic relationship. Several types of host are recognized in parasitology. A **definitive host** is one in which the parasite reaches maturity. **Intermediate hosts** are all hosts necessary to the development of a parasite other than the definitive host. It is required, to qualify as an intermediate host, that there be a distinct structural or metabolic development of the parasite within the host animal. A special type of intermediate host is the vector. **Vectors** of nematode diseases are specialized intermediate hosts that actively seek out the definitive host and during the process of feeding, generally as ectoparasites, transfer the nematode parasite to the definitive host. Therefore, in this section the intermediate host is viewed differently than a vector. The intermediate host, per se, is a passive transmitter of the parasite; as such, the intermediate host is sought out as a food source by the definitive host. This is in opposition to a vector which actively seeks the definitive host. Examples that illustrate these two concepts are found in some nematode parasites of freshwater salmonids and in dog heartworm. Mayfly nymphs are the intermediate hosts for several intestinal nematodes of trout. The trout (definitive host) actively seeks out mayfly nymphs (intermediate host) as preferential food. Whereas in dog heartworm the mosquito (vector) actively seeks out dogs and other definitive hosts for a blood meal. It is at this time the third stage infective larval nematodes that developed in the mosquito are transmitted.

Paratenic hosts (ecological hosts, intercalary hosts, carriers, or hotes d'attente) offer a useful ecological link between hosts or an evolutionary advantage to the parasite. No development occurs in a paratenic host, but generally, because of food size preference of the definitive host, the carriage by animals of larger size is advantageous. For instance, if the first intermediate host (where the nema develops to a third stage infective larva) is a grasshopper and the final stage a carnivore, the direct route to the definitive host is tenuous. However, by passage through insectivorous rodents (paratenic host), the third stage infective nematode becomes more available to a cat (definitive host) than it was while residing in a grass-hopper (intermediate host).

Reservoir hosts are definitive hosts in nature, either endemic natural hosts or introduced. Generally, they are wild hosts as opposed to domestic animals, though there are many exceptions. Trichinosis of man has pigs, bears, rats, and walruses as reservoir hosts. Heartworm of dog utilizes coyotes, wolves, fox, cats, and others as reservoir hosts.

The total body of information concerning disease in populations or communities is **epidemiology**. As a branch of medicine it includes both the data concerning infectious diseases in population groups, and also information resulting from anatomical deformities, genetics, malnutrition, aging, etc. Most information is from communicable diseases.

When communicable disease is at a relatively steady level in a human population it is said to be **endemic**. If there is a sharp increase in the incidence of a disease, or an outbreak of considerable intensity, it is considered to be **epidemic**. When communicable disease is disseminated over extensive areas of the world it produces a **pandemic**. Comparable terms referring to intensity or distribution of disease in lower animal communities are **enzootic**, **epizootic**, and **panzootic**. Diseases transmitted to man from lower vertebrates constitute a **zoonoses**.

B. Evolution of Vertebrate Parasitism

Vertebrate parasitism within Nemata evolved along at least three independent lines. In Secernentea there are two lines of development, one which shows a sequential transition from free-living bacterial feeders to obligate parasites of higher vertebrates (Rhabdtida: Ascaridida) and one which apparently also developed from bacterial feeders but through arthropod associations (Spiruria). In Adenophorea vertebrate parasitism may have had some of its development through associations with annelid worms. Nevertheless, in both classes of Nemata, parasitism has evolved in these divergent lines from free-living terrestrial nematodes. No known parasites of vertebrates have evolved from marine nematodes. This is an extremely important point to understanding the development of vertebrate parasitism. The nematode parasites of fish and marine mammals have their ancestral origins among terrestrial nematodes. The linking steps between free-living terrestrial nematodes and parasitic nematodes among Rhabditia are well established, especially among Ascaridida and Strongylida. In these groups early larval stages can only be distinguished from free-living bacterial forms by a trained nematologist or parasitologist. Spiruria, which are also secernenteans, are more difficult to link or relate to a group of free-living contemporary nemas because all known forms are parasitic. However, their general morphology (esophagus, excretory system, sense organs, female reproductive system, and male sexual characteristics) leaves no room for doubt that they are Secernentea and therefore of terrestrial origin. Only rarely, and then by secondary invasion, are members of Secernentea found in a marine or freshwater habitat. Since Secernentea embraces almost exclusively terrestrial nematodes they had no opportunity to evolve parasitism in the marine or freshwater environment. Such parasitism had to evolve terrestrially and was secondarily introduced to aquatic environments. Adenophorean parasites appear to be most closely related to the soil inhabiting dorylaims. This relationship is surmised by similar cuticular structuring and esophageal gland orifice placement, and in some instances confidence is heightened by early larval stage morphology.

These lines of development will be presented in more detail. The bacterial line in Secernentea is rather straightforward. Spiruria (Secernentea) and Adenophorea, on the other hand, indicate parallel evolution and some rather interesting implications in geological history.

C. Methodology for Formulating Hypotheses about the Evolution of Parasitism

Of the several approaches to parasite evolution, I believe the least satisfactory is one dependent on coevolution of host and parasite. Such schemes rely heavily on the theory of host specificity, which more and more appears to be a tenuous assumption that presents more problems than it solves. For example, the traditional listing of the Ascaridida parasites matched to their hosts in an evolutionary sequence serves to show that host evolution and parasite evolution need not follow. Indeed, they may seldom have followed each other.

Parasite	Host
(Acanthocheilidae)	
Acanthocheilinae	Chrondrichthyes
(Stomachidae)	
Raphidascaridinae	Osteichthyes
Stomachinae	Piscivorus Aves, marine Mammalia, marine Reptilia
(Agusticaecidae)	
Angusticaecinae	Amphibia and Reptilia
Ophidascaridinae	Reptilia
(Toxocaridae)	
Porrocaecinae	Aves
(Ascaridoidea)	
Ascarididae	Mammalia
Anisakidae	Marine Mammalia with intermediates

This coevolutionary scheme ignores a very significant fact: all nematode parasitism had its developmental origin terrestrially. One cannot have confidence in the taxonomic groupings of the parasites that appear to be based primarily on the hosts. On the basis of morphology, one might consider placing the Angusticaecinae, Porrocaecinae, and Ascarididae together. Since this development was terrestrial, then the further evolution of nematode parasitism among Ascarididae was most likely taken to the marine environment by marine Mammalia and aquatic Aves. This does not exclude the role that may have been played by anadromous fish that may acquire a parasitic burden during their freshwater existence. This does not, however, explain ascarid parasites in the marine environment. Sharks may have obtained their parasites from fish or, as Inglis suggested, they may have retained parasites acquired by freshwater ancestors; however, this would predate many terrestrial vertebrates. Parasite phylogeny inevitably leads to the conclusion that ascarid diseases were introduced into the aquatic environments from terrestrial vertebrate ancestors (probably mammalia) and not vice versa.

Others have the opinion that ascarid parasitism began in the aquatic environment and moved terrestrially. As evidence they cite the complex life cycles seen among Anisakinae in marine mammals. The implication is that these are newly acquired parasites from the marine environment, and that the transition from land to sea caused the normally terrestrial nematode parasites to become extinct. This assumption completely ignores the origin of ascarids from Rhabditia which are exclusively terrestrial.

Indirect life cycles with intermediate hosts occur within Ascarididae. Indeed it would be extremely difficult for anisakids or any vertebrate parasite to survive in the marine environment if the direct cycle existed exclusively. Even if the eggs did not require an incubation period, it would be very difficult for the eggs to be picked up directly by another definitive host. In view of this, it is a natural development for marine parasites to have indirect life cycles and it does not necessarily mean that their development was from already marine parasites and that the original terrestrial parasites were lost. The ability of parasites to adjust to new environments is denied by the above hypothesis. A further aspect apparently overlooked is the geological time span involved in the transition of nematode parasitism from terrestrial mammals to the marine environment. During the transition there certainly was sufficient time for the parasites to adapt. It is not unlikely that a time existed when the parasite was adapted to both a terrestrial and aquatic cycle.

Host–parasite coevolution may occur, but is insignificant to the overall development of parasitism by nematodes. Most such schemes, as the ascarid example, are complex and require assumptions unsubstantiated by contemporary forms. It cannot be denied that ecologically Secernentea are terrestrial nematodes (in soil–water) and only very rarely are they found in aquatic habitats (streams, lakes, etc.). Furthermore, that ascarids arose from bacterial feeding Rhabditia is easily confirmed by their larval stages and morphology. It is difficult to accept any scheme that jumps from terrestrial free-living bacterial feeders to the marine environment and sharks. Just because sharks are ancient does not mean their nematode parasites are ancient or that they originated with sharks. It is easy to understand sharks obtaining parasites from fish but how, in order to make the above scheme work, did fish acquire the parasite from sharks? Accepting the concept of a food chain leads to the opposite conclusion and one can understand how a carnivore such as the shark through time acquired parasites that eventually became adapted to their new host.

It seems to me more reasonable to base evolutionary schemes of the development of parasitism on ecology, taxonomy, and structure of the life cycle. Schemes that are proposed on these bases conform to taxonomic hierarchy and can be illustrated for the most part by contemporary examples.

Hypotheses based on the above ideas accept the premise that most nematodes are not markedly host specific and that the evolution of most groups of parasitic nematodes has been largely independent of the evolution of their hosts and tends to occur in host groups with similar feeding habits and ecological requirements.

The following hypotheses of the evolution of vertebrate parasitism are based on taxonomy, life cycle structures, feeding habits (of both parasite and host), and the ecological requirements of the host.

II. Adenophorean Parasites of Vertebrates

A. Evolutionary Sequence in the Life Cycle of Adenophorean Vertebrate Parasites

Among contemporary adenophorean parasites of vertebrates three types of life cycle are recognized: heteroxenous, simple direct, and complex direct.

1. Heteroxenous

A heteroxenous life cycle involves the utilization of intermediate or paratenic hosts; in Adenophorea these are generally annelids. The annelid ingests the nema eggs, usually deposited in feces, and develops in the annelid's body cavity. When the infected annelid is eaten by the definitive host the nema continues to adulthood in the alimentary canal.

2. Simple Direct

In the simple direct life cycle, the adult nemas are located in the intestine of the definitive host and eggs are passed in the host feces. These expulsed eggs are then ingested, through food contamination, by another or the same definitive host. Ingested eggs hatch and the released larvae develop to adulthood in the intestine.

3. Complex Direct

In the complex direct life cycle, the adult nemas are found in specific organs and deep tissues of the host; ejection to the exterior requires larval migration in order to attain the intestinal lumen and subsequent passage to the exterior with feces. The other avenue of escape from the definitive host requires cannibalism or carnivorous predation in order to effect the exchange to a new host. The best-known example is trichinosis.

In this sequence, evolutionary development is presented as progressing from the heteroxenous cycle to the complex direct cycle. Therefore, the use of intermediate or paratenic hosts is more ancestral than the derived cycle where only the egg is outside in the external environment (similar in this respect to the bacterial feeding line of development). The ultimate situation is where the parasite is never outside the host. In Adenophorea the infective stage is the first or rarely the second stage larva. Therefore, intermediate host of adenophorean parasites is not equivalent to intermediate host among Secernentea where the nematode must reach the third stage to become infective. In many instances the term paratenic host would seem more appropriate.

B. Examples of Adenophorean Parasites of Vertebrates

The sequence of examples follows the phylogenetic steps hypothesized in the early part of this chapter. The line of development of these life cycles is (1) heteroxenous, (2) simple direct, intestinal parasites, eggs passed in feces, and (3) complex direct, deep tissue parasites requiring larval migration or cannibalism for transmission.

1. Capillaria

a. Capillaria plica

Capillaria plica is a common parasite of the urinary bladder of carnivores, including dogs, foxes, and cats. These capillarids are relatively large worms, the males measuring 13–30 mm and the adult females 30–60 mm in length (Fig. 8.1). The adult females are found in the urinary bladder and when eggs are deposited they pass to the external environment with the urine of the host. When first laid the eggs are unembryonated, but under favorable external conditions the first stage larva develops in about one month. The eggs possess polar caps characteristic of the capillarids (Fig. 8.1A).

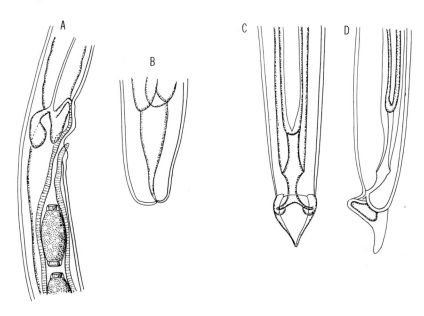

Figure 8.1. *Capillaria plica.* A. Vulval region of female. B. Female tail. C. Male tail ventral view. D. Male tail lateral view (redrawn from Rukhlyadeva).

Further development of the larva does not occur until the egg is ingested by an earthworm. Soon after the egg passes to the posterior two-thirds of the annelid's intestine, it hatches and the released nema larva burrows through the intestinal tract and into the connective tissue where it remains until the worm is eaten by the definitive host.

The route of migration within the vertebrate host is complex. (Figure 8.2 is for the reader's convenience in following body migration of this and subsequent parasites.) Upon being released from the earthworm the first stage larva molts and then burrows through the wall of the small intestine and molts to the third stage. This larva enters the hepatic portal system and proceeds through the liver and heart to the lungs. Larvae do not remain in the lungs, but pass through the pulmonary capillaries and return to the heart and enter the general circulation. When they reach the renal arteries they proceed to the kidneys and are carried in the blood vessels to the glomeruli. From the glomeruli they migrate into Bowman's capsule and through the tubular nephron to the pelvis of the kidney. Migration from the kidney to the urinary bladder is down the ureter. In the bladder they molt to the fourth and adult stages. The appearance of adult worms occurs some two months after acquisition of the infection.

These nemas are seldom pathogenic in small numbers. However, in heavy infestations there may be pneumonia and other bronchial complication.

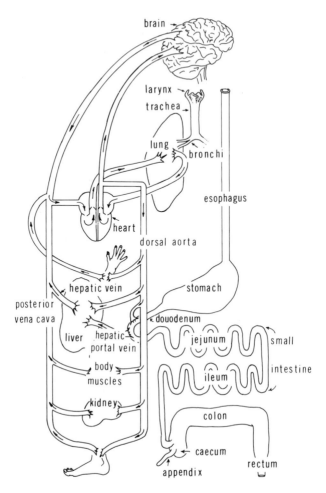

Figure 8.2. Schematic of human circulatory system and alimentary canal (original).

b. Capillaria philippinensis

Within the last twenty years a capillarid of man was discovered in the Philippines and was named *Capillaria philippinensis*. Man is the only known host and usually the syndrome ends with death. The adults are small, males measuring 2–3 mm and the females 2.5–4.5 mm. How the infection is acquired is unknown. The activity of the worms causes disruption of the mucosal lining of the intestine, degeneration of the lining epithelial cells, and inflammation of the mucosal layer. This extensive damage to the intestinal wall induces symptoms resembling malabsorption syndrome. The major symptom is intractable diarrhea that leads to rapid emaciation. There is some abdominal pain and distention, accompanied

by a low grade fever. In fatal cases the loss of nutrition normally terminates in shock, and death is due to this rather than specific tissue damage
attributable to the parasite. The mode of transmission is unknown, but
the first victim was known to eat a local preparation called "kilawan and
papaet," which is a food containing vital organs of various small mammals.

2. Trichuris trichiura, the Human Whipworm

Trichuris trichiura (Fig. 8.3) is one of the earliest described parasites of
man, it was first named by Linnaeus along with the giant intestinal worm
and human pinworm. The infection is worldwide, but more commonly it
is found in the moist warm regions of the world. In the southern United
States the incidence of infection may reach 20–25% but the intensity is
generally low. It has been estimated that nearly 400 million people throughout the world suffer from this worm.

   The adult females have a characteristic shape; anteriorly the first three-
fifths of the body is very attenuated and the last two-fifths is greatly
expanded (Fig. 8.3A). The males are similarly shaped, but the swollen
posterior is less dramatic (Fig. 8.3B). These nemas are relatively large,
males measuring 30–45 mm and females 35–50 mm. The expanded posterior region in both sexes houses the sacculate gonads. Gravid females
may contain up to 46,000 eggs.

Figure 8.3. *Trichuris trichiura.* A. Adult female. B. Adult male. C. Egg (modified from Jeffrey and Leach).

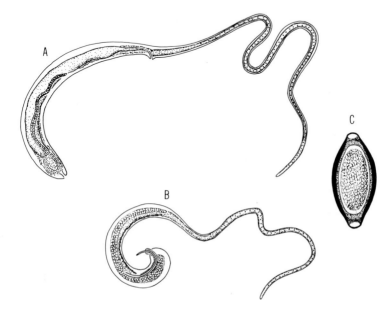

Males and females are found attached to the wall of the cecum and less frequently in the appendix, on the colon or posterior ileum. Eggs when deposited are voided with the feces. To become fully embryonated the egg must remain in the external environment at least 21 days. Humans acquire the infection when they swallow the fully embryonated egg, which then hatches in the small intestine. The larvae do not immediately proceed to the cecum, but seek harborage and nourishment in the crypts of the small intestine. After approximately 10 days the larvae begin to appear in the region of the cecum where they continue their development to adulthood. Full adulthood is achieved some three months after initial exposure.

The symptoms shown by victims of this nematode fall into two categories: allergic responses and traumatic responses. The allergic responses are manifest as inflammation of the colon (colitis) and rectal-anal region (proctitis). The symptoms are characterized by lower intestinal tract pain and diarrhea alternating with constipation. In heavy infections damage to the intestinal wall results in blood loss, blockage of the appendix, and marked irritation and inflammation of the cecum. The symptoms are very similar to those of hookworm and because of the chronic blood loss there is evidence of anemia. The most dramatic manifestation of this disease is prolapse of the rectum; on the everted rectum hundreds of worms can be easily seen attached to the wall. These worms are firmly attached to the mucosa and attempts to remove them by gentle traction leaves small spots of hemorrhage.

Diagnosis is based on egg identification (Fig. 8.3C). Unfortunately there are no completely satisfactory therapeutics that will eradicate all of the worms. A simple approach when the worms are in the colon and rectum is the use of high retention enemas, which wash out much of the worm burden. Any treatment must be followed by preventative measures to protect against reinfestation. Foremost in prevention is sanitation and sanitary disposal of feces.

3. Trichinella spiralis, the "Trichina" Worm

Trichinosis is probably the best known of all the nematode diseases. However, even though people are aware of the danger of raw pork, few actually realize that the culprit is a nematode. Trichinosis is not limited to man and pigs, it can occur in any carnivorous mammal. Especially dangerous is bear meat. The disease is unique in that no stage occurs outside the definitive host. Transmission occurs when humans or other meat eaters consume raw or rare meat contaminated with the cysts of *Trichinella.*

It is difficult to know where to begin describing the cycle because of its complexity in requiring two definitive hosts for continuity. When raw or rare meat containing cysts is consumed, the infective first stage larvae are released from their entombment by the action of gastric juices. Shortly after they are freed the larvae invade the duodenal and jejunal mucosa. In

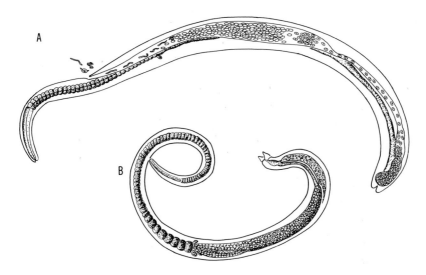

Figure 8.4. *Trichinella spiralis.* A. Adult female. B. Adult male (modified from Faust).

a few days males (1.4–1.6 mm) and females (3–4 mm) make their appearance (Fig. 8.4). Soon after impregnation of the females, the males die. The females subsequently increase to their maximum size and burrow deeper into the mucosa. Sometimes they go so far as to enter the peritoneum or mesenteric lymph glands.

Females deposit living young (Fig. 8.4A), the majority of which migrate into the intestinal lymphatics and mesenteric venules. Each female releases about 1500 larvae. From the lymphatics and venules they proceed to the heart and lungs and out into the arterial circulation. Upon reaching striated muscle they encyst; mainly they encyst in muscles poor in glycogen. Thus active muscles such as that of the diaphragm, larynx, tongue, abdomen, intercostal spaces, biceps, pectorals, deltoids, etc., are most heavily infected. In these muscles, generally near tendonous tissue, the larvae become encysted. Encystment is a host reaction and apparently in no way is it contributed to by the nematode. In six to nine months the fibrous capsule becomes calcified. Five cysts per gram of body weight can be lethal. Within the cysts, larvae may remain viable for more than five years. This is the termination of the cycle and the larvae must await the ingestion of this host if it is to continue. It has been shown that seal pups can acquire an infection through the mother's milk during the period of larval migration. Normally man is a dead end for this parasite. The list of normal hosts is very extensive and includes rats, pigs, bears, walruses, seals, and dogs. Humans normally become infected by breaking into the pig–rat–pig cycle.

In mild cases the symptoms do not differ greatly from common flu, i.e., mild rise in temperature accompanied by stomach upset and aches and pains. When the invasion is severe the syndrome is broken into three phases: invasion, migration, and encystment. Invasion is characterized by flu or food poisoning symptoms. Penetration of the gut by large numbers of larvae create symptoms such as vomiting, nausea, dysentery, and colic. Migration and initial larviposition in the muscles is manifest by difficulty in breathing, chewing, swallowing and speech; in the limbs there may be spastic paralysis. The conjunctiva of the eyes may appear jaundiced; there may also be a yellowing of the fingernails. Encystment is the critical third stage. Often nutritional stress and dehydration are evident. The pulse may at first be fast and strong and then it suddenly drops and cyanosis supervenes; as blood pressure falls the victim collapses. Prior to collapse, nervous disorders include visionary defects, loss of reflexes, disorientation, delirium, and encephalitis.

Diagnosis is by biopsy after the larvae reach the preferred muscle sites. Treatment is designed for symptomatic relief rather than destruction of the worms. Prognosis in mild cases is good, but heavy infestations may be fatal or permanently crippling.

Smoking, salting, and drying meat are not effective measures of prevention. The most effective measure is the thorough cooking of meat, especially of pork and bear. Freezing is also successful at $-15°C$ for 20 days; $-24°C$ for 10 days, or $-29°C$ for six days; quick freezing at $-37°C$ is also effective. These figures are based on a 15 cm block of meat.

Control consists of the destruction of all infected carcasses and viscera, extermination of rats and mice, and heat treating garbage fed to swine. Public education is an important part of any control program. In the United States federal law requires the cooking of garbage fed to swine, approved freeze treatment and inspection of prepared pork products and low temperature storage of pork; these regulations have led to a significant lowering of the incidence of the disease.

III. Secernentean Parasites of Vertebrates

A. Development of Vertebrate Parasitism from Bacterial Feeding Nematodes in Rhabditia

Bacterial feeding nemas among Secernentea exhibit a smooth transition from free-living soil inhabiting rhabdits to the complex obligate parasites in Ascaridida and Strongylida. The relationship between free-living and parasitic rhabdits is readily demonstrated through comparative morphology and biology. Among contemporary forms preadaptions to parasitism are seen in those bacterial feeding Rhabditia associated with fecal decomposition, decaying plants, or decaying animals. Nematodes in such substrates must be able to withstand variations in pH, oxygen tension, osmotic pressures, high temperatures, and moisture fluctuations. Nematodes with limited powers of dispersion compensate either by producing large numbers of eggs or through phoretic relations with other animals, generally insects or other invertebrates. The ability of Rhabditida to withstand desiccation and their behavioral adaptations favoring dispersion and transport to new environments are definite advantages not only to species survival, but to the development of parasitism.

Among the free-living taxa a phoretic association with insects and other invertebrates is important, but phoresis plays only a minor role in higher vertebrate parasitism among Rhabditia. However, it may well represent a developmental step to the utilization of intermediate hosts by certain ascarids and metastrongyles. Phoresis involves the specialization of the third stage larva to withstand the rigors of the external environment and apparently was the preadaptation to the development of an infective stage. The specialization of the third stage larva is an important adaptive feature because of penetrating ability, negative geotropism, or a behavioral stance that enhances either the phoretic relationship for dispersal or contact with the definitive host.

There are numerous examples of nematodes utilizing insects as transport (phoretic) host from one substrate to another. In many instances the nematode is a free-living bacterial feeder such as *Pelodera coarctata* and the transport host is a beetle. *P. coarctata* is free-living but has the ability to attach itself to the dung beetle, *Aphodius fimentarius*, and here it becomes a third stage resistant dauerlarva. The beetle then transports these resting nemas to fresh dung, at which time the nemas quit the insect and begin feeding and developing independently.

A closer example of the initiation of vertebrate parasitism by nematodes is the affiliation of *Rhabditis strongyloides* with phoretic hosts among rodents. Early larval stages of *R. strongyloides*, free-living bacterial feeders, are frequently encountered in the subcutaneous tissue of rodents or under their eyelids. Upon the death of the rodent transport host, the nematode quits the carcass either by boring directly through the skin or by departure from the eyes. Thus released to the environment the nematode proceeds to adulthood and any number of complete free cycles on saprobiotic substrates, one of which is often the decaying rodent.

The importance of behavioral activity of the third stage larva is illustrated by the relationship between *Dictyocaulus viviparus*, an intestinal parasite of cattle, and the dung pat fungus *Pilobolus*. The third stage nematode larvae are negatively geotropic, which causes them to climb to the tops of the fungal sporangia that extend above the surface of the dung pat. When the sporangiospores are discharged from the exploding sporangia the larvae of *D. viviparus* are catapulted away from the dung pat where cattle do not feed, to fresh grass, often three meters away, where cattle do feed.

The foregoing examples of bacterial feeding nematodes of the subclass Rhabditia are all associates of dung or decaying organic matter. This is a significant relationship, which preadapted these forms to the initiation of higher animal parasitism. In such substrates the nematode must be able to withstand great environmental fluctuation. The directional result of the development is to diminish the number of life stages of the parasite exposed to unfavorable conditions in the external environment. Among contemporary species we are able to document that behavioral and physiological adaptations essential to parasitism exists. If we accept that similar situations existed in geological history then steps can be outlined in the development among Secernentea of vertebrate parasitism from bacterial feeding nematodes:

1. The adaptation of bacterial feeders to a fluctuating environment.
2. The dispersion from the degenerating substrate to a fresh substrate by a transport host, or behavioral activity such as negative geotropism.
3. The development of an intimate association with the phoretic transport host and the establishment of a near necessary relationship while maintaining the ability to live entirely free; an example is *R. strongyloides* and rodents.
4. Initiation of actual parasitism. Among contemporary parasites this step is recognized as facultative parasitism, such as occurs with *Strongyloides stercoralis* (threadworms), which can complete several entire life cycles in the free-living state, feeding on bacteria or in the parasitic state or by a combination of both.
5. A refinement of the fourth step is illustrated by the various species of hookworm: only two larval stages can live free as bacterial feeders. The third stage, negatively geotropic larva, must find a vertebrate host to complete its life cycle. The third stage larva in both steps four and five enter the body of the definitive host by direct penetration, and make a blood system migration before settling down in a predesignated organ. This mode of entrance is enhanced by larval negative geotropism, which brings the larva above the height of the normal substrate, and therefore, in a better position to make host contact.
6. Indirect life cycle. The egg is in the external environment and either the egg is swallowed or hatched larvae penetrate a paratenic or an intermediate host (some ascarids and Strongylida). The development of the first three larval stages takes place in the intermediate host; little or no development occurs in a paratenic host. All other stages including egg laying occur in the definitive host. After transfer from the intermediate host and prior to settlement in the designated organ migration through the blood system is required.
7. A simple direct life cycle. Only the egg is in the external environment, but there is no intermediate host or vascular migration. This life cycle

is exemplified by human pinworm, *Enterobius vermicularis*. The egg is swallowed and then hatches and develops to adulthood in the intestine. Females migrate to the exterior to lay eggs.

The last two steps of vertebrate parasitism from rhabdit bacterial feeders are divergent: one utilizes an intermediate host or vascular system migration and the other does not. But the end result, in both instances, is convergent and culminates with only the egg exposed to the external environment.

1. Significance of Larval Migration in Definitive Host

In order to develop to adulthood in the lungs or intestinal tract many of the nematode parasites of vertebrates first migrate through the vascular system of the definitive host. There are several explanations for the necessity of this blood system migration. The simplest explanation for migration is that it is necessitated by the mode of entrance into the definitive host. For instance, both "threadworms" (*Strongyloides*) and "hookworms" (*Ancylostoma* or *Necator*) enter the definitive host by skin penetration. Therefore, migration through the blood system offers a convenient pathway to final establishment in the small intestine. The route is from the vascular system to the lungs, from the lungs over the larynx into the alimentary canal where they finally reach their preferred site for development, the duodenum (small intestine) (Fig. 8.2).

A second explanation is that migration substitutes for an ancestral development of the early larval stages in an intermediate host. The basis of this speculation is the need of the nema to develop through the third stage (ancestral infective stage) in an organ other than the intestine prior to the establishment of adulthood in the duodenum. In such instances migration is a mechanism of escape from conditions in the intestine or stomach that are detrimental to the survival of early larval stages that ancestrally developed in an intermediate host. The stomach environment is a known detriment to many nematode larvae and is an excellent barrier to parasitic development. Therefore, the circulatory system becomes an advantageous substitute when the intermediate host drops out of the cycle. This cycle of development is typical for the "giant intestinal worm" (*Ascaris*).

The third explanation proposes that the original or contemporary definitive organ could have been the lungs, in which case the migration is not representative of an intermediate host substitute, but merely the route to the definitive organ, the lungs. Such a vascular migration in the definitive host often occurs in nemas that still maintain the first larval stages in an intermediate host; an example of this is *Metastrongylus apri*, the "swine lung worm." The first three larval stages utilize an earthworm as an intermediate host; however, upon entrance into the definitive host, the third stage larva, released from the earthworm, immediately enters the blood system of the pig where it actively migrates until reaching the lungs where it continues development to the adult stage. Obviously in this case there is no substitution of the intermediate host, but merely a necessary use of the circulatory system to reach the lungs.

2. Examples of Rhabditia Parasites of Vertebrates

Only those species that are pathogenic to humans and other vertebrates will be discussed. Therefore, steps one through three in the evolutionary sequence previously presented will not be discussed further.

a. Strongyloides stercoralis, Threadworm

Though a facultative parasite, *S. stercoralis* (Fig. 8.5) is illustrative of an ancestral development. This parasite's life cycle is one of the most complex cycles involving vertebrates. This particular species is a parasite of humans, apes, and some monkeys in warm, moist areas of the world, as well as in the warmer temperate areas. It is most prevalent in Asia, tropical America, and Africa, but is also known from North America, U.S.S.R., and the Pacific Islands. Other species not transmissible to humans are encountered in sheep, goats, rabbits, rats, pigs, horses, etc.

Figure 8.5. *Strongyloides stercoralis.* A. Adult parasitic female. B. Adult free-living female. C. Filariform larva. D. Rhabditiform larva (modified from Faust).

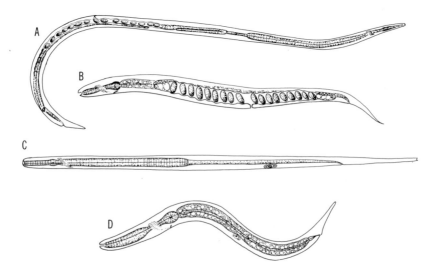

The life cycle of *S. stercoralis* is most easily understood when it is recognized that at each step the nematode has basically always two choices of direction to complete development. Fundamentally the life cycle is divided into a free-living and parasitic phase. However, because of the crossover between a free-living and parasitic existence the complete life cycle consists of one or any combination of three phases:

1. Indirect development, based for the most part on the free-living existence prior to a transformation to the parasitic phase, occurs primarily in the tropics.
2. Direct development, the most common means of human infection. The larvae passed in the feces immediately transform to the parasitic form and as the opportunity arises invade a new host.
3. Autoinfection, in which the infective larvae never really leave the host. Reinfection occurs in the perianal region or within the intestine.

The simplest way to begin the cycle is with the laying of eggs by the parasitic female (2.2 mm in length) in the definitive host. Normally oviposition takes place in the intestinal mucosa, but may abnormally occur in the columnar epithelium of the bronchi or trachea. The most common sites of intestinal infestation are the duodenum and the jejunum but all levels of the intestine from the pyloric wall of the stomach, appendix, and large intestine to the anus may be invaded. The eggs hatch within the tissue where they are layed. The escaping larvae migrate to the intestinal lumen as rhabditiform larvae. Rhabditiform larvae are so named because they resemble the larvae of free-living species in the genus *Rhabditis*. As the larvae pass down the intestine two avenues of development are presented: the larvae may remain rhabditiform and pass to the exterior environment in the host feces, or they may transform to the infective filariform larvae while still in the intestine. The infective filariform larvae (Fig. 8.5C) differ morphologically from the rhabditiform larvae: they are longer, thinner, and do not possess the valved rhabdit esophagus. Their esophagus is cylindrical without any divisions, bulbs, or valves. These filariform larvae are ready for immediate reinfestation either in the intestine or in the perianal region (autoinfection). More will be said about this phase later. The unchanged rhabditiform larvae (Fig. 8.5D) that pass to the exterior have two choices in the external environment. They may remain rhabditiform and proceed to develop through a series of free-living life cycles feeding on bacteria in the host feces, or they may transform immediately to the filariform infective stage and await the appearance of a new host. Those choosing the free-living existence continue so until the substrate deteriorates. At this time they do not proceed to a rhabdit third stage, but metamorphose to the filariform stage. The filariform larvae, whether transforming immediately in feces or after one or more free-living generations, are the infective stage for the disease continuance between hosts.

Invasion of the definitive host occurs when skin contact is made with feces infested with third stage filariform larvae. The larvae can penetrate directly through the skin or more easily, through a hair follicle. Once in the subcutaneous blood system they migrate into the veins of the peripheral blood system. In the venous blood (Fig. 8.2) system they are carried through the right heart to the lungs where they break from the pulmonary capillaries into the lung alveoli. From here the larvae ascend the respiratory tree to the epiglottis and are swallowed. Upon reaching the small intestine, normally in the duodenal and jejunal regions, the female larvae burrow into the mucosa and grow to adulthood. Males may or may not be present; fertilization is not required for successful oviposition because the females are capable of parthenogenetic reproduction. If present the males must copulate with the female prior to her burrowing into the mucosa. Males are not tissue parasites and are evacuated in a short time from the intestine with the feces.

Those larvae that underwent the change to filariforms in the intestine (autoinfection) must either penetrate the intestine or penetrate the skin in the anal region in order to gain access to the venal circulatory system. They cannot develop in situ; they must make the circulatory system migration.

The pathological effects of *Strongyloides* begin with penetration and continue for as long as the worms persist in the body. Abdominal symptoms generally commence about one month after the initial invasion. At the point of invasion small pinpoint hemorrhage spots (petechiae) are produced. These areas produce an intense itching (pruritus) and some swelling of the surrounding tissues. No further inflammation occurs unless accompanied by other infectious organisms such as bacteria. When the migrating larvae break out of the pulmonary capillaries into the alveoli both hemorrhage and cellular infiltration into the air sacs and bronchials occurs. The latter can slow the passage of the larvae up the respiratory tree and may be a partial explanation for the occasional development seen in the bronchia and trachae. This passage through the lungs produces symptoms of bronchopneumonia, with unproductive coughing and an intrathoracic burning sensation. After invasion of the intestinal wall the females continue to burrow in the mucosa. In heavy infestations the mucosal wall becomes honeycombed by the burrowing females and migrating larvae, which often results in the sloughing of extensive patches of the mucosa. When internal autoinfection is excessive there may be secondary bacterial invasion, often resulting in fatal strongyloidiasis.

The symptoms of intestinal invasion are watery mucous diarrhea, alternating with constipation and dehydration, which unattended can result in death. As with almost all parasitic disease eosinophilia is characteristic and thus indicative but not diagnostic. Confirmation of the disease is by recovery of the larvae in the stool or by duodenal drainage; the latter is by far the more accurate determination.

The most important preventative measures are sanitation and education. Care must be taken to avoid contact with contaminated feces. Once the infection is recognized, anthelmintic treatments should be taken. During treatment the anal sphincter should be carefully cleaned in order to avoid, as much as possible, autoinfection.

b. Ancylostoma and Necator, Hookworms

Hookworm is one of the major diseases of mankind in the warm moist climates of the world. The two important human hookworms are the "Old World hookworm," *Ancylostoma duodenale* (adult female 10–13 mm, adult males 8–11 mm) and the "New World hookworm," *Necator americanus* (adult female 9–11 mm, adult males 7–9 mm) (Fig. 8.6). These two hookworms are characterized by the presence of oral cutting organs, consisting of toothlike processes in *Ancylostoma* spp. and of semilunar plates in *Necator* spp. (Fig. 8.6A, D). The common names are misleading because the two overlap in much of their range, especially throughout Asia and the Indonesian archipelago. Europe and the Mediterranean basin are primarily subjected to *Ancylostoma* while Africa, South America (except for the western coastline), and North America are primarily subjected to

Figure 8.6. A. *Ancylostoma duodenale*; female head. B. *A. braziliense*; female head. C. *A. caninum*; female head. D. *Necator americanus*; female head. E. *A. duodenale*; male tail. F. *A. braziliense*; male tail. G. *A. caninum*; male tail. H. *N. americanus*; male tail (redrawn from Jeffrey and Leach).

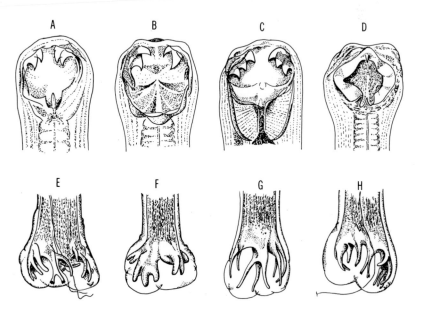

Necator. Because the life cycle, symptomatology, epidemiology, and prevention are so similar, the two will be discussed together.

The life cycle has similarities to that of *Strongyloides*, but is simpler because there is no complete free-living cycle. An important difference is that hookworm passes to the exterior environment in the egg stage. In regions where both occur this is important in initial diagnosis since *Strongyloides* is passed in the larval stage. Normally the egg is not fully embryonated at the time of passage with the feces. Stool consistency influences the rate of development, being much slower in constipated stools. In favorable stools development to hatching occurs in 24 to 48 hours. Death of embryos is rapid in stools diluted with urine. Unusual as it may be, the first stage larva emerges from the egg, not the second stage, as is common among Secernentea. These larvae are rhabditiform and by experienced personnel can be distinguished from larvae of *Strongyloides* and other *Rhabditis*. After feeding freely on bacteria for approximately one week they have molted once and are ready to metamorphose to the third stage filariform larva, whose esophagus is narrow anteriorly and expands to a pyriform, valveless bulb posteriorly.

These larvae are negatively geotropic and raise themselves up on microelevations to await the passage of a suitable host. Upon contact with human skin they penetrate hair follicles, interdigital spaces, or under scaling fragments of the epidermis. Once they have invaded the bloodstream the path followed is the same for *Strongyloides*, i.e., right heart, lungs, up the respiratory tree over the epiglottis, and then swallowed. Their final destination is the jejunum. Five weeks after exposure the females begin oviposition and each female is capable of laying 9000 eggs daily. In the jejunum both males and females are parasitic and they use their huge bulbose stomas with cutting teeth or plates to attach to the mucosa and suck blood. The blood sucking is continuous even during copulation and egg laying. As long as the adults are attached, blood can be seen oozing from the worm's anal opening. Even after the worms are removed, the wounds continue to bleed for sometime; indicating that some anticoagulant may be injected into the feeding site. Each worm is capable of passing 0.67 ml of blood through its gut daily so that in heavy infestations of 250–400 worms, an infested person may loose as much as 268 ml of blood per day. Though in the absence of reinfestation the worms begin diminishing in numbers in about one year, the average life span is five years and in one experimental case egg laying continued for fifteen years.

The three primary sites of pathological damage are the skin, lungs, and intestine. The hemorrhagic areas of the skin, the itching, and the swelling are more severe with hookworm than threadworm. At the actual site of entry there is little damage, but as the larvae burrow deeper in the blood capillary beds considerable local tissue reaction occurs. The common name given to these skin symptoms is "ground itch." Lung symptoms are characteristic of any parasite breaking through the capillaries into the alveoli, i.e., unproductive coughing, pneumonialike symptoms, etc., as in threadworm. The most damage attributable to hookworm occurs in the intestine. Here they produce deep ulcerations at the site of attachment where the intestinal villi sucked into their mouth capsule are digested. The degree of pathological change and symptoms produced in a victim are dependent on (1) the worm burden, (2) how long the infection has persisted, (3) resistance of the victim, (4) host nutrition, and (5) reinfection.

In severe cases there is diarrhea, constipation, poor digestion, and anemia. The patients skin becomes dry and there is a yellow pallor in light-skinned victims. The patient feels cold even in a hot climate. Though digestion is poor, victims crave bulky materials in an attempt to relieve the intense abdominal pain. Swelling of the face is noted, particularly around the eyes. The stomach in severe cases becomes distended and "pot belly" is typical in children; this also occurs with malnutrition. In advanced cases there is mental dullness that is physiological and not just due to anemia.

The clinical condition called "creeping eruption" is caused by cutaneous larval migrations of hookworm larvae of an origin generally from dogs and cats. The commonly encountered species is *Ancylostoma braziliense*, which has strains in cats and dogs (Fig. 8.5B, F). The filariform larvae of this species are unable to negotiate passage through the deeper skin strata of humans. The larval tunnels, which can reach several millimeters to a few centimeters daily, are reddened and raised. As advance continues, the abandoned areas become dry and crusty. The invaded areas give rise to intense itching. Without reinfestation the larvae are lost with the sloughing of the epidermis. Medically supervised freezing has been shown to be effective; however, in cases of multiple lesions the treatment is often worse than the infection.

There are several anthelmintic preparations for relief from human hookworm. In almost all cases because of protein deficiency the patients nutritional status must be built up. Generally prognosis is good.

Prevention and control of human hookworm falls into two categories: (1) individual prophylaxis and (2) mass control. The number one preventative measure is individual and public education about sanitation. Educating people in sanitary methods for the disposal of human excrement is extremely important. Practical solutions are still lacking in countries where human feces, "night soil," is used for crop fertilization. In heavily infested areas all members of the community should undergo anthelmintic treatment. Some of the efforts that can be made to avoid the disease are to (1) avoid contact with soil, easily achieved by wearing shoes, (2) encourage the use of toilets and sanitary latrines, and (3) raise, where possible, the level of nutrition; well nourished individuals are the least susceptible to the establishment of the worms.

In the case of "creeping eruption" people should be educated to the dangers of pet diseases to humans. Laws should be enacted to prevent pet owners from allowing their animals to defecate in public parks or public beaches.

c. Miscellaneous Strongylida

Strongyles usually undergo a direct life cycle with the first three larval stages developing in the exterior environment. In some species of strongyles there is passage through a paratenic host, often erroneously called an intermediate host, this passage is never accompanied by larval development. There are many examples among Strongylida such as the fatal human parasites in *Angiostrongylus* that require intermediate hosts, generally molluscs or annelids. These nematodes are extremely important veterinary parasites throughout the world.

c1. Horse Strongyles   *Strongylus vulgaris, S. equinus,* and *S. edentatus* are all serious parasites of horses. All three have direct life cycles and an infestation is acquired through the ingestion of contaminated feed. These three species also illustrate a transition away from the circulatory system migration so common among strongyles. *S. vulgaris* (Fig. 8.7A, D) migrates in the blood, *S. equinus* (Fig. 8.7B, E) migrates in the abdominal coelom and *S. edentatus* (Fig. 8.7C, F) limits migration to the intestinal wall.

c1a. Strongylus vulgaris   *S. vulgaris* is a small-toothed strongyle: (male 14–16 mm; female 20–24 mm) (Fig. 8.7A, D). The ingested egg hatches in the intestine and develops to the third stage larva. When the larvae reach the colon or cecum they burrow through the intestinal wall and make their way into the venules of the hepatic portal system and are carried to the lungs by way of the liver and heart. After breaking through the pulmonary capillaries into the lung alveoli they proceed up the respiratory tract over the epiglottis and are swallowed. Upon reaching the colon and cecum they proceed to develop to full maturity. Those larvae not breaking into the lung alveoli are carried back to the heart and then to the general circulation. When these larvae reach the mesenteric artery they attach to the intima and grow. The attachment and feeding in the artery cause thrombi and aneurysms.

This nematode is the most serious and pathogenic of horse strongyles, especially in foals. Parasitism by these nemas is almost always fatal. The pathogenic syndrome is characterized by (1) increased body temperature, (2) loss of appetite, (3) rapid loss of weight, (4) marked depression and decline in physical activity, (5) colic, (6) diarrhea and constipation, and (7) death.

c1b. Strongylus equinus   *S. equinus* is among the so-called large strongyles (males 26–35 mm; females 38–47 mm) of horses, and is commonly

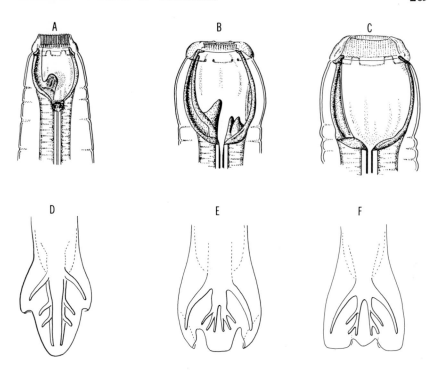

Figure 8.7. Horse strongyles. A. *Strongylus vulgaris*; female head. B. *S. equinus*; female head. C. *S. edentatus*; female head. D. *S. vulgaris*; male tail. E. *S. equinus*; male tail. F. *S. edentatus*; male tail (heads original, tails redrawn from Olsen).

referred to as the "large-toothed strongyle" (Fig. 8.7B, E). The blood system migration has been foresaken by this nematode and replaced by migratory sojourns in the abdominal coelom. When the infective larvae reach the colon or cecum, after ingestion, the cuticle is shed and the larvae immediately burrow into the intestinal serosa where nodules are induced. The larvae remain with the nodules for approximately 11 days. At this time they quit the nodule and enter the body cavity and make their way to the liver where they reside for six to seven weeks. The larvae leave the liver after a short period of residence and return to the body cavity, and by this route enter the pancreas. They, most likely, leave the pancreas by way of the pancreatic duct and thus return to the intestinal lumen and cecum where full maturity is reached.

c1c. Strongylus edentatus *S. edentatus* (Fig. 8.7C, F) is a medium-sized toothless strongyle (males 23–44 mm; females 33–44 mm) that restricts migration to the intestine. There is no sojourn in the circulatory system or body cavity of the horse. Upon reaching the large intestine the third stage larva frees itself from the retained second stage cuticle and burrows through the intestinal wall to the outer layer of connective tissue. In this tissue nodules are formed and the contained larva remains for about three months. At the end of this time the larva leaves this nodule and migrates back into the intestinal wall and induces yet another nodule. Eventually this nodule is also left and the larva returns to the intestinal lumen and cecum where development to full maturity is attained in about 11 months.

Just as the migratory habit declines among these three species, so also does the pathogenicity. Therefore, *S. vulgaris* is the least derived parasite and *S. edentatus* is the most derived.

c2. Metastrongylus apri, Lungworm of Swine   *M. apri* is common in the bronchi and bronchioles of swine throughout the world. The males are moderate sized and range up to 25 mm; the females on the other hand, often attain a length of 85 mm (Fig. 8.8).

The fully mature female oviposits her eggs in the bronchi and bronchioles. These eggs are coughed up, swallowed, and then voided with the host feces. The eggs are thick shelled and very resistant to adverse conditions. Viability, in pasture situations, can be maintained for up to 13 months; they can even survive five or more months of freezing. In order to com-

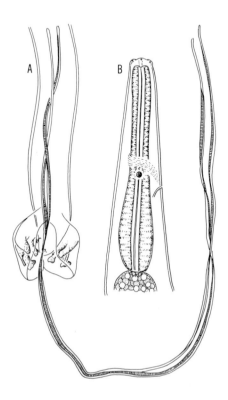

Figure 8.8. *Metastrongylus apri.* A. Male tail. B. Female anterior (modified from Shul'ts and Kaminsky).

plete the life cycle the egg must be ingested by an earthworm. Several species of earthworms throughout the world can serve this purpose. Within the earthworm's alimentary tract the eggs hatch. It is believed that gastric secretions by the earthworm aid the process by weakening the heavy outer shell. The liberated larvae make their way into the tissues of the earthworm's crop. Here they are found primarily in the lamellar sinuses of the calciferous gland. The third stage infective larvae appear in 10 to 30 days. From the sinus lamellae they enter the circulatory system and accumulate in the hearts and wall of the esophagus. There is no further development until the earthworm is ingested by a pig. Longevity is limited only by the life span of the earthworm. When a rooting pig ingests an earthworm the nemas are released and they immediately invade the intestinal mucosa. From here they travel by way of lacteals to the lymph nodes where they molt to the fourth stage. These new fourth stage larvae proceed by way of the lymph vessels to the heart and thence to the lungs. Breaking through the pulmonary capillaries they enter the alveoli; from here they proceed to the bronchi and bronchioles where they mature, mate, and with oviposition renew the cycle. This worm is the alleged vector of the swine influenza virus.

c3. Angiostrongylus cantonensis, Rodent Lungworm   *A. cantonensis* is another example of a strongyle that utilizes intermediate hosts. Normally, the intermediate is a snail, but apparently almost any invertebrate such as freshwater prawns, oysters, slugs, and landcrabs can harbor and allow larval development to the third stage. The normal definitive host is a rat; however, it is discussed because it can also infect man and is usually fatal.

   A. cantonensis is a moderate size worm, the males measure between 16–19 mm and the females are between 21–25 mm (Fig. 8.9). The eggs, which are voided in the feces of the definitive host, must be ingested by an intermediate host. The hatched larvae develop in about two weeks to the

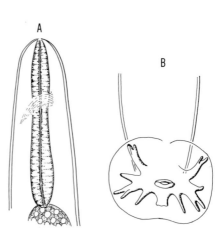

Figure 8.9. *Angiostrongylus cantonensis.* A. Female head. B. Male tail (modified from Chen).

infective third stage. Ingestion of the intermediate is only one source of infestation. In the case of human infection this is accomplished when the mollusc or prawn is eaten raw. Slugs, for instance, which serve as intermediates when eaten raw are believed by some Orientals to improve the voice; as such this is a common habit among orators and singers in the Orient. The other common mode of ingestion is the swallowing of the third stage larvae attached to leafy vegetation. This comes about when infested snails or slugs crawl over vegetables. The infective larva leave the mollusc and are, for some time, protected in the mucous slime trail of the slug or snail. This is believed to be a common mode of transfer in Malaysia. Therefore, the mode of infection in man changes with the prevailing social customs, i.e., vegetables in Malaysia, slugs in Japan, raw prawns in the South Pacific Islands. Upon gaining entrance into a suitable definitive host the larvae penetrate the intestinal wall and proceed on a circulatory sojourn to the heart, and from there to the lungs. However, in this disease cycle they do not enter the lungs, but stay in the blood system where they return to the heart and then proceed to the brain. Here they feed on the meninges and develop to the adult stage. When oviposition is near the females migrate back to the heart and via the pulmonary arteries they make their way into the lungs where oviposition takes place. Eggs make their way to the exterior when they are coughed up and swallowed and subsequently voided in the host's feces.

In many parts of the Orient and south Asia many cases of eosinophilic meningoencephalitis of unknown origin may well be the result of human exposure to *A. cantonensis*. Anthelmintics can be successful against the early invasive stages but rarely are these recognized in time; as such, prognosis is poor and death inevitable. Some people apparently show immunity when invasion occurs at low levels. There has been some speculation that clinical symptoms require a certain threshold of larval invasion. However, there are those that develop the characteristic syndrome and the development can be unbelievably rapid. It can begin with a low grade fever that persists for a few weeks; this may be followed by urinary problems. Almost attendant on the latter symptom can be unconsciousness followed by death within 24 hours. In others the early symptoms are similar and then begins a syndrome of steady mental deterioration, which becomes increasingly serious. Generally death comes within two years. Autopsy reveals that the worms invade the vascular and perivascular tissues of the brain as well as blood vessels of the meninges. In addition there are small necrotic areas in the brain tissue along with areas of inflammation; this is believed to result from the wanderings of the worm. Granulomatous tissues may also surround the worms, encasing them in a dense exudate. Vascular congestion, however, seems to be the prominent feature.

Diagnosis at this time is less than satisfactory. Most often recognition comes with autopsy.

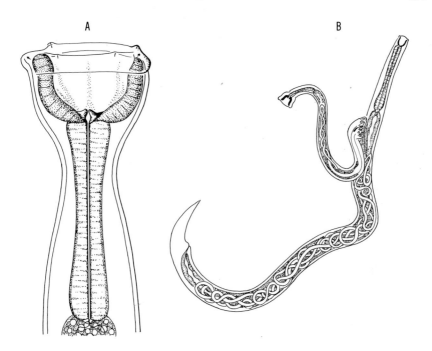

Figure 8.10. *Syngamus trachea.* A. Male anterior. B. Male and female in copula (original).

c4. Syngamus trachea, Gapeworm of Galliform Birds   *Syngamus trachea*

(Fig. 8.10) is an example of a strongyle utilizing a paratenic host. The common name derives from the fact that young chickens, pheasants, turkeys, etc., show symptoms of gapes due to their labored breathing resulting from nemas occluding the trachea.

There is a considerable difference in size between the adult male and female (Fig. 8.10B). The male is generally between 2 and 6 mm, while females attain lengths up to 20 mm. These males and females are in continuous copula while attached to the bird's trachea. The attachment of numerous pairs of nemas induces the birds to gape and cough, thereby bringing the eggs up the trachea only to be swallowed and voided with the feces. The eggs quickly develop in moist shady soil at temperatures between 24 and 30°C. Apparently they do not feed during their soil residence because reportedly all cuticles and larvae are maintained within the egg. The egg is also the overwintering stage because the males and females, which are short-lived, are eliminated during the course of winter. The egg (carrying the third stage larva) and the hatched larva are both infective to susceptible birds. However, when the larvae are eaten by dung infesting earthworms, slugs, snails, or fly maggots they make their way to the body cavity and become encysted. Within molluscs they can remain viable for several years. Birds eating any of these paratenic hosts become infested when the encysted larvae are released by gastric juices in the bird's intestinal tract. Whether the larva is ingested directly or through the auspices of a paratenic host the subsequent sequence of events is the same. The larvae burrow into the wall of the bird's intestine and make their way to the hepatic portal system; from here they are carried through the liver to heart and finally they reach the lungs. In the lungs they leave the circulatory system and enter the alveoli. Here they develop into immatures and adults who then migrate upward into the respiratory tree and trachea. The cycle from ingestion to egg laying takes about three weeks.

Other members of the genus *Syngamus* attack cattle and sheep. Death in chickens and other galliform birds infested with this disease often occurs due to asphyxiation.

d. Ascaris lumbricoides, The Giant Intestinal Roundworm

Ascaris is one of the oldest known species of animal parasitic nematodes. There are several related genera in the order Ascaridida that are also important to men and other animals; reference will be made to these at the appropriate time.

Ascaris lumbricoides, the giant intestinal worm of humans, is the most widespread and common helminth of humans. Nearly 700×10^6 people are afflicted with these worms. Infections are common in warm moist climates or in warm moist regions with temperate climates. A World Health Organization survey (1967) disclosed that throughout the Mediterranean basin approximately 30% of the students carried infestations. The range was from a low of 2.5% of the college students in Pisa, Italy, to a high of 97% of the school children in restricted areas of Yugoslavia and Czechoslovakia. Foci of high incidence still persist in the United States of America. In the rural population of South Carolina the prevalence among children in the six to eleven age group was 63.7%.

These worms are extremely large. The adult females are 20–35 cm in length, some reaching 49 cm with a diameter of 3–6 mm (Fig. 8.11). Adult males are slightly smaller, measuring 15–31 cm in length and 2–4 mm in diameter. The ovaries in a mature female are extensively developed (Figs. 5.3A, 5.4) and when dissected out and measured are often more than one meter in length. A female may contain at any given time 27,000,000 eggs and daily lay 200,000 eggs. The life cycle, unlike related genera in marine mammals, fish, and birds, is simple and direct.

The fertilized eggs are distinctive, the shape is broadly oval, and the shell is thickened and tuberculate (Fig. 8.11D). The eggs are not fully embryonated when discharged in the host's feces. In nine to thirteen days, in a moist shady habitat at temperatures between 22° and 33°C the eggs achieve full infectivity and contain second stage larvae. The eggs are

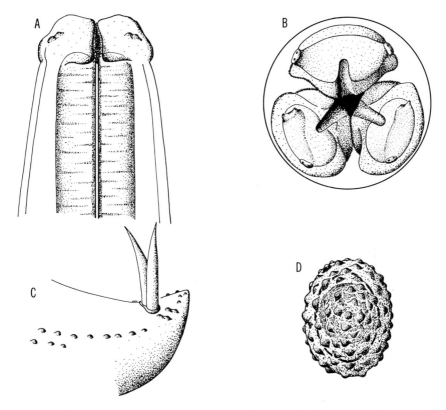

Figure 8.11. *Ascaris lumbricoides.* A. Female head. B. Face view, female. C. Male tail. D. Egg (original).

resistant to desiccation, low temperatures, putrefaction of the substrate, and several strong chemicals; however, under these conditions they remain practically dormant. Eggs in contaminated soil can remain viable for five or more years. When eggs are exposed to the direct heat and rays of the sun or moist temperatures above 70°C they are rapidly killed. Eggs have never been known to hatch in soil, no matter what conditions prevail. Infection takes place when the infective egg is swallowed.

Infection is acquired by ingesting eggs accidently from contaminated soil or from contaminated food or drink. When the embryonated egg reaches the intestine the shell is weakened and the encased larva is stimulated to activity by the intestinal juices. Upon release from the egg the second stage larvae (0.2–0.3 mm in length) immediately penetrate the wall of the small intestine. Larvae that do not penetrate the small intestine are soon killed by the intestinal juices. After penetrating the duodenal wall the larvae migrate to the mesenteric lymphatics or mesenteric portal hepatic venules by which they proceed to the right heart and lungs.

In the lungs the larvae molt twice, each interim being approximately five days. At 10 days they are fourth stage preadults (1.5 mm in length) that have acquired resistance to the gastric juices of the stomach. Having reached this stage in the lungs they now migrate up the respiratory tree to the epiglottis and are swallowed; on reaching the small intestine they proceed to develop to adult males and females. The first eggs appear 60 to 75 days after the initial exposure.

When exposure is extremely heavy, many of the larvae may make a migration across the pulmonary capillaries and be carried to the left side of the heart from whence they enter the systemic circulation. Here they become filtered out in the lymph nodes, thyroid, spleen, brain, and spinal cord. This is extremely serious because these larvae provoke acute tissue reactions. On rare occasions they pass the placental filter of pregnant mothers and reach the developing fetus.

In passage through the liver and lung capillaries the robust larvae cause minute hemorrhages at each site of rupture. The resulting symptoms of these penetrations resemble hepatitis and pneumonia. In the intestine the adults cause traumatic or toxic symptoms. The traumatic damage is especially serious in heavy infestations for the confined nemas may penetrate the bowel and begin a migration through the body cavity and its contained organs; in lieu of this, blockage of the tract by large knots of worms may occur. Toxic reactions increase with repeated infections and are believed to be caused by the systemic absorption of the by-products of the living and dead worms.

When infestations are light there are no remarkable manifestations. Most commonly there is some abdominal discomfort which may be accompanied by diarrhea and acute inflammation of the bowel. In children these symptoms are usually accompanied by a mild fever.

Wandering ascarids cause the most problems and can cause serious damage, the least of which is the initiation of appendicitis. In one survey of almost 1400 individuals, the larval distribution was 32 incidences in the head region, which included 2 findings in the brain, and 19 in the ear; 23 in the thoracic cavity and organs; 788 in the liver and bile ducts; 73 in the pancreas and 356 in the appendix.

Diagnosis is usually by examination of feces for eggs. Prevention centers around sanitation and the avoidance of eating raw vegetables in areas where the incidence of the disease is established. The eggs are almost exclusively of human origin though other ascarids can cause serious problems even though they do not develop to adulthood.

d1. Miscellaneous Ascaris   One of the most serious conditions within humans, as a result of exposure to ascarids of lower animals, is **visceral larva migrans**. The most common agent involved in this zoonosis is the dog ascarid, *Toxocara canis*; however, *Toxocara cati* of cats may also be involved. Just as with normal human *Ascaris* migration to ectopic foci results in extremely serious and often fatal conditions, especially to small children. This grim picture should be of more concern because the parasite occurs worldwide in dogs. In human tissues such as the liver capillaries, lungs, brain, eye, viscera, or musculature the larvae become encapsulated by granulomatous tissue. Prognosis varies with the number of larvae invading and the tissue invaded. With granulomatous involvement of the eye surgical removal is indicated. Though in the majority of cases eventual recovery without lasting symptoms occurs, it must be emphasized that fatality is not uncommon. Children are most commonly affected because their play habits often get contaminated soil into their mouths. The best prevention is to keep dogs from defecating in play areas and to maintain a schedule of dog worming.

Other species of ascarids that can cause problems to humans are in the genera *Anisakis* and *Contracaecum*. These are generally ingested while eating raw fish, a common custom in many areas of the world. Of the two, *Anisakis* is probably the more serious because the normal definitive hosts are found among marine mammals. The symptoms generally are of an acute abdominal syndrome. Thorough cooking or prolonged freezing eliminates the danger.

e. Enterobius vermicularis, Human Pinworm or Seatworm

The oxyurid *E. vermicularis* (Fig. 8.12) represents the most derived life cycle among parasites whose ancestry was directly from bacterial feeding nematodes. In this nematode the life cycle is simple and only the egg is found in the exterior environment; all other stages occur and develop within the human body. Though worldwide it is more common in temperate and cold climates than in the warm regions of the world. In part this is attributed to less frequent bathing and confinement of clothes characteristic of colder climates. Incidence among school children in cooler regions of the world is often 100%. In the United States it is estimated that nearly 10% of the population suffer from this parasite and those suffering are primarily school children.

Some authors consider the simplicity of the pinworm life cycle an indication of primitivity. However, we cannot always relate what appears to be complex with advancement and what appears to be simple with primitive. The mechanics of a life cycle and the adaptation to a host are not synonomous. True, the mechanics of the pinworm cycle are simple, but the adaptations, such as man being the only natural host, and only the egg occurring in the external environment where it can persist only days, are evidence of complex physiological requirements. I do not believe this is an ancestral cycle. This appearance of simplicity was hard won through evolution and is highly derived.

Within the intestinal tract the males and females occur together in the cecum, appendix, and adjacent portions of the ascending colon and ileum,

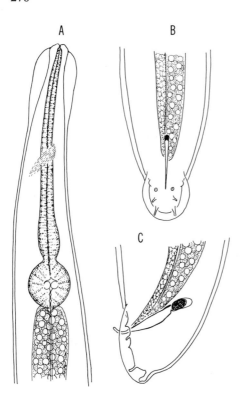

Figure 8.12. *Enterobius vermicularis.* A. Adult female anterior. B. Male tail, ventral view. C. Male tail, lateral view (original).

with their heads embedded in the intestinal mucosa. The males are less than half the size of the females (males 2–5 mm; females 8–13 mm).

The gravid female body, because of greatly distended uteri, becomes filled with eggs. When this condition occurs, the female releases herself from the mucosa and migrates voluntarily down the colon and out the anus. Eggs are deposited as the female randomly crawls on the perianal and perineal region. Generally this occurs during the night hours when the victim is asleep. Not uncommonly, in female patients, the worms enter the vagina. A single female lays between 4500 and 17,000 eggs during her external trek that usually ends with death. Each egg is in the "tadpole" stage of development and cannot proceed further without free oxygen of the exterior environment which is nearly absent in the bowel. Once deposited the eggs rapidly mature and usually this is accomplished within six hours at body temperature. Under conditions of cool moist temperatures, eggs may remain viable for 13 days.

There is disagreement about larval development in the egg. Some authors claim no molt occurs in the egg, so that the first stage is the infective stage. Others such as Alicata and Chitwood claim a molt occurs in the egg. The issue is further confused by the uncertainty of molts that occur in the host's intestinal tract subsequent to ingestion. The reports vary from two molts to three. Both reports on intestinal molts indicate that there must be one or two molts in the egg since four molts is considered universal among Nemata. In sifting these conflicting reports I conclude one molt occurs in the egg, two molts occur subsequent to hatching in the intestine, and the last molt, from fourth to adult, occurs when a suitable site of attachment is found in the intestine. Fifteen days after ingestion the females are fully gravid and ready to begin the voluntary migration and egg laying.

An infection with *Enterobius vermicularis* is technically called enterobiasis and is more common within families or asylums than in the population at large and is more common among North American whites than blacks. Additionally, it is more common in children than adults and is an occupational hazard among laundry workers. The most common methods of transmission are fourfold. The most common means is direct anus to mouth transmission by finger contamination. Anus to mouth transmission is facilitated when people draw their sleeping garments over their heads. The second most common means is among persons sleeping in the same bed or bedroom with carriers. Such persons become exposed to viable eggs on bed linens and other contaminated objects in the room, for example, doorknobs and other objects from which hands can become contaminated. A third form of transmission allows a relatively few worms to infect a large group of people by airborne eggs, which either get into the mouth or are inhaled and swallowed. The fourth method provides for reinfestation of the same individual. In moist climates the eggs may hatch on the anal mucosa and the released larvae are then free to migrate up into the bowel, thus perpetuating self-infection. This mode of infection is called autoinfection.

The damage done directly by the nematode is not great; however, secondary pathogens invading the minute ulcerations that develop at the site of attachment result in open ulcers or submucosal ulcers. The most distressing effect is the intense itching in the anal region produced by the trail of eggs layed by wandering females. This pruritus of the perianal, perineal region is nearly intolerable. The intensity of the itching invariably leads to scratching which in turn provides, through wounds, infection courts for secondary invaders such as basteria. Internally the fourth stage larvae and females may incite appendicitis by secondary bacteria entering the inflamed walls of this organ. A Russian report proclaims that of 685 surgically removed appendices, 411 showed evidence of pinworm damage.

When the worms enter the reproductive tract in women, symptoms and discomfort may continue years after the infestation is eliminated. In a mild infection there may be an increased mucoid vaginal discharge. There are numerous records of the encystment of the female nema in the fallopian tubes or peritineal cavity. In the tubules they may induce an inflammation that can persist for years.

Behavior symptoms associated with this disease are nail biting, nose picking, grinding of the teeth at night, inattention, and poor cooperation. These symptoms are often not considered as part of pinworm etiology, so the infection may not be diagnosed correctly. This syndrome is often coincident with nervousness, insomnia, nightmares, and even convulsions.

Diagnosis is by the recovery of the eggs or the capture of adult females. Eggs are generally collected with an N.I.H. swab, or its tongue depressor modification. Both systems are good, but the latter allows immediate microscopic examination. The tongue depressor swab is a simple procedure that can be performed at home and taken to a doctor for confirmation. Cellulose adhesive tape held sticky side out over the rounded end of a tongue depressor by the thumb and index finger is swabbed around the anal region. The tape is then transferred, sticky side down, with a drop of toluene onto a glass microscope slide and directly examined.

There are a number of chemical treatments available for the evacuation of the worm burden. The difficulty is reinfestation from contaminated objects within the household or institution. A common source of reinfestation of children occurs in the cloak room at school. Cleanliness cannot be overemphasized; personal hygiene must be carefully attended to. Bed linens, bed clothes, and underclothing should be sterilized by boiling daily. Draperies and furniture should be carefully vacuumed. Fingernails should be kept short and clean, especially before meals. Toilet seats should be regularly scrubbed and sterilized. This must all be done concomitant with chemotherapy. If one member of the family has enterobiasis it is a good preventative measure for all to take therapy. It is only through persistent efforts that an infection can be eradicated once it becomes established.

B. Spiruria

1. Evolutionary Sequence among Spiruria

There are no known free-living members of Spiruria. They are the only wholly parasitic subclass in Nemata. All the vertebrate parasites among Spiruria utilize intermediate hosts, either among insects or crustaceans. Experimentally, annelids have been shown to be capable of acting as intermediate hosts, but have not been found doing so in nature. Annelids and molluscs are, however, definitive hosts for spirurids in the superfamily Drilonematoidea. Whether annelids were, in geological times, intermediates for vertebrate hosts is unknown.

The evolutionary sequence of parasitism in Spiruria is not so much reflected in variations of the life cycles, as in other Secernenteans, but in the intermediate hosts utilized and the ecological niche of the definitive host. When terrestrial insects such as cockroaches, grasshoppers, or beetles serve as intermediate hosts the most common definitive hosts are found among galliform birds (chickens, pheasants, quail, etc.), herbivores, or a few insectivorous mammals. It appears that secondarily, through geological time, these parasites may have found their way into raptors or carnivores. In some instances it may even be that animals that now serve as paratenic (ecologically advantageous hosts in contemporary times) may be in the process of becoming definitive hosts. The most derived group of spirurids are in Filarioidea; unlike other spirurids these are actively vectored by Diptera. In addition, they are subcutaneous tissue parasites while other spirurids are almost exclusively gut parasites.

It seems most plausible that those spirurids for which fish are the definitive hosts were introduced into the aquatic environment along with aquatic insects. Circumstantial evidence for this speculation is found in the life cycle of spirurid parasites of salmonids. Mayflies (Ephemeroptera), which are considered ancient insects, generally act as the intermediate hosts for freshwater salmonids such as trout. Interestingly, trout are recognized as ancient Osteichthyes (bony fishes). In addition, these parasites have, more often than not, subjected their reproductive cycle to controls emanating from the fish. This is further evidence that the parasitic relationship is one of long standing.

The spirurids that utilize crustacea such as copepods generally have their definitive hosts among Anseriformes (ducks, geese, etc.), fish, or utilize fish as paratenic hosts in which case the definitive hosts may be fish-eating birds, or fish-eating mammals. Many spirurids utilizing Crustacea as intermediates directly infect vertebrates that drink from infested waters.

Crustaceans may have become involved in the life cycle of spirurids in at least two ways. One way is via the avenue of substituting for aquatic insects in environments where the suitable insect intermediate did not occur. This is a likely explanation for the parasites of fish that utilize crustaceans for intermediate hosts. The second possibility is the presentation throughout geological history of the parasites as eggs or larvae through the demise of terrestrial insects in an aquatic environment, or as now occurs, when infected terrestrial vertebrates enter water and eggs or larvae are released through cutaneous ulcers. The cycle in the latter is completed when the potential host drinks water containing infested Crustacea harboring the infective third stage larva.

In any event, these spirurid cycles that utilize arthropod intermediates as passive transmitters in the food chain are never very complex. The life cycle begins with the deposition of eggs or larvae from an infected definitive host into an environment in which the intermediate host lives. Infestation of the intermediate host is generally through ingestion of the nematode egg or larva. Within the intermediate host the larva develops to the infective third stage and upon ingestion of the intermediate by a definitive host the cycle is completed. With the exception of drinking in Crustacea the intermediate host is most often a food source for the definitive host. This is also true of many infections utilizing Crustacea, for example, Crustacea are a food source of young fish as well as filter feeding birds as some ducks, spoonbills, and flamingoes.

Spirurids are a fascinating group of parasites, not only because of the mystery of their introduction to parasitism in geological history, but because they exhibit one other unique characteristic in Nemata, i.e., the utilization of insects as active vectors that seek out the definitive host. Insect vectors of spirurid parasites are largely limited to the Diptera and are especially common among mosquitoes and biting flies. The nematodes vectored are for the most part limited to the Filarioidea and the disease induced is called filariasis. Probably the best-known example is "elephantiasis" or possibly because of recent notoriety in West Africa, "onchocerciasis," commonly called "river blindness."

Though vectorism is limited to Filarioidea there are closely related parasites that lend insight into how this unique situation of active vectors evolved. Two genera *Draschia* and *Habronema* illustrate in their life cycles the probable intermediate step from typical spirurid parasitism to that of the filarids. Both genera are parasites of horses: *Draschia* (Fig. 8.13) is "vectored" by the housefly *Musca domestica* and *Habronema* by stable flies in the genus *Stomoxys*. Nematode eggs are evacuated in the definitive host feces. Upon hatching the larvae are ingested by fly maggots. Within

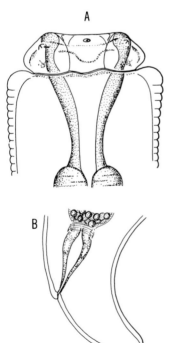

Figure 8.13. *Draschia megastoma.* A. Female head. B. Female tail (original).

the maggot they develop to the third infective stage and are carried through the insect's metamorphosis to adulthood. When adult flies alight on the moist surface of the horse, such as nostrils or lips, the nemas quit the fly by way of the mouthparts. These larvae are swallowed or inhaled. Those swallowed develop to adulthood in the stomach of the horse. Those nemas that enter the skin or lungs fail to develop, whereas in filariasis the development occurs subcutaneously in the lymph system, etc., but not in the intestinal tract. It is significant that the larvae of *Habronema* and *Draschia* both attempt subcutaneous development.

In filariasis both entrance and exit from the definitive host are via the skin. Details of the cycle will be presented later with specific examples. Unlike many suppose, transfer from the vector is not by direct injection, in the case of mosquitoes, but by the nematode crawling down the insect's proboscis and subsequent skin invasion while the insect feeds or after the insect has fed. Invasion of the definitive host is through hair follicles or the wound created by the insect.

a. Spiruria Parasites Possible Links to Dinosaur Parasitism

The solution to the mystery of the development of parasitism in Spiruria may lie within the reign of the dinosaurs, and as such may always remain a secret lost at the time of the late cretaceous disaster that nearly finalized the chapter of dinosours in the history of life on earth. Yet recent discoveries and hypotheses may allow a glimpse of parasitism among the dinosaurs. Two important facts that aid any extrapolation are that dinosaurs did not become extinct without descendants, and the convergence in shape among the dinosaurs, mammals, and birds reveals that their ecological requirements and feeding habits may have been similar to contemporary counterparts. Few today question the belief that birds are descendants of the dinosaur. It is equally evident that dinosaurs not only occupied the niches of contemporary animals, but because of their niche they superficially resembled today's mammalian counterparts. In the upper Triassic *Aetosaurus* and *Stagonolepis* were dinosaur relatives within the swine niche that possessed piglike snouts for rooting through the soil; *Struthiomimus* was a cretaceous dinosaur that not only occupied the niche of modern ostriches, they even looked like ostriches. Link these facts with contemporary theories that at least some of the dinosaurs were warm-blooded and a setting for the development of warm-blooded vertebrate parasitism is created.

Another premise to be considered is that most nematodes are not markedly host specific, and the evolution of most groups of parasitic nematodes has been largely independent of the evolution of their host and has tended to occur in groups of hosts with similar ecological requirements and feeding habits. This implies that the development and persistence of parasitism were in conjunction with host's feeding habits and ecological requirements throughout geological history regardless of what host occupies the niche in contemporary times. Therefore, it is probable that some of the nematode parasites of dinosaurs have become adapted to mammals. From contemporary parasitism one would expect, in addition to spirurids, these parasites to be among the ascarid–strongyle groups that normally exhibit a direct life cycle with passage to the exterior as eggs in the host excrement, with reinfestation being by food contamination. Obviously the species need not be the same as contemporary parasites because morphological changes and speciation were likely to have been coincident with several host changes passed through in diverse taxonomic groups with time.

Some parasites undoubtedly became extinct concomitant with the extinction of various dinosaurs or subsequent substitute hosts. This would account for gaps in the developmental steps of vertebrate parasitism, especially among spirurids. It might also be an answer to our inability to, with conviction, relate spirurid parasites with nonparasitic counterparts. It may be that these are very ancient parasites whose nonparasitic history is lost in antiquity. On the other hand, the history of parasitism among spirurids may have been preserved through the developmental line of coelurosaurian theropods (small bipedal dinosaurs), *Archaeopteryx*, and modern birds.

The term preserved is important because spirurids contemporarily are a taxonomic island, for all intents and purposes, isolated from other free-living or parasitic Secernentea. It can be pointed out that today they do occur in most vertebrates including fish but the great majority occur in birds and among birds they are predominantly found in Galliformes and Anseriformes. Their occurrence in other animals is limited relative to common encounters with Rhabditia types: Ascaridida, Oxyurida, Strongylida, etc. It could be concluded that the thread through time that preserved Spiruria was the lineage of birds. The taxonomic isolation of Spiruria would indicate that their period of great radiation antedates mammalia and may well have been among the dinosaurs. Some might raise the question of Reptilia, but here again they are not common parasites of reptiles in modern times.

Arguments can be put forth that would discount a dinosaur–mammalian lineage in the development of parasitic nematodes of the ascarid–strongyle types. First, the development from free-living rhabdits to parasitic species appears rather clear and documented by contemporary forms principally among mammalia. From this it could be concluded that the ascarid–strongyle development of parasitism is recent and radiated with the speciation of mammalia. It cannot be discounted completely that they too occurred in dinosaurs and many forms among them may have experienced extinction with their host as is here being proposed for the Spiruria. Critical proposals extrapolated from contemporary parasites are that there was a strong dependence, in geological time, of Spiruria on the dinosaurs, whereas Rhabditia were likely to have been developing and radiating among mammalia. Mammalian life style in the Mesozoic placed mammals in a position to be introduced to the preadapted bacterial-feeding nematodes of Rhabditia associated with feces and the carrion of reptiles, dinosaurs, and other mammals. If in fact these two propositions occurred, then upon the extinction of most dinosaurs Spiruria would remain a taxonomic anomaly, which they are, but the Rhabditia, having radiated principally among mammalia, would present an almost complete story of their development of parasitism, which apparently they do.

Extinction as such is a continuously occurring natural phenomenon, however, it is among selected groups and species. It is not normally a natural occurrence involving whole orders or families at one time. If the hosts of spirurids underwent "expected" extinction there would be gaps in their story, but not complete discontinuity. One "host" group presents itself for "widespread" or "mass" extinction which subsequently was represented by a limited line of descendants and that group would be the dinosaurs. The descendants are birds; their progenitor is *Archeopteryx* which was preceded by coelurosaurians. This could be an explanation for the uniqueness of spirurids, if one accepts that parasitic nematodes occur in groups of hosts with similar ecological requirements.

In perusing the host lists of spirurid parasites one is struck by the overwhelming occurrence of these nematodes in birds, particularly Galliformes and Anseriformes. For the moment let us continue to focus attention on the spirurids and Galliformes. Birds placed in this group are found in a large variety of habitats from alpine to grasslands. No matter the habitat, be it a coniferous forest, deciduous forest, brush, or grass, feeding habits among Galliformes remain remarkably similar and independent of species. Primarily they are ground birds seeking aboreal refuge normally only during resting periods or to escape terrestrial enemies. Food habits in Galliformes are diverse; they are omnivores whose food consists of seeds, insects and small animals; they are even at times cannabalistic. Among invertebrates their food is commonly annelids or insects such as beetles, cockroaches, and grasshoppers. It should not be surprising then that these insects and annelids act as intermediate hosts for the transfer of spirurids to the definitive hosts (Galliformes and Anseriformes). It is true these parasites are found in other vertebrates but nowhere to the extent that they are encountered in these two families of birds. As might be expected, some of the strongyles that utilize insects as intermediates are also found in these same birds; however, their occurrence and speciation is limited.

The contemporary parasitism by spirurids has significance to vertebrate evolution specifically birds. That spirurids evolved with Galliformes is unlikely because these birds are not considered ancient or ancestral in bird evolution. If coevolution of spirurids and Galliformes were true then, as is accepted with other nematode parasites with presumed recent adaptations to parasitism, we should have knowledge of free-living counterparts. Spirurids are, as already stated, rather unique among Nemata and without known free-living counterparts. Their morphology (open stoma, two-part esophagus) and biology (totally parasitic and totally dependent on an invertebrate intermediate host) indicate an ancient origin. Further evidence of antiquity could be assumed from the fact that spirurids include among their members the only nematodes that utilize annelids as definitive hosts.

Accepting, as a working hypothesis, that for the most part, contemporary Spiruria were preserved through a "Galliformes niche" then throughout time occupants of the niche would have the opportunity to act as definitive host. In the Cretaceous this niche was occupied by small bipedal dinosaurs, which are classed as coelurosaurians, and *Archaeopteryx*, the precursor to birds. A major difference between these two types of dinosaurs is that *Archaeopteryx* possessed feathers and the coelurosaurians did not. In those dinosaurs there was, as is true of modern birds, already independent actions and functions of the hind and forequarters. Ostrom suggests, "that the unique and highly diverse nature of avian flight was possible only because it was preceded by uncompromised, obligatory bipedal posture and locomotion that allowed the forequarters in a proto-*Archaeopteryx* stage to be adapted for other entirely separate and unrelated activities." It is also evident that in *Archaeopteryx* the running apparatus of the hind limbs was far more advanced than was the flight apparatus of the forelimbs and chest.

A major question is whether birds originated flight from the ground up or from the trees down. Ostrom believes the former to be true because even though *Archaeopteryx* had feathers it was not anatomically suited to power flight. I believe spirurid parasitism provides corroborating evidence to Ostroms' hypothesis. The hind feet of *Archaeopteryx* were adapted from coelurosaurians for a ground dwelling habit. Like modern birds the feet consisted of three main toes directed forward and the first toe (hallux) was directed posteriorly in a fully opposable position to the second or middle toe. Even more important is the nature of the claws of these toes. The claws of the third toe hallux of *Archaeopteryx* are nearly identical to those found in modern Galliformes (Fig. 8.14). The hallux in both is higher on the metatarsus and lacks the well-developed flux or tubercle for muscle attachment seen in Passeriformes (perching birds) (Fig. 8.14). As such the feet of *Archaeopteryx* were adapted to ground scratching and not to perching. Ground scratching and flushing low vegetation insects is an important aspect to spirurid parasitism. If *Archaeopteryx* had the ability of power flight and were arboreal the types of insects available would probably not have been satisfactory intermediate hosts for the ancestral spirurids found in Galliformes, i.e., they would have been of the types utilized to transmit the more derived spirurids that are important to the development of filariasis, the most sophisticated form of parasitism

third toe   hallux   third toe   hallux

predatory bird: eagle   perching bird: crow

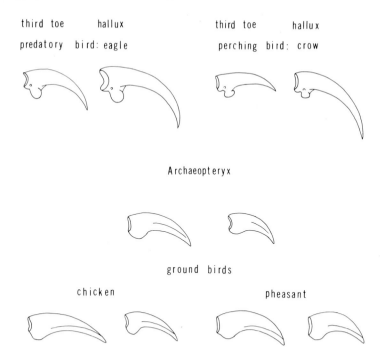

Archaeopteryx

ground birds

chicken   pheasant

Figure 8.14. Diagrams comparing third toe and hallux of predatory birds, perching birds, *Archaeopteryx*, and ground birds (redrawn from Ostrom).

known among the subclass. In modern perching birds the type of spirurids found exhibit life cycles preadapted to filariasis. Their intermediate hosts are generally in Diptera (flies), not in the more ancient Orthroptera (grasshopper, cockroach) used by most Galliforme parasites.

If flight originated from the trees down then the chain of development of spirurid parasites in grasshoppers, cockroaches, and ground beetles would have been broken because these insects seldom inhabit trees. It is difficult to postulate how a cycle most certain to have existed among coelurosaurians could have been rejuvenated if *Archaeopteryx* were first arboreal and secondarily terrestrial. The burden of carrying these spirurids through time prior to the advent of Galliformes would have to have been the responsibility of remnant coelurosaurians. That they persisted this long is questionable. It would seem that if *Archaeopteryx* broke the terrestrial life cycle in originating flight arboreally, these particular spirurids would not have been available when Galliformes entered the scene as ground scratchers and flushers. Bear in mind that few insectivorous mammals house related parasites so they were not likely preservers of spirurids.

The arboreal origin of powered flight seemingly requires a tortuous logic uncorroborated by contemporary life cycles in Spiruria or the anatomical adaptations available to *Archaeopteryx*. On the other hand the life cycles, intermediate hosts, and definitive host of spirurids would lead one to conclude that the line of development was unbroken from coelurosaurians through *Archaeopteryx* to modern birds. In order to be unbroken, however, *Archaeopteryx* had to originate flight from the ground up.

The foregoing hypotheses and the sequence of the development of the parasitic habit among Nemata will now be illustrated by examples. The function here is not to teach parasitology or the identification of parasites. The purpose is to acquaint the reader with some of the more important nematode parasites as well as to convince the reader that there is a logical sequence in the evolution of nematode parasitism.

2. Spirurid Diseases

Spirurids are common parasites found in a wide variety of animals in all parts of the world. Two groups in which they figure prominently are birds and fish. They are also serious parasites of man where they induce some of the most insufferable of parasitic diseases such as "guinea worm," "elephantiasis," and "river blindness." Most spirurid diseases, as mentioned earlier, are not host specific but tend to occur in groups of hosts with similar ecological requirements and food habits. The examples presented here will be from fish, birds, and mammals including humans.

a. Sterliadochona

Sterliadochona (Figs. 3.7, 8.15) is a common genus of parasites in the esophagus and stomach of freshwater salmonids. There is only one group of insects, i.e., the various species of mayflies (Ephemeroptera), that act as intermediate hosts for this genus of nematode and some other closely related nema genera. The life cycle of the nematode is closely allied with the activities of the fish at different times of the year. It is not uncommon to find the incidence of disease to be 100% in trout over 10 cm in length.

There are, in this example, two generations of parasites throughout the year. The selective advantage of this will be discussed later. The intermediate mayfly hosts will be discussed as mayfly 1 and mayfly 2, because the species differ from stream to stream. One generation of the parasite occurs in spring and one occurs in autumn. In May and July the fully gravid female nematodes release their eggs in the trout's intestinal tract; these are then voided with the trout feces. These eggs are ingested by mayfly 1 feeding on the stream bottom in spring and summer. Inside mayfly 1 the larvae hatch, and make their way through the gut and into the body cavity where within one month they develop into third stage infective larvae. Mayfly 1 plays only a small role in the life cycle of the parasite on this particular stream because the trout at this time are feeding primarily on terrestrial insects. During August and September the

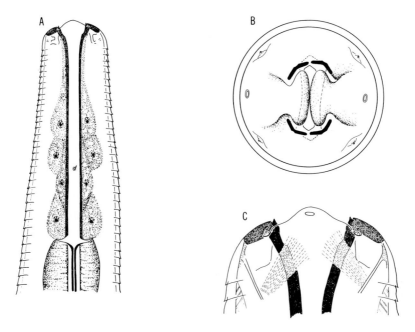

Figure 8.15. *Sterliadochona pedispicula.* A. Female head. B. Female face view showing lateral lips and cheilorhabdions. C. Female head showing cheilorhabdions in lateral view (redrawn from Maggenti).

played-out adult nemas quit the trout leaving only juveniles behind. In autumn these juveniles develop into adults and begin producing eggs in September and October. These eggs are ingested by the second species of mayfly which is the normal intermediate on this stream. By November this mayfly 2 is again infective and during these months is the main source of food for the trout. During this same period the adult nemas again leave the trout. Throughout the winter months only juvenile nematodes remain in the gut of the trout. On this hypothetical stream trout spawn from November to January. By February the nemas are mature but do not contain embryonated eggs. Eggs belonging to the spring generation of parasites become fully embryonated beginning in May.

There are four factors prevailing in this example: (1) the fall spawning cycle of the trout; (2) the two species of mayfly; (3) the species of trout, and (4) the time when trout change their food size preference (10 cm). Keeping in mind that spirurids parasitize hosts in similar environments and with similar feeding habits then the use of what appears to be a dead-end intermediate (mayfly 1) can be understood. On another stream where the trout spawn in spring or early summer then mayfly 1 would be the preferred intermediate for 10 cm developing trout which would begin feeding on mayflies in August and September, while adult fish are feeding on terrestrial insects. On the stream in the example, 10 cm trout are developed in March and April when mayfly 2 contains infective nemas. On any given stream there may be dead ends as mayfly 1, but over the range of streams, trout, and mayfly distribution it is this flexibility that serves to preserve the parasite in any freshwater environment containing trout. Though unconfirmed there may also be some selective advantage on streams containing several species of mayfly and more than one species of trout whose feeding habits may differ in order to avoid excessive competition.

b. Oxyspirura mansoni, Eyeworm of Galliformes

Oxyspirura mansoni (Fig. 8.16) is a common eyeworm among chickens, turkeys, pheasants, and several birds in most of the warm parts of the world.

The life cycle is interesting because of the preferred site selected by the adult worms under the nictitating membrane of the eye. Eggs layed by females wandering in this environment are washed down the nasolacrimal ducts into the pharynx, swallowed, and subsequently voided in the feces.

Figure 8.16. *Oxyspirura mansoni.* A. Female head. B. Female face view. C. Male tail. D. Female tail (modified from Ransom).

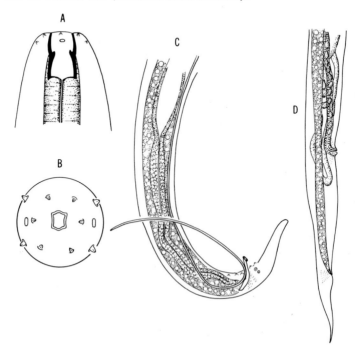

In the exterior environment no further development occurs until the egg is ingested by the woods or burrowing cockroach *Pycnoscellus surinamensis*. Forty-eight hours after reaching the insect gut the eggs begin hatching. Eight days after the initiation of hatching, the larvae begin to show up in the hemocoel of the cockroach. The insect reacts to this invasion by sometimes encysting the nema; however, this does not interfere with the nematode's development. The third stage larvae, either free or encysted, are infective to birds, which consume the cockroach intermediate, in the bird's crop. The larval nematodes now reverse the route followed by the egg, i.e., they make their way up the esophagus into the nasolacrimal ducts and thereby gain entrance to the bird's eyes. Sexual maturity is reached in about 30 days.

c. Tetrameres americanus and T. crami

Tetrameres americanus and *T. crami* are discussed together because the former parasitizes Galliformes and the latter parasitizes Anseriformes. Interest is generated in these two *Tetrameres* not only because of the marked sexual dimorphism between the vermiform male and globose female (Fig. 8.17), but also because this genus utilizes crustacea to perpetuate the life cycle in Anseriformes and terrestrial insects to perpetuate the cycle in Galliformes. In both Anseriformes and Galliformes *Tetrameres* are parasites in the proventriculus.

The life cycles differ only by the intermediate host. The parasites' eggs are fully developed and contain first stage larvae when laid and voided with the bird's feces. When ingested by the appropriate intermediate the eggs hatch. For Anseriformes the common intermediates are freshwater gammarids in the genera *Hyallela* and *Gammarus*; with Galliformes the intermediates are grasshoppers and cockroaches. Within the intermediate the eggs hatch and the larvae penetrate the gut and thus gain entrance to

Figure 8.17. *Tetrameres fissipina*; nongravid female (modified from Travassos).

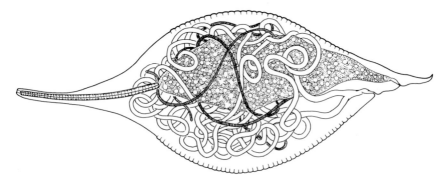

the body cavity. These larvae may or may not encyst; in either case they remain infective. Infection of the definitive host occurs when the intermediate is eaten. The nematode escapes from the intermediate in the host crop from where they make their way to the proventriculus. Males (2.9–4.1 mm) develop in about 30 days. Females which are globular to spindle-shaped (1.5–3.25 mm long by 1.2–2.2 mm wide) develop over a slightly longer period, sometimes extending into two months.

Tetrameres are also found in meadowlarks, canaries, and pigeons, but the intermediates are always grasshoppers or cockroaches.

d. Philonema oncorhynchi, Dracunculoid of Anadromous Salmonids

Philonema oncorhynchi is a body cavity parasite of anadromous salmonids, especially the sockeye salmon, *Oncorhynchus nerka*. Sockeye salmon are among that group of anadromous salmonids that die after spawning. This routine has become not only an important facet to the life cycle of the salmon but also the parasite. Four years may elapse from spawning to spawning, and a nematode, in order to survive, must be closely attuned to the reproductive cycle of the fish.

When the sockeye salmon return to the northern lakes to spawn in September they harbor mature female worms filled with embryonated eggs. Mature females can exceed 42 mm and the males range around 18 mm (Fig. 8.18). When the salmon releases its roe the female worms are expelled with it; within one minute of exposure to freshwater the female worms burst, releasing thousands (500,000 eggs/female) of first stage larvae into the environment. The released larvae are active and capable of keeping themselves suspended in water.

When the adult salmon dies its decaying body becomes an ideal provider of nourishment for freshwater copepods. In the immediate environs of these dead salmon there is a phenomenal bloom of these crustacea. This bloom is important for two reasons: (1) the copepod is the intermediate host for the parasite and (2) the copepod is an important food source for the young salmon fry which become infected between January and February. Thus this infection, which may remain with the fish for the four years of its life, is acquired at a very early age. Within the copepod it takes 17 days at 12°C for the nematode to become a third stage infective larva (Fig. 8.18B). Once infected the copepod can maintain the parasite up to seven months. When the young salmon fry ingest the copepod the nema is released.

The salmon begin their seaward migration in April. The nematode is still in the third stage and about 1 mm long. In this early stage of infection they are located in the wall of the swimbladder and in connective tissues of the parietal peritoneum. No development of the nematode is noted until the salmon is 18 to 34 months old. The nema now begins to develop

Figure 8.18. *Philonema oncorhynchi.* A. Male tail ventral view. B. Third stage larva anterior. C. Second stage larva anterior (redrawn from Platzer and Adams).

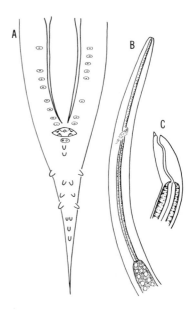

rapidly and in the fourth stage the young females are 4.8 mm and the male larvae are only slightly smaller at 3.9 mm. Growth from this time on is rapid and soon the adult females and males appear; the females now measure 42 mm and the males 18 mm in length. The salmon is also undergoing rapid changes and development preparatory to its return to the freshwater spawning grounds. At the age of 32 months the salmon undergoes maturation and as the salmon's gonads increase in size so also do those of the nema. By the time the salmon reaches freshwater the nematode eggs contain morulae, i.e., they are in the "tadpole" stage. When the salmon reaches the spawning grounds their eggs are mature and ready for fertilization. The female nematode at this time is fully gravid with active first stage larvae.

Noticeable pathology is noted in the salmon during the period when the nematode is migrating through the coelom. Most noticeable are the adhesions with encysted male nematodes. These adhesions occur on the pyloric caeca, liver, stomach, gonads, and between the intestine and body wall. Female nematodes have never been found in these encystments. The adhesions themselves are transitory and are not seen when the female salmon is in the spawning grounds.

There seems to be little doubt that the cycle of this nema is controlled by internal factors of the fish. The advantages of being responsive to the fish are clear. The nematode requires the freshwater copepod for its intermediate host, and if development of the nema occurred at sea the cycle would be broken and the parasite lost. The slowing of the nematode's development also reduces pathological damage to the fish. If the nematode developed rapidly during the early days and months of the fishes life, salmon fry probably could not abide the nutritional loss or damage caused during the worm's migration. If the nemas completed their cycle while the fry were in freshwater, death would be inevitable because of the nutritional drain, i.e., theoretically when the fish were 80 mm long the nema would be 40 mm, and obviously a worm burden of 30–90 worms could not be tolerated.

The relationship between *Philonema* and *Oncorhynchus* is an ideal example of host–parasite adaptation that has some benefit to both: the host, which is not killed by the parasite, and the parasite, which is allowed to sustain a four year life cycle. It is interesting that life begins and ends for the host and the parasite in the spawning grounds.

e. Dracunculus medinensis, Guinea Worm

It is the species *Dracunculus medinensis* that is thought by many to be the fiery serpent of Moses. It certainly is among the oldest known animal parasites, having been studied and recorded by the early Egyptians. Human involvement is still common in much of Africa, the Middle East, and Southeast Asia. In North America it is reported from several fur bearing mammals, i.e., fox, racoon, and mink; additionally it has become established in a few restricted foci of tropical America.

The adult worms (females 70–120 cm long; males 40 mm long) develop in various body cavities and retroesophageal tissues, and when the females reach full maturity they migrate into the subcutaneous connective tissues generally of the lower extremities. When the head end of the adult female (Fig. 8.19) approaches the skin of the host a papule is formed that soon develops into an enlarged blister that eventually ruptures. Most commonly these blisters are found on areas of the body most likely to come in contact with water; however, these papules may also be found on the abdomen or back. When contact is made with freshwater the anterior portion of the female bursts open and a loop of the uterus, which has prolapsed through the ruptured female, is extruded out of the perforation in the blister. The uterus in contact with water bursts and thousands of larvae are discharged into the water. Larval discharge will repeatedly occur each time the ulcer comes in contact with water. Unless the worm is removed from the body successive discharges of larvae will occur until all are evacuated.

The released larvae are wiry and measure 0.5–0.7 mm. Though the larvae are motile, their movement is stiff. The means whereby these larvae gain entrance into the intermediate crustacean host, usually a *Cyclops*, is by ingestion. The larvae have no mechanism for direct entry. Once ingested the larvae penetrate the *Cyclops'* intestine to gain entrance to the body cavity. Generally, if five or more worms are ingested the *Cyclops* dies. Within the cavity the larva molts, losing its striated cuticle. Later a

Figure 8.19. *Dracunculus medinensis.* A.
Male anterior. B. Female tail. C. Female
tail tip (modified from Moorthy).

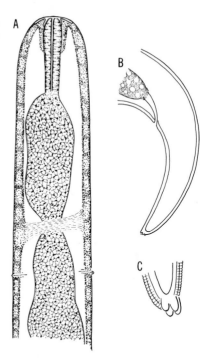

delicate sheath develops about the nematode. Whether this is formed by
the nema or is a weak attempt on the part of the *Cyclops* to encyst the
larva is unknown.

Ingestion of water containing infective *Cyclops* is the only known source
of human or other mammalian infections. In the gastric juices of the
duodenum the larvae are freed from the *Cyclops*. Immediately they make
their way through the wall of the intestine and reach the loose connective
tissue of the peritoneal lining of the abdominal cavity and viscera. In
these tissues they develop into adulthood in eight months to one year.
The role of males is unclear, but their scarcity suggests that they are not
necessary for reproduction. When the female migrates to the subcutaneous
tissues the ovaries are postfunctional and the uteri are distended with the
active larvae. No vulva has been found on these gravid females and this
is why the anterior extremity must burst in order to release the larvae to
the exterior environment.

The incidence of infection is high wherever people depend on step
wells, either natural or rain cisterns, for drinking water. Such wells require
that the lower extremities of the body be immersed in the water in order
for a person to fill hand carried water vessels. Areas of high infection are
often in areas where religious ablutions require rinsing of the mouth
during the act of being "purified" by water. In these areas the incidence
is often higher among men than women.

The only notable pathology occurs when the female worm migrates to the subcutaneous tissues of the skin. The worm causes an allergic reaction because of the large amounts of toxic by-products released into the host's system. The situation is improved upon the bursting of the blister. The syndrome of allergic symptoms preceding the bursting of the blister consist of flushed skin (erythema), rashlike eruptions (urticaria) with intense itching accompanied by nausea, vomiting, diarrhea, respiratory impairment, giddiness, and a drop in blood pressure (syncope). The rupture of the blister usually alleviates these symptoms but the open ulcer is an invitation to secondary disease agents and septicemia.

Diagnosis is only possible with the onset of symptoms followed by the formation of the cutaneous lesion. Prognosis is usually good unless the symptoms are complicated by secondary septicemia. Surgical removal is called for once the papule has formed or the so-called "Indian barber" technique can also be used if the worm has not yet burst. This technique consists of rolling the adult female out of the ulcer and onto a stick.

Prevention involves (1) keeping infected people from contaminating the water and (2) not drinking suspect water from step wells, rain cisterns, etc.

3. Filariasis and Related Infections

Filariasis encompasses some of the most devastating parasitic infections of humans. The results of filarid infections range from blindness to gross malformation of the body.

Filarial disease differs from usual nematiasis by not being intestinal. Normally filariasis is adapted to a habitat in the subcutaneous and deeper tissues of the vertebrate body, including the circulatory, lymphatic, muscular, and connective tissue layers, or serous body cavities of all vertebrates but fish. Filarids are the only group of nematodes actively vectored.

The name filariasis is derived from the stage of the nematode that is picked up by the vector in subcutaneous capillaries or other cutaneous tissues. This stage is a motile embryo, which collectively are called microfilariae. In some species the flexible eggshell remains with the motile embryo and the microfilaria is said to be "sheathed"; when the embryos are not still in the eggshell the microfilariae are described as "unsheathed." Within the arthropod vector the microfilariae develop to the third stage; it is this stage that is returned to the definitive host during the feeding operation of the vector. The infective larvae are not injected into the definitive host. They actively crawl down the mouthparts and then make their way into the definitive host either directly, through wounds, or via hair follicles.

The following examples are presented in a sequence that is designed to further corroborate the earlier hypothesis that birds learned to fly from the ground up. One point presented in that discussion was that many of the spirurid parasites of birds appear to have a taxonomic affinity to the filarids which are for the most part vectored by Diptera. The first example is of a "filarid-like" infection that has perching birds as definitive hosts, but utilizes a grasshopper as an intermediate host. Again the direction appears to be from the ground up and secondarily introduced to "flying" or arboreal insects, in the order Diptera, and there primarily limited to blood sucking or biting flies. Once introduced to these Diptera fecal associations are no longer necessary.

a. Diplotriaenidae, Air Sac Parasites of Birds and Reptiles

Diplotriaenidae (Fig. 8.20) are oviparous and the eggs hatch after being voided in the feces of the bird or reptile and swallowed by mandibulate insects such as grasshoppers. Their life cycles differ from the general ovoviviparous filaroids whose motile embryos are transmitted by biting or blood sucking arthropods.

Eggs containing first stage larvae are laid in the air sacs and lungs of the definitive host. After deposition they are conveyed up the trachea to the pharynx, swallowed, and passed to the external environment in the feces. For further development to occur they must be swallowed by an insect. In the insect's midgut the eggs hatch and the released larvae burrow through the gut into the hemocoel where they develop in the fat bodies. Third stage infective larvae appear, encysted or free, in the insect hemocoel some 30 days after ingestion.

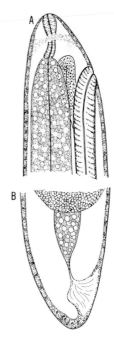

Figure 8.20. *Diplotriaena translucidens.* A. Female anterior showing anterior placement of the vulva. B. Female tail (modified from Anderson).

The route taken in the vertebrate host is not certain, but evidence gained from studies with thrushes may shed light on the matter. It appears that once the larvae are freed in the gut of the definitive host they burrow into the intestine in order to gain access to the blood or lymph system. The migration is then to the lungs via the heart. The adults then mature in the lungs; this takes from two months to one year.

The prime interest in this group is that arboreal birds that feed on ground dwelling insects may have introduced filarial type infections to flying insects such as Diptera. This indicates that filarid diseases are secondary to the more common spirurid diseases of ground birds.

b. Habronema and Draschia, Stomach Ulcers of Horses

Habronema and *Draschia* contemporarily illustrate how filariasis of mammals may have developed ancestrally. The adult females (Fig. 8.13) are located in ulcerated regions of the horse's stomach. Their eggs when deposited pass on with feces. These embryonated eggs are ingested by fly maggots. Upon hatching, the larvae enter the Malpighian tubules where they molt to a sausagelike stage reminiscent of the second stage larvae of filarids. The sausage stage persists until the fly pupates; they then molt, leave the tubules, and make their way to the head of the fly as third stage infective larvae.

When the fly visits the moist mucous membranes of the lips, nostrils, or open wounds, the nemas leave the fly by way of the labellum. Those that penetrate subcutaneously die; however, those that are swallowed proceed to the stomach. In this preferred site they presumably enter the glands of mucosa where ulcers containing the adult worms are induced.

Of more interest here from a phylogenetic standpoint are not those that undergo development, but those entering the cutaneous tissues. Though they cannot develop in wounds or nasal membranes they can sustain themselves for several months. Their activities are such that the continuous initiation, especially in wounds, induces granuloma. These large pulpy tissues persist as long as the larvae. The significance is that these may be reiterations of ancestral attempts of filaroids to invade the cutaneous tissues of mammals.

c. Wuchereria bancrofti, Elephantiasis

Elephantiasis is indigenous throughout most warmer regions of the world from 41° N to 28° S. In Europe the infection occurs in scattered areas surrounding the Mediterranean. No longer is it known from the mainland Australia or in the United States. Until the late 1930s or early 1940s it was endemic in an area near Charlestown, South Carolina. More than one genus and species is involved in elephantiasis throughout the world. *W. bancrofti* is chosen as illustrative of the group (Fig. 8.21). All forms of

Figure 8.21. *Wuchereria bancrofti.* A. Female anterior showing anterior placement of the vulva. B. Female posterior (modified from Yorke and Maplestone).

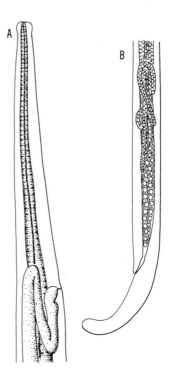

elephantiasis are transmitted by mosquitoes and it has been demonstrated that over 70 species are efficient vectors.

The third stage infective larvae quit the mosquito by crawling down its proboscis during the act of feeding. At the feeding site the mosquito deposits a small droplet of fluid. This fluid is protective to the nematode and prevents drying during the time they are penetrating the host's skin, either directly or through a hair follicle or the bite wound left by the mosquito. These larvae, after penetrating the skin, pass through the peripheral lymphatics, in which some growth occurs, to nearby lymph vessels or to the lymph nodes where development to full maturation occurs (female length, 80–100 mm; male 40 mm). Females do not deposit eggs as such but motile embryos called microfilariae, which in this species retains the flexible eggshell or "sheath." The microfilariae are only 0.24–0.29 mm in length and in the blood or lymph their movement is graceful and snakelike.

Microfilarial periodicity refers to the surfacing of large numbers of these motile embryos in the peripheral blood either nocturnally or diurnally. These circadian rhythms differ according to local mosquito vector flight times and species or biotypes of the nematode. In this species microfilaria are found in the blood nocturnally during the hours between 10 P.M. and 2 A.M. During the daylight hours it is unusual to find microfilariae in the peripheral blood. In some regions *W. bancrofti* does not undergo any noticeable periodicity.

The microfilariae are picked up by the vector mosquito during the act of taking a blood meal. When they reach the insect's stomach the sheath is discarded. The freed "larvae" work their way through the wall of the mosquito's proventriculus or cardiac portion of the midgut. In the course of several hours they make their way to the insect's thoracic muscles. Here they metamorphose to the sausage-shaped stage; they then proceed through two molts to become filariform third stage larvae. Under optimal conditions this stage is reached in about 11 days. From the thoracic muscles they migrate to the head region to await the mosquito's next blood meal when they will crawl down the proboscis and start the cycle again.

How severe the pathological effects of elephantiasis are depends on (1) the infected individual's tolerance to the parasite, (2) the number of invading third stage larvae, (3) the number of times the individual is bitten by infective mosquitoes, (4) the site within the body where the nemas choose to develop, and (5) whether the infection is complicated by secondary pathogens.

Like most filaroid diseases the incubation period is an extended one often lasting a year or more. During this period there may be transient symptoms of inflamed lymphatics or other allergic responses to the initial invasion. In some individuals swelling of the limbs may occur after periods of exertion, but this generally subsides with rest. In the acute stages swelling of limbs is accompanied by extreme pain and tenderness; during this period "filarial fever" is usually conspicuous. These symptoms are recurrent and may manifest themselves monthly. In the absence of additional complications the attacks become less severe and will eventually dissipate. Repeated infections inevitably lead to the classical symptoms associated with the chronic stage: elephantoid extremities, and in males enlargement of the scrotum. It is from these gross enlargements that the common name elephantiasis is derived. The enlargements are a physical and mental burden to the victim who is often ostracized from society. The elephantoid growth consists of lymph and fat in a fibrous matrix; the covering skin is thickened and resembles elephant hide.

The disease is diagnosed by recovery of circulating microfilariae. In mild cases the prognosis is good, but depends on the number of infective larvae that have gained entrance and the potential for reinfection as well as individual sensitivity. Chemotherapeutics are available that eliminate the circulating microfilariae and sterilize the adult females. In subchronic cases elephantoid tissue can be surgically removed, with mild cases pressure bandaging reduces the swelling.

Prevention involves protection from mosquito bites, mosquito control, and more recently the experimental treatment of whole populations in endemic areas with diethylcarbamazine. A test program in Brazil adds this chemical to table salt.

Figure 8.22. *Onchocerca volvulus.* A. Female face view. B. Male tail (modified from Sandground).

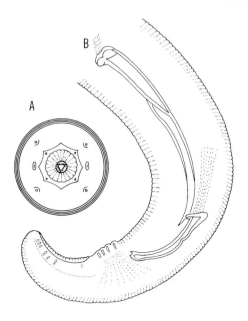

d. Onchocerca volvulus, River Blindness

Another important filariasis of man is onchocerciasis or "river blindness" caused by *Onchocerca volvulus* (Fig. 8.22). This disease is endemic to tropical Africa, but was introduced with slavery to the sugarcane and coffee growing areas of Central and upper South America. The life cycle is similar to elephantiasis, but this disease is vectored by black flies of the family Simulidae. Instead of blocking the lymphatics, the adult worms (female length 33–50 cm; male 19–42 mm) are located in subcutaneous tumors. Though the tumors may be uncomfortable, the most serious damage is caused by the migration of microfilariae into the eyeball. This invasion induces lesions causing opacities or keratitis of the cornea; complete blindness occurs when the microfilariae invade the optic nerve. Adults can be removed by simple surgical procedures. The use of chemotherapeutics eliminates the circulating microfilariae and greatly reduces the chances of blindness.

e. Loa loa, Loa Loa Disease

Loa loa (female length 50–70 mm; males 30–34 mm) is vectored by biting tabanid flies of the genus *Chrysops*. The disease loa loa is endemic to the rain forest belt of West Africa and Equatorial Sudan; it was introduced to the West Indies with slavery. The adult nemas (Fig. 8.23) do not settle in tumors, but continuously wander subcutaneously provoking temporary swellings and inflammation. Because of their wandering habit the best time for their removal is when they are migrating through the

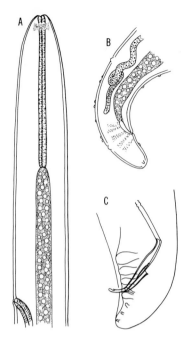

Figure 8.23. *Loa loa.* A. Female anterior showing anterior intestine and vulva opening. B. Female posterior showing cuticular bossing. C. Male tail (modified from Yorke and Maplestone).

corneal conjunctiva or across the bridge of the nose as they migrate from one eye to the other. During their migrations they will temporarily settle down. At this time they provoke temporary inflammations called "fugitive" swelling or "Calabar" swelling. Though generally subcutaneous they may be found penetrating the viscera or even the kidney.

Prevention consists of three phases: (1) protection from the bite of the *Chrysops* fly; (2) insecticidal campaigns against the larval flies, and (3) detection and treatment of human victims.

f. Dirofilaria immitis, Heartworm of Dog

Heartworm of dog, *Dirofilaria immitis* (Fig. 8.24), is a worldwide infection. Over 90% of the dogs in Hawaii are infected. Only within the last 10 years has the disease been known from the western United States, but it is rapidly spreading and nearly 30% of the dogs in the foothills of Sacramento, California, are positive for this nematode. The adult males measure 12–16 cm and the females are 25–30 cm long. Both adult males and females are found in the right chamber of the heart. This filarial is mosquito transmitted and on rare occasions can develop in man.

Dirofilaria immitis develops to the third infective stage within the Malpighian tubules of the vector mosquito. Once developed, they make their way to the insect's head region to await its taking of a blood meal.

Figure 8.24. *Dirofilaria immitis.* A. Female anterior showing micro-filaria in the uterus. B. Male tail (modified from Olsen).

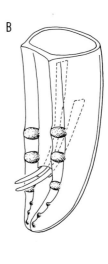

At this time they make their way down the mosquito's proboscis into the droplet of serum around the feeding site. They then proceed to penetrate the host. Having penetrated the host's skin, the larval nemas make their way to the submuscular membranes, subcutaneous tissues, subserosa, adipose tissues, and sometimes are found within muscle tissue throughout the body. In these tissues they develop to a length of 25 to 110 mm. Subsequent to this development they make their way to the venous system and are carried to the heart. This period of development and movement to the heart takes approximately three months. Microfilariae begin showing up in the host's blood some six months after exposure and may continue for more than two years.

Though the disease can be extremely serious, dogs may harbor 25–50 adult worms with symptoms varying from none to moderate. The first obvious symptoms include tiring easily and shortness of breath but with recovery that makes the dog appear normal.

In heavy infestations (50–100 worms) there is generally a dilation of the right side of the heart as well as a pathological involvement of the lungs, liver, and kidneys. Skin lesions have also been reported.

Little can be physically done to protect dogs in endemic areas; however, chemotherapeutics are available, generally at rather nominal cost, which prevent the development of the infective stages. They also kill the microfilariae and sterilize the adult worms; this can be dangerous because a heavily infected dog may die of anaphylactic shock. Mosquito control is another factor in limiting the disease. In California the nemas apparently can be vectored by *Aedes sierriensis*, the "tree hole" mosquito. Two important factors with this mosquito are that it can, and apparently has, vectored the nematode to wild reservoir hosts, and that it breeds in water anywhere, in tree holes, tin cans, bottles, etc., and therefore is virtually impossible to control.

The use of mosquito repellents has not proved practical; therefore screening, mosquito abatement, and preventative chemotherapeutics are the means of control and prevention.

Chapter 9
Classification of Nemata

The classification of Nemata presented here gives characters only to the superfamily level. The reader should bear in mind that I believe many of the superfamilies listed are likely to be families when the concepts of nematology mature to those of other zoological sciences. This classification follows the skeletal structure proposed by Filipjev in 1934. It further incorporates the internal framework of Chitwood. The higher categories include the proposals I made in 1982.

We cannot be complacent about classification, it is not merely a system of categorizing groups for identification. The greatest contribution of a classification is predictability. A close study of a sound classification reveals a great deal about the included animals, not only their relationships to each other but also the evolution of the group. Proper placement allows us to predict likely life histories as well as biological niches.

The soundness of nematode classification weakens at the lower levels, especially at the family level. There are perhaps 15,000 nominal species described and these are generally arranged into some 250–300 families, of which nearly 25% are monotypic. Additionally 30% of the proposed subfamilies and 35% of the genera are monotypic. Contrast this with some quarter of a million beetles in less than 100 families.

The point is not that we should mirror an entomological classification, but heed the concepts inherent in the "family" designation. What is implied in the nematology family arrangements is that nematodes have undergone widespread extinction and therefore the family group is left as islands clear-cut from each other as in a Pacific atoll. This, however, is not the case. The characteristics between families among nematodes are so slight as to defy identification, i.e., the characters are few and often meaningless. The blending of characters from one group to another seldom allows for clear distinctions characteristic of a family group to be made. This also indicates that little extinction has occurred, with the possible exception of the subclass Spiruria. Oftentimes what is designated a family in reality probably represents no more than a genus.

Too often the proposal of a family is based on emotion. The family is a subjective division but it is supposed to have meaning as stated above. It is not the size of a group that determines if it is a family; it is the contained information, morphological or biological, that determines family or generic status.

For an expanded classification with characters through the family level see references Chapter 9, Maggenti; 1981.

I. Classification of Nemata

Phylum: Nemata   313
  Class I. Adenophorea   313
    Subclass A. Enoplia   314
      Order 1. Enoplida   314
        Suborder a. Enoplina   315
          Superfamily 1. Enoploidea   315
            Family: Enoplidae
            Family: Lauratonematidae
            Family: Leptosomatidae
            Family: Phanodermatidae
            Family: Thoracostomopsidae
          Superfamily 2. Oxystominoidea   315
            Family: Paroxystominidae
            Family: Oxystominidae
            Family: Alaimidae
        Suborder b. Oncholaimina   316
          Superfamily 1. Oncholaimoidea
            Family: Oncholaimidae
            Family: Eurystominidae
            Family: Symplocostomatidae
        Suborder c. Tripylina   316
          Superfamily 1. Tripyloidea   316
            Family: Tripylidae
            Family: Prismatolaimidae
          Superfamily 2. Ironoidea   317
            Family: Ironidae
      Order 2. Isolaimida   317
          Superfamily 1. Isolaimoidea
            Family: Isolaimiidae

Phylum: Nemata

Nematodes are bilaterally symmetrical unsegmented pseudocoelomates, body generally elongate, cylindrical (wormlike) covered by cuticle secreted by hypodermal cells. Among parasitic species body form may be spindle-shaped, pear-shaped, lemon-shaped, or variations of saccate. Anterior end usually bluntly rounded; oral aperture terminal and surrounded by lips. Anterior body characteristically bears sixteen setiform or papilliform sensory organs and two chemoreceptors (amphids). Oral opening followed by stoma, esophagus, intestine, and rectum that opens through a subterminal anus. Excretory system when present opens through an anterior ventromedian pore. Sexes usually separate. Female reproductive system tubular opening through a ventromedian vulva. Tubular male reproductive system joins posteriorly with digestive system to form a cloaca that opens ventromedially through the "anus". Males generally possess paired secondary cuticular sex organs called spicules. Nervous system consists of a ganglionated circumesophageal nerve ring and longitudinal nerves; main nerve ventral and ganglionated. Body musculature limited to longitudinally oriented fibers. Organs of respiration and circulation absent.

Eggs with determinate cleavage, oviparous or ovoviviparous. Life cycle generally direct: egg, four larval (juvenile) stages, adult. External and internal cuticle shed at each larval molt.

Class I. Adenophorea

Adenophorea, a class of nematodes divisible into two subclasses: Enoplia and Chromadoria. Amphids postlabial, variable in shape: porelike, pocketlike, circular, or spiral. Cephalic sensory organs (sixteen), setose to papilloid, postlabial and/or labial in position. Somatic setae and hypodermal glands commonly present; somatic papillae usually (always?) present. Hypodermal cells uninucleate; cuticle four layered: epicuticle, exocuticle, mesocuticle, and endocuticle. Cuticle surface generally smooth, but may have transverse or longitudinal striations. Excretory organ, when present, single celled, ventral, and without collecting tubules; excretory duct usually noncuticularized. Pseudocoelomocytes six or more in number. Caudal glands (three) generally present; absent in most Dorylaimida, Mermithida, and Trichocephalida. Males with two testes (some exceptions) with single ventral series of preanal supplements, papilloid or tuboid. Male tail rarely with caudal alae.

This class includes free-living marine, freshwater, and terrestrial inhabitants, as well as parasites of plants, invertebrates, and vertebrates. Marine nematodes with rare exceptions are exclusively limited to the Adenophorea.

Subclass A. Enoplia

When amphids are subcuticularly pouchlike, the external opening is transverse (cyathiform). When the internal pouch is tubiform, external opening porelike, ellipsoid or greatly elongated. Cephalic sensory organs are papilliform or setiform and somatic setae may be absent or present. Caudal glands may be present or absent, generally present among marine forms. Subventral esophageal glands often open into buccal cavity through teeth or at anterior esophagus; in others all glands open posteriorly. Esophageal glands generally five or more. Esophagus cylindrical, conoid or divisible into a narrow anterior portion and an expanded glandular posterior region; animal parasites have a stichosome esophagus. Cuticle generally smooth, but both distinct transverse and/or longitudinal markings occur.

As now recognized the subclass contains seven orders, one *incertae sedis*: Enoplida, Isolaimida, Monochida, Dorylaimida, Trichocephalida, Mermithida, Muspiceida (*incertae sedis*).

As currently structured, this is the largest subclass within the phylum, over 3000 nominal species are assigned to Enoplia. Members of this group occupy every ecological habitat and niche available to nematodes.

Order 1. Enoplida (Figs. 3.15A, B; 4.9A, B, C; 4.10A)

Amphids pouchlike, postlabial, with slitlike or ellipsoid aperture. Only in the family Oxystominidae is the external aperture greatly elongated longitudinally. Cephalic sensory organs in three whorls. The first whorl: six papilliform circumoral sensilla sense organs; whorl two: six sensilla, and whorl three: four sensilla. Circlets two and three generally setiform. Esophagus cylindrical-conoid, widening gradually toward the base. Five esophageal glands: one dorsal, four subventrals. Dorsal and anterior two subventral glands open into or near the stomatal region. Posterior subventral glands open anterior to the level of the nerve ring. Females with one or two gonads. Male typically with paired testes; spicules paired (rarely 1 or 0). Ductus ejaculatorius heavily muscled; male supplementary organs when present, often elaborate, either papilloid or tuboid in a single ventromedian row. Excretory system, when present, confined to a single cell. Lateral somatic glands present; generally three caudal glands open through terminal cuticular spinneret.

The known members of this order are found in marine, brackish waters or freshwaters where they feed as carnivores or feed on diatoms, algae, and/or organic detritus. The order is divided into three suborders: Enoplina, Oncholaimina, and Tripylina.

Suborder a. Enoplina (Figs. 3.15A, B; 4.9A, C)

This suborder characteristically has the cephalic cuticle doubled (=helmet or cap). Helmet formed by infolding of cheilostomal cuticle over anterior portion of esophagus. Stoma may or may not possess armature (teeth or movable jaws). Movable armature of cheilostomal origin. Esophagus well developed, more or less cylindrical. Esophago-intestinal valve well developed. Five esophageal glands; one dorsal and two anterior subventral gland orifices located near anterior end of esophagus or open through teeth. Remaining two subventral glands open anterior to nerve ring. Amphids either pouchlike with exterior opening through transverse slit or pouch may be greatly enlarged and external opening oval. Cephalic sensilla generally in two whorls. Inner whorl: six papillae; outer whorl: ten setae (sensilla trichodea). Second whorl may be separated in two whorls of six and four setae. Males with or without medioventral supplementary organs. Spicules two with supporting gubernaculum. Caudal glands and spinneret present. Single simple medioventral excretory gland usually present.

The suborder contains two superfamilies: Enoploidea and Oxystominoidea. The species placed in this group are principally marine, rarely in soil and are presumed to be predaceous.

Superfamily 1. Enoploidea (Figs. 3.15B, 4.9C)

Helmet punctate, smooth or penetrated by numerous fenestrae. Stoma sclerotized and surrounded by esophageal tissue; usually armed with teeth or jaws. When teeth present usually three in number, sometimes only one tooth in stoma. Esophagus cylindrical with or without crenations. Amphids pouchlike external opening transverse slit. Cephalic sensilla in two whorls, one circumoral and papilliform; the other ten postlabial setiform sensilla. Dorsal and subventral esophageal gland orifices near anterior end of esophagus or through teeth. Males with one or two tubiform preanal supplementary organs.

Almost exclusively marine in habitat, only a few species reported from brackish water or the soil environment.

Superfamily 2. Oxystominoidea (Figs. 3.15A, 4.9A)

Lip region (=cephalic region) not set off by a groove. The unarmed stoma is weakly developed and surrounded by esophageal tissue. Cephalic sensory organs in three separated whorls: six circumoral papillae or setae. Only known group of nematodes to have the circumoral sensilla setiform. Second whorl: six papilliform or setiform sensilla; third whorl: four submedian setae. Posterior whorls generally well separated. Amphids generally elongated with variable manifestations of the external opening. In some the opening is a transverse slit or porelike; in others the pouch is extraordinarily expanded and elongate, the opening oval; still others have the pouch very expanded and elongate, the external opening is longitudinally elongated. Esophagus greatly elongated expanding gradually posteriorly. Males supplementary organs generally absent; when present conspicuous or minute and setose. Females with one or two ovaries; males with one or two testes.

Suborder b. Oncholaimina

This suborder contains a single superfamily Oncholaimoidea. Amphids pocketlike; external aperture oval or widened ellipsoid. The buccal cavity or stoma generally vaselike with thick cuticular walls. Stoma armed with one dorsal and two subventral teeth, may also be armed with transverse rows of small denticles. Stoma divisible into cheilostome and esophastome (posterior portion of buccal cavity surrounded by esophageal tissue). Adult male stoma may be indistinct or collapsed. Lips six in number and amalgamated. Cephalic sensilla: one circlet six circumoral papillae; the second circlet ten setae combined from circlet two (six setae) and three (four setae). In some sensilla papilliform. Esophagus muscular, cylindrical or conoid, sometimes with a series of muscular bulbs posteriorly. Cuticle generally smooth, with scattered somatic setae and papillae.

Primarily marine and brackish water nematodes with alleged predaceous and carnivorous habits.

Suborder c. Tripylina (Figs. 4.9B, 4.10A)

Cephalic cuticle simple, not duplicated. Somatic cuticle smooth sometimes superficially annulated. Cephalic sensilla, circumorally papilliform; external whorls variable, in some they are sensilla chaetica (bristlelike), sensilla trichodea (hairlike), or sensilla basiconica (conelike). Amphids pouchlike, external opening inconspicuous or transversally oval. Stoma generally simple, collapsed, funnel-shaped, or cylindrical, armed or unarmed and surrounded by anterior esophagus; when expanded both cheilostome and esophastome recognizable. Esophagus cylindrical-conoid. Dorsal esophageal gland always opens anterior to nerve ring, subventrals generally open anterior to nerve ring. Male supplementary organs generally three in number, sometimes more; gubernaculum present. Caudal glands generally present.

Primarily found in freshwaters and soils throughout the world, few marine forms. Those with armed stomas presumed predaceous.

Superfamily 1. Tripyloidea (Fig. 4.9B)

Cuticular annulation well developed. Stoma of two types: collapsed and primarily esophastome, or globular and composed of two distinct parts. The dorsal gland and two of the subventral glands open in anterior extremity of esophagus. When dorsal tooth present gland may open through it. Cephalic sensory sensilla in outer whorls are sensilla chaetica, sensilla basiconica, in a few the sensilla trichodea are jointed. Amphidial opening porelike or oval. Esophagus muscular, more or less cylindrical; esophago-intestinal valve well developed often with denticulated pads. Males with preanal supplementary organs either papilloid or vesiculate. Caudal glands and spinneret generally present.

Primarily found in freshwater and moist soils, some marine and brackish water forms. Food consists of small microfauna including nematodes.

Superfamily 2. Ironoidea (Fig. 4.10A)

Cephalic sensory organs in two circlets. Inner circlet six papillae; outer circlet either ten sensilla trichodea or ten sensilla basiconica. Stoma elongate, cylindrical armed either anteriorly with three large teeth or posteriorly with small teeth. Anterior teeth eversible. Stomatal walls well cuticularized. Esophagus cylindrical, esophageal glands open anteriorly near the stomal cavity. Esophago-intestinal valve small. Females with one or two ovaries. Males when present have paired, arcuate spicules and a feeble gubernaculum. Male supplementary organs papillose or setose. Caudal glands and spinneret present or absent.

Primarily found in freshwater and moist soil. Feeding habit carnivorous.

Order 2. Isolaimida (Fig. 9.1A, B)

Elongate and cylindrical nema ranging in length from 3–6 mm which is 60–70 times the body width. Anteriorly cuticle may be annulated. Cuticle may be ornamented with punctations most obvious on tail in longitudinal and transverse rows. Oral opening surrounded by six hollow tubes and two whorls of sensilla. The six papillae of each whorl large and recessed. Amphids apparently absent or represented by dorsolateral papillae. Buccal cavity elongate, triradiate, usually with thickened walls anteriorly. Esophagus clavate. Females with two ovaries sometimes with flexures. Male testes paired. Male spicules thick, cephalate or noncephalate, with or without longitudinal medial strengthening bars. Gubernaculum small but with two dorsal apophyses. Male supplements papilloid, preanal and midventral; in addition, large paired caudal papillae on male tail and small paired papillae on female tail. Tails in both sexes short, conoid bluntly rounded with pointed tips.

The known species are soil inhabitants.

Order 3. Mononchida (Fig. 4.7A)

Amphids small and cuplike; located just posterior to the lateral lips; external aperture slitlike or ellipsoidal. Cephalic sensilla in two whorls: circumoral circlet consists of six sensilla and the second circlet (external) has ten sensilla. All are sensilla coeloconica. Stoma strongly cuticularized bearing one or more massive teeth, may be opposed by denticles in either transverse or longitudinal rows. Esophagus cylindrical conoid lumen

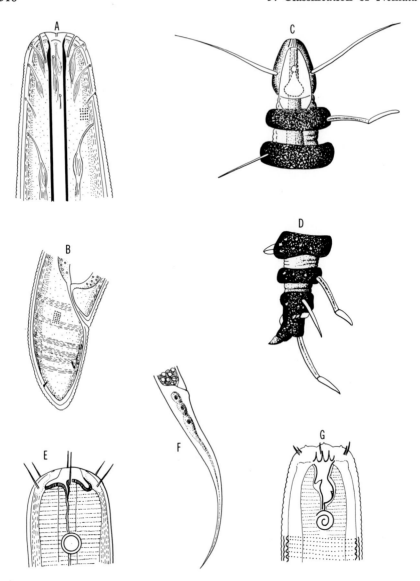

Figure 9.1. A. *Isolaimium stictochroum*; female head. B. *I. stictochroum*; female tail. C. *Desmoscolex longisetosus*; female head. D. *D. longisetosus*; female tail. E. *Monhystera vulgaris*; female head. F. *M. vulgaris*; female tail. G. *Achromadora ruricola*; female head. (A–D redrawn from Timm; E–G redrawn from de Man).

lining greatly thickened with special thickenings (apodemes) on the radial arms. Five esophageal glands are uninucleate, one is dorsal and four are subventral. Gland ducts short and open posterior to nerve ring. Excretory system degenerate. Males with ventromedian supplements and two equal spicules with or without lateral accessory pieces, gubernaculum sometimes present. Females have one or two ovaries. Caudal glands and the spinneret usually present but may be degenerate or absent.

All known mononchs are predaceous on small soil microorganisms and are even cannibalistic. They are found in soil and freshwaters throughout the world. No marine forms are known. There are two suborders: Mononchina and Bathyodontina.

Suborder a. Mononchina (Fig. 4.7A)

Lip region somewhat expanded; lips generally angular and distinct. Cephalic sensilla of external whorl terminal on each lip. Stoma strongly cuticularized and primarily cheilostomal in origin; barrel-shaped or globular and very large. Stoma armed with a large immovable dorsal tooth; subventral teeth may be present or absent; if present, may be as large as dorsal tooth or in form of denticles. Basal plates of the stoma surrounded by esophageal tissue. Esophagus nearly cylindrical; triradiate lumen heavily cuticularized with apodemes on radial arms. Esophageal intestinal valve large and well developed. Females have one or two reflexed ovaries. Male spicules slender, gubernaculum well developed with prominent lateral accessory pieces. Preanal supplements papilloid and in a single ventromedian row. Caudal glands and spinneret usually present.

This suborder contains a single superfamily Mononchoidea.

Suborder b. Bathyodontina

Lip region rounded or slightly expanded and rounded. Cephalic sensilla all sensilla coeloconica. Stoma compartmented primarily esophastome. Stomal cavity tubular posteriorly. Esophastome armed with a ventrosublateral tooth of varying length, sometimes a small denticle also present. Esophagus cylindrical or nearly so, five esophageal glands open posterior to nerve ring. Esophago-intestinal valve well developed. Females have one or two gonads. Males rare, when found spicules are simple and paired, without an associated gubernaculum. Preanal supplements papilloid in a single ventromedian row. In both sexes the tail is usually short and blunt with caudal glands and a spinneret.

This suborder is divided into two superfamilies: Bathyodontoidea and Mononchuloidea. These nemas are found in soil and freshwater where they are mainly predators on small microorganisms.

Superfamily 1. Bathyodontoidea

Lip region high, usually well developed. Oral opening hexaradiate with cuticularized walls. Main stoma divided into two sections: anterior tubular, posterior triradiate with elliptical radial arms. Left ventrosublateral wall with a minute tooth. Esophagus nearly cylindrical. Excretory pore conspicuous. Eight longitudinal rows of body pores. Caudal glands and spinneret are present in both sexes.

Inhabitants of soil or freshwater; predaceous.

Superfamily 2. Mononchuloidea

Lip region generally rounded; cephalic sensilla three circlets of sensilla coeloconica; two circlets on lips; third circlet, four sensilla, located just posterior to lips. Amphids small, pocketlike with narrow inconspicuous slitlike apertures. Stoma divided into two sections: short anterior region, slightly tapered, wall thickened; posterior region long and thin walled. Large ventrosublateral mural tooth located at the junction of the two stomal regions; minute denticles may also be present. Females with one or two ovaries; generally one, vulva post equatorial. Males have paired testes, paired spicules, but no gubernaculum. Tail in both sexes short and rounded or moderately elongate, caudal glands and spinneret present.

Order 4. Dorylaimida (Figs. 3.15D, 4.10B, 4.11A–B, 6.1–6.3)

Lip region may be set off by a constriction. Lips usually well developed but many forms exhibit a smoothly rounded anterior. Cephalic sensory sensilla: sensilla coeloconica, located on lips. Amphids postlabial. Amphidial pouch-shaped like an inverted stirrup, external opening ellipsoidal or transverse slit. Stoma is armed with a movable mural tooth or a hollow axial spear. Esophagus divided into slender muscular anterior, an elongated or pyriform glandular portion. Five esophageal glands all open posterior to the nerve ring. In some forms only three esophageal glands, and rarely, seven. Esophago-intestinal valve well developed. Prerectum common. Females with one or two reflexed ovaries. Males with two equal spicules, rarely with gubernaculum. Males often have paired adanal supplements preceded by a ventromedian row of papilloid supplements. Caudal glands and spinneret absent.

The order is subdivided into three suborders: Dorylaimina, Diphtherophorina, and Nygolaimina. All the nemas in this group are found in soil or freshwater; some forms are subterranean ectoparasites of plants.

Suborder a. Dorylaimina (Figs. 3.15D, 4.11A, 6.1, 6.2)

Lips generally separate and bearing cephalic sensilla. Amphids cyathiform with slitlike or elongate-ellipsoidal apertures. Two part protrusible spear: anterior odontostyle, posterior odontophore. Occasionally anterior to onchiostyle there may be accessory denticles or other cuticular structures. Esophagus two parts: a narrow muscular corpus followed by a swollen and enlarged muscular-glandular postcorpus with 3–5 glands. Postcorpus may be surrounded by special muscles. Excretory cell and pore absent or rudimentary. Prerectal region usually distinguishable. Ovaries are usually paired. Males have paired testes, and spicules with lateral accessory pieces, gubernaculum sometimes present. Paired adanal supplements preceded by a ventromedian series.

Members of this suborder are generally terrestrial, but may be found in freshwater. Also included in this suborder are important plant parasites that can act as vectors of plant viruses.

Superfamily 1. Dorylaimoidea (Figs. 3.15D, 4.11A, 6.1, 6.2)
Anterior stoma is straight and tubular, into this vestibule projects the hollow axial odontostyle. The odontophore may have tripartite extensions or flanges posteriorly. Body cuticle smooth, or may be ornamented with longitudinal ridges. Esophagus with slender anterior part and an elongate expanded posterior part at least one-third of the total length of the esophagus. Esophago-intestinal valve well developed. Female gonads paired or unpaired. Male supplements a single adanal pair preceded by a ventro-median series. Gubernaculum usually absent.

Algal feeders in soil and freshwater environments. Only the family Longidoridae contains representatives that are important plant parasites and vectors of plant viruses.

Superfamily 2. Actinolaimoidea
Axial spear typical of dorylaimids, but walls of stoma (cheilostome and esophastome) may be heavily cuticularized or vestibule may be in form of a cuticular basket. These variations may be accompanied by large teeth or denticles. Lip region usually rounded but lips may be well developed and separated from neck by constriction. Cuticle may have minute transverse striations or longitudinal ridges. Esophagus almost equally divided between narrow corpus and expanded postcorpus. Females have paired ovaries that are opposed and reflexed.

Superfamily 3. Belondiroidea
Principal characteristic: a thick sheath of spiral muscles that enclose basal portion (postcorpus) of esophagus.

The lip region in most narrow and varies from rounded to angular. Amphid apertures large, nearly as wide as the head. Odontostyle short, normally measuring less than the width of the lip region. Odontophore rarely more than cylindrical; in one genus, flanged. Usually tails of both sexes similar. Females have one or two ovaries, if one ovary present, then posteriorly directed.

Superfamily 4. Encholaimoidea
Cephalic sensilla in two whorls. Inner whorl, six papilliform circumoral sensilla, one on each raised lip. External whorl, six large and four small sensilla basiconica. Amphids pouchlike with slitlike openings. Stoma armed with axial spear, flanges on the odontophore well developed. Esophagus: elongated cylindrical corpus and small pyriform postcorpus. External cuticle with widely spaced transverse annulation and widely spaced longitudinal annulations, body has corncob appearance. Female one ovary posteriorly directed and reflexed; vulva in the anterior half of body. Males have tuberculate supplements.

Suborder b. Diphtherophorina (Figs. 4.10B, 6.3)
Lip region more or less typical for dorylaims, only slightly expanded and bearing cephalic sensory organs. Amphids inverted stirrup-shape with

ellipsoidal apertures. Stomatal armature variable and complex. When spear complete, cheilorhabdions and esopharhabdions are most strongly developed dorsally. In some forms subventral walls are thin and membraneous; thickened dorsal wall then functions in manner of a movable dorsal tooth. Esophagi are narrow anteriorly and expanded posteriorly into a short pyriform bulb. An anterior ventromedian excretory pore present. Females, paired reflexed ovaries; males, only a single testis. This is a most unusual character among males in the subclass Enoplia. Male spicules paired, sometimes accompanied by gubernaculum. In a few forms a weakly developed bursa also present on male tail. Where present, preanal ventromedian supplements are papilloid.

The suborder contains confirmed plant parasites and others, though unconfirmed, are often associated with plant roots.

Superfamily 1. Diphtherophoroidea

Stomatal armature composed of an elaborate series of peripheral plates. Plates originate from cheilorhabdions and esopharhabdions. Anteriorly stoma is distinguished by an archlike vestibule that joins the spear extension. The dorsal and ventral portions of "spear" not joined. Thickened dorsal section (esophastome) bear a toothlike structure of cheilostomal origin. This "tooth" which is short, pointed and curved is formed within esophageal tissues and migrates to anterior placement at time of molting. Posterior portion of spear has basal swellings (apodemes) for muscle attachments. Esophagus narrow anteriorly and ends in plain, pyriform, or elongate-conoid postcorpus. Female ovaries paired, opposed and reflexed. Male spicules slightly arcuate, gubernaculum reduced. Preanal ventromedian supplements are reduced or vestigial.

These nemas are all soil forms and though their feeding habits are unknown, they are normally found in association with plant roots.

Superfamily 2. Trichodoroidea (Figs. 4.10B, 6.3)

This superfamily distinguished by unusual stomatal armature formed as elaboration of dorsal half of "spear" seen in Diphtherophoroidea. Anterior tip a dorsal tooth formed by cell in anterior esophagus. Anterior half solid but posterior extremity hollow. Posterior portion (esopharhabdion) part of and fused to dorsal wall of anterior esophageal lumen. Esophagus narrow and cylindrical in corpus and expanded posteriorly. Expanded postcorpus and contained glands may overlap anterior intestine slightly. Females have one or two ovaries. Males have single telogonic testis, anteriorly outstretched. Spicules simple, slightly curved, proximal end slightly cephalated. Gubernaculum not only cups spicules dorsally but possesses a cuneus that projects between spicules. Inconspicuous bursa may be present. Preanal supplements in a ventromedian row.

These nematodes are important plant parasites whose feeding upon roots results in a symptom called "stubby root." They also transmit plant viruses.

Suborder c. Nygolaimina (Fig. 4.11B)

Primary characteristic distinguishing the suborder: eversible stoma with protrusible subventral mural tooth. Esophagi variable: some of Dorylaim two-part type, some bibulbar and a few with distinct posterior bulb appearing valved. Disagreement exists over the esophago-intestinal glands. Some claim three glands are universal, others claim disclike glands in one group. It seems likely that three glands are characteristic but degree of development varies. Prerectum present or poorly defined, sometimes absent.

The nemas encompassed by this suborder are all presumed predaceous.

Order 5. Trichocephalida (Figs. 8.1, 8.3, 8.4)

Trichocephalida possess a protrusible axial spear in early larval stages, not found in adults. Cephalic sensory organs are sensilla coeloconica; amphids just off the lip region. Esophageal glands characteristic: posterior glands lie in one or two rows along lumen of esophagus, not enclosed by esophageal tissue (= stichosome). Stichosome and individual gland orifices are posterior to nerve ring. Reproductive system in males and females unique among Nemata: germinal zone extends entire length of gonads and forms a serial germinal area on one side or around the gonoduct; no rachis. Both males and females have a single gonad. Males have either one or no spicule. Eggs operculate.

The life cycles exhibited by these vertebrate parasites are either direct, often requiring cannibalism, or indirect, in which case they utilize arthropods or annelids as intermediate hosts.

Superfamily 1. Trichuroidea (Figs. 8.1, 8.3)

Stichosome of adults limited to a single row on one side of esophagus. In early larval development there are two rows. Body in both sexes divisible into two regions: elongated, narrow anterior body houses esophagus, with stichosome; posterior half with reproductive system begins at esophago-intestinal junction; bacillary band occurs laterally; glandular tissue opens to exterior through cuticular pores. Males and females have single gonads, reflexed. Males with one spicule. Eggs operculate; females oviparous. In some species males small and degenerate, and live within uterus of female.

This superfamily includes parasites of man and other mammals. Their life cycle is either direct or indirect.

Superfamily 2. Trichinelloidea (Fig. 8.4)

Stichosome formed of a single short row of stichocytes. Two distinct body regions not clearly developed. Taxa included in this superfamily lack bacillary band. Female genital opening far anterior in region of stichosome; ovary posterior to stichosome. Males with single gonad but no spicule. Females viviparous.

Only one genus is known: *Trichinella*. This genus is an important parasite of man and other mammals.

Superfamily 3. Cystoopsoidea

Stichosome doubled in larvae and adults. Body not distinguishable into two parts. Vulva anterior in region of esophagus, anterior to stichosome. Digestive tract ends blindly. Males and females have single gonads. Males with single spicule. Females oviparous and eggs operculate.

This nema is known from a single species parasitic in sturgeons.

Order 6. Mermithida (Figs. 7.3–7.7)

Characteristically exceptionally long and slender; achieving lengths of over 30 cm. Hypodermis forms eight chords; one dorsal, one ventral, two lateral, and four submedian. Anteriorly chords nucleated. External amphid either porelike or pocketlike. Oral opening normally terminal but may be located subterminally. Stichocytes in two rows. Intestine degenerate in adults, forms a storage organ (trophosome) connected to esophagus by bifurcating or ramifying canal. Anterior portion of trophosome or intestine overlaps posterior esophageal region. Gonads of both sexes generally paired but occasionally single. Males generally with two spicules, taxa with a single spicule do occur. Eggs modified not operculate.

All known forms are parasites of insects and other invertebrates.

Superfamily 1. Mermithoidea (Figs. 7.3–7.5)

Long and slender nemas with smooth cuticles. Anterior extremity or lip region with four submedian and two lateral sensilla coeloconica Amphids behind lip region, in aquatic taxa amphidial pouch very large. Mouth and esophagus nonfunctional in adults. Larval stages possess piercing tooth in stoma. Larval and adult esophagi similar, stichosome in two rows always exceeding four cells. Intestine replaced by trophosome in adults. Both sexes have paired gonads. Female vulva equatorial, leads to muscular S-shaped or barrel-shaped vagina. Some taxa with eggs ornamented with polar filaments (byssi). Males have one or two spicules, papilliform supplements are numerous in three rows both post- and preanal.

The nemas are parasitic in a wide variety of insects, slugs, and snails.

Superfamily 2. Tetradonematoidea (Figs. 7.6, 7.7)

Larval esophagus reminiscent of Dorylaimida: narrow anteriorly and swollen posteriorly. In adults stichosome evident; composed of three or four cells; anteriorly esophagus reduced to simple hollow tube. Juveniles with small axial stylet. Adult males and females gonads paired. Males with one spicule; preanal supplements may be present.

Species generally attack insects of the order Diptera, both terrestrially and in the aquatic environment.

Order 7. Muspiceida (*incertae sedis*)

Neurosensory structures greatly reduced. Neither amphids nor cephalic papillae observed except in one species. Males unknown. Among known females digestive tube reduced. Females appear to be didelphic, amphidelphic, and viviparous.

All known forms are parasites of higher vertebrates or fish. The type of damage they cause is tumorous and cancerlike in nature.

Subclass B. Chromadoria

Cephalic and/or labial sensilla generally in three rows, two external rows may be combined. Circumoral sense organs always sensilla coeloconica; outer circlets may be sensilla trichodea, sensilla basiconica, or sensilla coeloconica or combinations. External amphid may be spiral, circular, vesiculate, or forms derivable from these. When stoma is well developed may contain a large dorsal tooth, three jaws or six inwardly acting teeth. Esophagi with three uninucleate glands. Dorsal gland orifice anterior to nerve ring, subventral gland orifices at base of corpus. Shape of esophagus variable from cylindrical to subdividable into corpus, isthmus, and post-corporal bulb. Basal bulb often valved. Esophago-intestinal valve usually well developed. Gonads in both sexes single or paired. In males muscles of ductus ejaculatorius only moderately developed. Caudal glands nearly always present. No parasitic forms are known. Generally they are microbivorous feeding on algae, diatoms or bacteria; some may be predaceous.

Order 1. Araeolaimida (Figs. 2.7, 2.8, 3.15C)

Amphids simple spirals which may appear as elongated loops, shepherd's crooks, question marks, or circular. Cephalic sensory organs often separated into three whorls, circumoral circlet and first external circlet are sensilla coeloconica or the first external are sensilla basiconica; the third whorl (second external) normally composed of four sensilla trichodea; rarely are external whorls combined to a circlet of ten.

Body annulation simple without punctations. Stoma anteriorly funnel-shaped; followed by tubular esophastome, only rarely are denticles found in stoma. Esophagus generally ends in terminal bulbous swelling, which may or may not be valved. Females: paired ovaries, except a few taxa. Male supplements tubular, rarely papilliform.

Most often encountered in marine habitat or in brackish waters. Feeding habits are unknown.

Suborder a. Araeolaimina (Figs. 2.7, 2.8, 3.15C)

Cephalic sensory organs in three whorls. Circumoral circlet and first external circlet sensilla coeloconica. Third whorl four sensilla trichodea or sensilla basiconica. Stoma shape varies, most often a simple tube slightly funnel-shaped anteriorly. Amphids diverse but always derived from simple spirals, sometimes appearing circular. Esophagi distinguished by tuboid endings on esophageal radii of anterior corpus. Basal bulb may or may not be valved.

Found in marine and freshwater environments, rarely in soil. Some known bacterial feeders, others unknown.

Superfamily 1. Araeolaimoidea

Amphids generally simple spirals, elongated loops, or hook-shaped. Cephalic sense organs in three whorls, generally only four organs of third

circle are setiform. Stoma not well developed, cheilostome thinly cuticularized and esophastome tubular or funnel-shaped. Esophagus cylindrical or
externally divisible into corpus, isthmus and swollen valveless terminal
bulb. Female gonads paired, outstretched with only 20–25 oocytes. Males
lack preanal tubular supplements.

Found mainly in marine environment, although many forms have been
collected from freshwater and soil.

Superfamily 2. Axonolaimoidea

Amphids diverse, generally a single turn loop of a wide sausage-shape.
Elongated arm loop dorsal. Ventral portion of loop returns to almost
touch dorsal arm, therefore amphid often appears circular with a doubled
outline. Cephalic sense organs of first two circlets papilliform, third external circlet of four sensilla are elongate setae. Stoma divisible into two
parts, cheilostome is armed with eversible rhabdions misnamed teeth;
esophastome is funnel-shaped, generally with heavily cuticularized walls.
Excretory pore often opens at base, or just posterior to stoma. Females:
generally paired outstretched ovaries.

Found in marine or brackish waters. Feeding habits unknown.

Superfamily 3. Plectoidea (Figs. 2.7, 2.8, 3.15C)

Papilloid cephalic sensory organs except for outer circlet, four of which
may be setiform or elaborated into winglike lamellae. Amphids simple,
may appear as circles or shepherd's crooks. In one taxon amphidial opening is a transverse oval. Stoma tubular or expanded anteriorly, followed
by shortened cylindrical portion. Esophagi cylindrical anteriorly, with or
without definite isthmus, followed by expanded muscular bulb with valve.
Tuboid endings on radii of anterior esophagus. Female ovaries paired,
reflexed. Male supplements vary according to genus from none to many.

Most often found in freshwater, soil and brackish water environments.
Microbivorous in feeding habits.

Superfamily 4. Camacolaimoidea

Amphid simple unispiral at level of outer whorl of cephalic sense organs.
Sensilla of first two are papilliform, third whorl four setiform sensilla.
Dorsal stomal wall developed into an onchium or protrusible spear.
Esophagus expanded slightly at postcorpus which is principally glandular.
If male supplements present, they are papilliform.

Found in marine or brackish environments. Feeding habits unknown.

Suborder b. Tripyloidina

Lips amalgamated to three. Second and third whorls of cephalic sensilla
combined into a single circlet of six long and four short setae. In one
genus setae appear jointed. Amphids vary from circular to spiral. Stoma
almost always developed and divided into a series of chambers, normally
armed with denticles or small teeth in posterior chamber. Esophagus cylindrical and without noticeable swelling of postcorpus.

Found in marine or brackish environments. Feeding habits unknown.

Order 2. Chromadorida (Fig. 9.1G)

Amphids simple spirals, reniform, transverse elongate loops or multiple spirals. Cephalic sensory organs at extreme anterior in one or two whorls. Cuticle in all taxa show ornamentation generally manifested as punctations. Punctations seen whether cuticle annulated or apparently smooth. Stoma when developed primarily esophastome and normally armed by dorsal tooth, jaws or protrusible rugae. Esophagus almost always cylindrical in corpus and expanded in postcorpus. Muscular postcorpus has a heavily cuticularized lumen forming so-called crescentric valve. Esophago-intestinal valve triradiate or flattened. Females almost always have paired reflexed ovaries.

Found in marine waters, freshwaters, or soil.

Suborder a. Chromadorina

Amphids variable: transversly elongate, oval to loop-shaped or circularly spiral. Usually amphids located just posterior to external circlet of cephalic sensilla but sometimes found between four setae of third sensilla whorl. Stoma cylindrical or funnel-shaped and cheilostomal vestibule generally possesses weakly to well developed rugae (cheilorhabdions). Main stomatal chamber usually with three teeth. Teeth may be hollow and unmoveable or solid and protrusible. Esophagi may be slightly swollen posteriorly or postcorpus may be a well developed bulb. Cuticle almost always ornamented with transverse rows of punctations or other intracuticular designs, except in families Spirinidae and Microlaimidae. Males may or may not have cuticularized, knoblike preanal supplements.

Though mainly marine, they are found in freshwater and soil.

Suborder b. Cyatholaimina

Distinguished by the tightly looped multispiral amphids. Body almost always annulated; each annule with one or more rows of punctations. Stoma generally armed with three well developed teeth, one dorsal and two subventral. Vestibule may or may not have rugae. Esophagi either cylindrical or with well developed posterior bulb. Males may or may not have preanal supplements.

Found in marine, freshwater, and soil habitats. Feeding habits unknown.

Superfamily 1. Cyatholaimoidea

Amphids tightly coiled multispirals located a short distance posterior to cephalic sensilla. Cephalic sensilla in two whorls; circumoral whorl, six papilliform sensilla; outer whorl combined external circlets of ten setiform sensilla. Stomatal vestibule lined with twelve riblike cheilorhabdions. Esophastome funnel-shaped to collapsed, armed anteriorly with a strong dorsal tooth; in other taxa esophastome guarded by minute denticles. Esophagus ends in slight swelling or well developed bulb. Males may or may not have preanal supplements.

Found in marine, terrestrial, and freshwater environments. Food habits unknown.

Superfamily 2. Choanolaimoidea

Stoma structure complex and not well understood. Cheilostome variable consisting of flaps or rugae. In some taxa these are converted into jaws or moveable teeth. Esophastome sometimes divided into two sections with immobile teeth between sections. Head region blunt and not well set off. Commonly cephalic sensilla are in form of short cones, in some instances they are setose. Amphids circular, spiral or multispiral. Cuticle ornamented with heterogeneous punctations and pores. Preanal male supplements may be large and cuticularized or just a series of papillae.

This group is entirely marine and nothing is known of the food habits; some are allegedly predaceous.

Superfamily 3. Comesomatoidea

Amphids wide multispirals making at least two turns; situated just posterior to the outer whorl of cephalic sensilla of either four or ten sensilla trichodea. Body cuticle smooth, but ornamented with punctations that simulate annules. Stoma shallow and bowl-shaped and generally at anterior of the esophastome there are small denticles. Male spicules may be greatly elongated and accompanied by a simple gubernaculum, or short and complex in two consecutive arcs. In other taxa spicules are long and accompanied by a gubernaculum with a caudal apophysis.

Found in marine habitats. Feeding habits unknown.

Order 3. Desmoscolecida (Fig. 9.1C, D)

Distinguished by conspicuously annulated bodies. Annulations not simple but may be covered by concretion rings, or cuticle may be ornamented with scales, warts, or bristles. Though not confirmed, cephalic sensory organs seem to be reduced in number; only second and third whorls remain. Second whorl papilliform; four sensilla of third whorl conspicuous and setiform. Amphids vesiculate and circular, or oval and occupy much of head region. Body setae tubular, often distal end elaborate and open. Esophagus slightly swollen posteriorly, does not overlap intestine. Pigment spots or ocelli generally present just posterior to esophageal base. Females amphidelphic, normally no ovarial flexures. Males: single testis, usually without flexures along its length. Male spicules paired, usually accompanied by gubernaculum. Tail of both sexes contain three caudal glands and spinneret.

Typically marine, but have been found in freshwater and soil habitats. Feeding habits unknown.

Superfamily 1. Desmoscolecoidea (Fig. 9.1C, D)

Body annulation prominent, ringlike with granular concretions on main body annules. Annulation heterogeneous, main encrusted annules separated from each other by varying numbers of small nonencrusted annules. Scattered elongated sensilla trichodea occur along body.

Primarily marine but have been recorded from freshwater and soil. Feeding habits unknown.

Superfamily 2. Greeffielloidea

Prominent annulation homogeneous, not interspersed with smaller annules. Each annule bears a corolla or ring of elongate spines or short scales that do not pass through cuticle. Large subdorsal and subventral tubular setae occur along the body.

Primarily in marine habitat, rarely in brackish estuarine waters. Feeding habits unknown.

Order 4. Desmodorida (Fig. 3.7A)

Amphids vary from reniform, to elongate loops, to simple or multiple spirals. Cephalic sensory organs in three successive whorls, sometimes intermixed with numerous cervical setae. Rarely second and third circlets of sensilla combine into a single whorl of ten sensilla. Distinguished by having a cephalic helmet (additional cuticular layers in the head capsule) and prominent body annulation. Cuticle annulated but no punctations except on terminal tail section of Draconematina. In some groups anterior and posterior adhesion tubes occur that are utilized in locomotion.

Principally marine but are sometimes found in brackish and freshwater environments. Feeding habits unknown.

Suborder a. Desmodorina (Fig. 3.7A)

External amphid varies from elongated hook to simple or multiple spirals. Head capsule with subcuticular thickening (helmet). Cephalic sensilla, generally in three whorls, in exceptional instances two circlets combined of ten sensilla. Stoma may be armed. Generally esophagus terminates in bulb with well developed valve. Cuticle annulation varies from fine to coarse. Often longitudinal rows of setae along body, numerous setae in cervical region.

Found in the marine, brackish water, and freshwater environments. Feeding habits unknown.

Superfamily 1. Desmodoroidea

Development of the helmet within the cephalic capsule striking. Body strongly and distinctly annulated except for head capsule. Additional to normal complement of cephalic sense organs there may be numerous cervical setae on head capsule. Amphids always simple spirals never exceeding 1½ turns. Stoma generally with a well developed dorsal tooth; however, development of subventral teeth variable. Esophagus variable, but almost always postcorpus swollen, either oval or elongate. Usually valvelike cuticular lining well developed. Ovaries paired, reflexed.

Found in the marine environment although some have been collected from brackish water. Feeding habits unknown.

Superfamily 2. Ceramonematoidea (Fig. 3.7A)

Cuticular ornamentation of crested annules tilelike, or longitudinal rows of tablets. Head capsule free of annulation, and bears 16 sensilla in either two or three circles. Outer whorls always setiform; second whorl may be conelike sensilla basiconica or papilliform. Circumoral whorl very reduced.

Amphids simple loops either elongated and hooklike or they turn back to touch the dorsal arm and therefore appear circular. Amphids either near junction of capsule and general body or near middle of capsule. When at junction, cephalic sensory setae are on anterior half of capsule. When amphids are in midcapsule region, cephalic sensory setae are located far anteriorly and are combined into single circlet of 10 or four. Stoma not armed with a dorsal tooth or small subventral teeth.

Marine environment. Feeding habits unknown.

Superfamily 3. Monoposthioidea

Cuticle distinctly annulated, annules have longitudinal rows of basically spinelike ornamentation. Spines may be setose, simple, or bifurcated. Head region variable, may appear as an enlarged body annule distinguished by presence of cephalic sensory setae or it may be capsulelike bearing simple spiral amphids. When there is no capsule amphid position is within annulated area below lip region and posterior to sensory setae. Stoma may or may not be armed with a well developed dorsal tooth opposed by small subventrals. Males may have either one or two spicules.

Marine habitat. Feeding habits unknown.

Suborder b. Draconematina

Well developed helmet always present. Lip region variable, ranges from lips obscure to well developed. Cephalic sense organs have a 6+ (6+4) arrangement. However, in most genera numerous anteriorly placed cervical setae obscure symmetry of outer whorls. External amphid varies from simple elongate loop to a loop intersecting itself and appearing as a single spiral. Cephalic adhesion tubes may or may not be present. Annulated cuticle may be ornamented with spines, ridges, granules, or internal inflations. Elongate somatic setae scattered over body. Posteriorly, adhesion tubes or stilt setae generally present. Females amphidelphic, didelphic and with ovaries reflexed. Males have single testis outstretched and cephalated spicules accompanied by a gubernaculum.

All species are marine and have been collected from intertidal zones. Feeding habits unknown.

Superfamily 1. Draconematoidea

Length varies from 0.3 to 3.0 mm. When relaxed body configuration usually dorsally and ventrally arched into shallow sigmoid. Greatest body width occurs in esophageal and midbody region. Inner circle of cephalic sensilla papilloid or setose. Outer circles always setiform but obscured by numerous cervical setae. Amphids elongate loops or simple unispirals, located laterally or dorsolaterally on nonannulated rostrum. Lips obscure or conspicuous and supported internally by cheilorhabdionic framework. Cephalic adhesion tubes located either dorsally on rostrum or just posterior to it. Stoma obscure or conspicuous and correspondingly armed or unarmed. Esophagus variable, may have anterior and posterior swellings separated by constriction or may have cylindrical corpus and swollen post-corpal bulb, which may or may not be valved. Cuticle except for rostrum and tail terminus conspicuously annulated. Posterior tubular adhesion tubes connected to glands; stilt setae absent. Posterior adhesion tubes occur in four or more rows extending over posterior third of body. Sometimes there are rows of sublateral adhesion tubes on posterior third of body. Sometimes there are rows of sublateral adhesion tubes extending posterior to anus. Tails, in both sexes, conoid to elongate cylindrical with non-annulated but punctate terminal region.

All are exclusively marine. Feeding habits unknown.

Superfamily 2. Epsilonematoidea

Small nemas seldom exceed 0.5 mm. Body when relaxed arched dorsally and ventrally in a sigmoid manner. Greatest body width occurs in anterior and posterior regions. Sensory organs arrangement typical of order. In some paramphidial setae present. Obscure lips apparently not supported by thickened cheilorhabdions. Spiral amphids usually located dorso-laterally. No cephalic adhesion tubes. Stoma inconspicuous rarely developed. In a few species stoma armed with a dorsal tooth. Esophagus divided into cylindrical corpus and valved posterior bulb. Body annulation may be ornamented. Posterior adhesion tubes arranged in four rows anterior to anus, none posterior to anus. In both sexes tails are cylindrical-conoid.

Marine habitat. Feeding habits unknown.

Order 5. Monhysterida (Fig. 9.1E, F)

In general stoma funnel-shaped with lightly cuticularized walls; others have spacious stomas with heavy walls and protrusible teeth. Amphids variable from simple spiral to circular. Generally cephalic sensory organs combined, second and third circlets form one whorl of ten setae; in some only third circlet of four setae remain. Cephalic end often bears many cervical setae (subcephalic) confusing normal pattern of distribution. Esophagi generally cylindrical but sometimes postcorpus swollen. Body cuticle smooth or with more or less distinct annulation or ornamentation. When annulation distinct somatic setae may be long and arranged in four or eight longitudinal rows. Female ovaries outstretched, generally singular, when paired they are still outstretched.

These nemas occur in all environments: marine waters, brackish waters, freshwaters, and soil.

Superfamily 1. Monhysteroidea

Stoma variable, generally funnel-shaped and not strongly cuticularized. Sometimes, walls are well enforced and stoma spacious. In a few, stoma cylindrical and very elongated, this generally is associated with a greatly attenuated anterior end. Cephalic capsule often with subcephalic setae. Cuticle often ornate because longitudinal striae interrupt transverse annulation. Amphids, generally circular but some simple spirals. Two external circlets of cephalic sensilla often combined into one of ten setae. Esophagi cylindrical and generally without a terminal swelling or bulb. Radii of esophagus converge and do not form terminal tubuli in region of corpus.

Can be found in all environments from marine waters, to freshwaters, and in soils. Feeding habits unknown.

Superfamily 2. Linhomoeoidea

Stoma variable: cavity indistinguishable, or cavity in form of truncated cone, either armed or unarmed, or the stomal cavity may be long and straight with two denticles at junction of cheilostome and esophastome. Esophagus generally has a terminal swelling; often esophagus divisible into corpus, isthmus, and postcorporal bulb. Amphids are circular, generally with a light refractive spot in center of circle. Esophageal musculature concentered and with special attachment points.

Can be found in marine, freshwater, and soil environments. Feeding habits unknown.

Superfamily 3. Siphonolaimoidea

Stoma narrow tube or in form of spear. Esophagus often divisible into corpus, isthmus, and a well developed posterior bulb. Amphids, large and circular; often occupy 30–40% of body width at their level. Musculature of esophagus concentered without specialized cuticular attachment points.

All known forms are from marine waters. Food habits unknown.

Class II. Secernentea

External amphidial apertures most often are porelike and located dorsolaterally on lateral lips or anterior extremity. In some instances they are oval, cleftlike, slitlike, or located postlabially. In all cases cephalic sensilla are labial and porelike or papilliform. Generally, 16 are present in two circles: a circumoral circle of six and outer circle of 10. In parasitic groups cephalic sensilla may be reduced. Deirids generally occur cervically at the level of the nerve ring. Comparable sense organs located caudally: phasmids. Hypodermis uninucleate or multinucleate. Cuticle varies from four layers to two layers, almost always transversally striated, and laterally modified into a "wing" area marked by longitudinal striae or ridges, generally raised slightly above body contour. In some parasitic forms lateral alae may extend out a distance equal to body diameter.

Esophagi variable but most have three esophageal glands, one dorsal and two subventral. Dorsal gland always opens anteriorly either in procorpus or in anterior metacorpus. Subventral glands open into posterior metacorpus. Excretory system opens ventromedially through a cuticularized duct connected to collecting tubules either on both sides of body or limited to one side. Body cavity with four to six coelomocytes are reported. Somatic setae or papillae absent on females; however, caudal papillae may be present on males. Male preanal supplements paired and often elaborate. In some males there is a medioventral preanal papilloid supplement. Males commonly possess caudal alae sometimes referred to as a bursa copulatrix.

Members of this class are almost exclusively terrestrial and only rarely are they encountered in the freshwater habitat.

Subclass A. Rhabitida

Subclass distinguished by form of esophagus which, in at least larval stages, is divisible into corpus, isthmus, and valved postcorporal bulb. In parasitic forms adult esophagus may be clavate or cylindrical but second stage larvae have rhabditoid form or are reminiscent of it. Valve in posterior bulb distinctive and occurs nowhere among Nemata except in this subclass. Lumen of posterior bulb expanded into a trilobed reservoir with three muscular lobes lined with cuticle. Muscular contraction rotates lobes posteriorly, moving food from the trilobed antichamber posteriorly toward intestine. Stoma without movable armature and composed of two parts: cheilostome and esophastome. Each region may be further subdivided into two or more sections. Males generally have well developed bursae supported by papillae or cuticle rays.

Order 1. Rhabditia (Figs. 3.4B; 4.8A, B; 7.2; 7.13; 7.14)

Lip region varies from a full complement of six lips to three or two or none. Stoma generally tubular but may be separated into five or more sections. Esophagus divided into corpus, isthmus, and postcorporal bulb; always muscular and generally possesses rhabditoid valve. Terminal excretory duct cuticularly lined and has paired, lateral collecting tubules running posteriorly. Female reproductive system with one or two ovaries, if one then vulva preanal and gonad antepudendic. Intestinal cells may be uni-, bi-, or tetranucleate. Hypodermis may be coenocytic. In males when caudal alae present it contains papillae rather than supporting rays accompanied by musculature.

These nematodes are freeliving and microbivorous in soil, or parasites of invertebrates and vertebrates.

Suborder a. Rhabditina (Figs. 4.8B, 7.2, 7.13, 7.14)

Stoma commonly cylindrical without distinct separation of rhabdions, generally two or more times as long as wide. In most instances lips distinct and bear all papilloid cephalic sensilla and porelike amphids. Esophagus clearly divided into corpus (procorpus and metacorpus) and postcorpus (isthmus and valved muscular posterior bulb). Females may have one or two ovaries. Males generally have paired spicules accompanied by a gubernaculum. Caudal alae common, but absent in some families.

Microbivorous forms as well as parasites of invertebrates and vertebrates.

Superfamily 1. Rhabditoidea (Figs. 4.8B, 7.13, 7.14)

Superfamily distinguished by generally possessing a well developed cylindrical stoma. Number of lips varies from two to six. Esophagus generally possesses, at least in larval forms, muscular posterior bulb with a rhabditoid valve. Males have caudal alae supported by 5–9 papilloid supplements.

Free-living microbivorous forms and parasites of invertebrates and vertebrates.

Superfamily 2. Alloionematoidea (Fig. 7.2)
Anterior extremity bears either none or six small amalgamated lips which en face view show only six papilliform sensilla and porelike amphids. Stoma infundibuliform, main section cheilostomal. Esophagus irregularly cylindrical and muscular in corporal region. Distinct tissue change at narrow isthmus which then expands to posterior bulb with rhabdiform, but not well developed, valve. Females have paired, opposed, reflexed ovaries. Males have preanal and caudal papillae but no bursa copulatrix.

Superfamily 3. Bunonematoidea
Body strikingly asymmetrical. Asymmetry includes labia and distribution of cephalic sensilla. Asymmetry involves elaborate ornamentation limited to right side of body; left side maintains an almost normal appearance. Stoma and valved esophagus typical rhabditoid. Females have two ovaries and males a leptoderan but often asymmetrical bursa. Asymmetry appears to be a manifestation of only cuticle since internal organs maintain normal nema symmetry.

Reportedly these nemas are bacterial feeders or sometimes fungus feeders.

Suborder B. Cephalobina (Figs. 3.4B, 4.8A)
Cheilostome almost always expanded and two regions recognized. Anteriorly cheilostome vestibule thin walled and followed by heavy walled chamber of essentially same width. Esophastome narrower than cheilostome and either funnel-shaped or tubular, composed of several consecutive sections, commonly four: named pro-, meso-, meta-, and telorhabdions. Esophagus divisible into three parts; posterior bulb always contains rhabditoid valve. Females with rare exceptions have one antepudendic ovary, sometimes with several flexures. Males lack bursa copulatrix, but do have patterned genital papillae. Paired spicules accompanied by gubernaculum.

These nemas are all free-living forms, commonly found in high moisture soils, decomposing vegetation, or as insect associates.

Superfamily 1. Cephaloboidea (Figs. 3.4B, 4.8A)
Superfamily distinguished by elaboration of labial region. These modified labia are called probolae. Labia may be fringelike, lobelike, flaplike, or liplike. Probolae often divisible into an internal circle of three called labial probolae and external circle of three called cephalic probolae. Stoma distinctly divided into two parts: an anterior expanded cheilostome and a narrow (collapsed) esophastome composed of several jointed rhabdions. Females ovary unpaired but generally vestigial posterior ovary present.

Unpaired reproductive system anteriorly directed to junction of ovary and oviduct; it then turns posteriorly and extends past vulva where it generally possesses two flexures. Vulva usually near or just posterior to equatorial. Because the single ovary usually extends anterior and posterior to the vulva, cursory examination gives the impression of paired ovaries.

Generally collected from moist soils, decomposing vegetation, or from soil about plant roots.

Superfamily 2. Panagrolaimoidea

Cheilostome a broad thick walled chamber as long or slightly longer than broad. Esophastome funnel-shaped, short, and lined with small rhabdionic plates. Esophagus sometimes has a distinct muscular metacorpus. Single female gonad anteriorly directed then reflexes back past vulva usually with no further flexures. Vestigial posterior uterus acts as a seminal receptacle.

All free-living and microbivorous.

Superfamily 3. Robertioidea

Parasite with very specialized body form. Anteriorly body smoothly rounded, i.e., distinct lips not visible. Stoma tubular but not heavily cuticularized. Anterior body to level of vulva almost cylindrical. In region of vulva to anus body is bulblike and about four times wider than preceding neck region. Just posterior to anus body constricts rapidly and projects a long setiform tail. Elongate esophagus cylindrical throughout corpus and isthmus, postcorporal bulb distinct but apparently lacks valve. Female ovary typical of Cephalobina, it is single and extends both anteriorly and posteriorly.

Superfamily 4. Chambersieiloidea

Oral opening heavily cuticularized and this cuticularization may take form of six alleged mandibles. Immediately surrounding oral opening, in one subfamily, are six tufted labial cirri. Each tuft of main cirri has five cirri arising from a common stem (fernlike). External to cirri six large sensilla basiconica and four papilliform sensilla make a combined circle of 10. These sensilla basiconica intermixed with sensilla coeloconica common to both subfamilies. Posterior are found the distinct postlabial oval amphids located at base of stoma. Stoma, well developed inarching cheilorhabdions, followed by broad spacious esophastome with well developed walls. Esophagus in three distinct parts: corpus, isthmus, and posterior bulb. Terminal bulb valvated. Vulva near equatorial; gonad antepudendic and reflexed almost back to anal opening. Male tail adorned with tubercle-like papillae.

Found in a variety of moist situations and presumed microbivorous.

Superfamily 4. Elaphonematoidea

Body cuticle transverse annulation is interrupted laterally by three incisures without transverse interruptions. Labial region highly complex; "face" ventrally directed and oral opening subterminal. On either side of oral opening are delicate membranous finlike appendages. Each fin strengthened by three ribs. Proximally fins narrow but at about their middle broaden out and become truncated in a hatchet-blade appearance. Dorsal to oral opening is another truncate appendage. Complicated lobes are located between dorsal and lateral appendages. Ventral to oral opening are two small roundish lobes, each with a single papilla. Stoma collapsed and bordered by rasplike denticles. Esophagus anteriorly cylindrical with little distinction between corpus and isthmus. Posterior bulb well developed and valvated. Single female ovary elongated, anteriorly directed, spermatheca at reflexure of ovary and oviduct. Ovary proceeds posteriorly past vulva and about half way to anus without further flexures. Female also possesses a postuterine sac. Six pairs of caudal papillae on male tail and three pairs preanal.

Feeding habits are unknown, but original nemas collected from cultivated soil.

Order 2. Strongylida (Figs. 8.5–8.10)

Labial region variable may consist of three or six lips or may be replaced by corona radiata. Stoma variable, may be well developed or rudimentary but never collapsed and unobtrusive. When well developed principally formed of cheilostome, only basal portion of capacious cup surrounded by esophageal tissue. When reduced stoma represented as a cheilostomal vestibule. In larval forms esophagus typical rhabdiform (corpus, isthmus, posterior bulb) and posterior bulb contains typical trilobed rhabdiform valve. In adults, esophagus cylindrical to clavate. Excretory system paired lateral canals and paired secondarily associated subventral glands. Females have one or two ovaries and heavily muscled uterina. Males distinguished by bursa copulatrix which usually contains muscles; this form of bursa is known only among members of this order. Paired genital papillae in muscled rays have the following pattern: ventroventral, lateroventral, extenolateral, mediolateral, posterolateral, externodorsal, dorsoventral, and terminodorsal. Males also have paired equal spicules.

All are parasites of vertebrates in adult stage; early larval stages bacterial feeders or parasites of annelids or molluscs.

Superfamily 1. Strongyloidea (Figs. 8.5, 8.7–8.10)

Stoma extremely variable, but always well developed and large. Stoma may be hexagonal in cross section or globular, cylindrical, or infundibuliform, never with mandibles. Oral opening hexagonal or surrounded by six small lips or a corona radiata. Never is mouth guarded by teeth or cutting plates. Massive stoma primarily cheilostomal in origin, only base of globe normally surrounded by esophageal tissue. Esophagus of hatched larvae always rhabditiform. Adult esophagus cylindrical or clavate. Males have well developed bursae, rays not fused.

Obligate parasites of reptiles, birds, or mammals. Early larval stages may be free-living and microbivorous.

Superfamily 2. Diaphanocephaloidea

Nematodes distinguished from all other Strongylida by modification of stoma into two massive lateral jaws and absence of corona radiata or lips. Cephalic (labial) sensilla of outer circle separate: submedians not fused. Bursa copulatrix bell-like or trilobed.

Obligate parasites in the intestinal tract of reptiles and amphibians. Early larval stages free-living and microbivorous.

Superfamily 3. Ancylostomatoidea (Fig. 8.6)

Distinguishing characteristic is constituted principally of cheilostome. Capacious stoma thick walled, globose, and either unarmed or armed anteriorly with teeth or cutting plates; neither lips nor a corona radiata present. In addition, dorsal rays of bursa copulatrix have greatly reduced branches.

Obligate parasites of intestinal tract of mammals. First two larval stages are free-living, usually fecal associates. Commonly referred to as hookworms.

Superfamily 4. Trichostrongyloidea

Oral opening surrounded by three or six inconspicuous lips; in some taxa lips completely wanting. No evidence of corona radiata. Stoma reduced or collapsed. Cuticle of the cephalic region commonly inflated. Somatic cuticle thick and often has several longitudinal ridges, which accounts for these nemas being described as "wiry." Esophagus of hatched larva rhabditoid but posterior bulb may not have a rhabditoid valve. Dorsal rays in male bursa copulatrix may be atrophied, but lateral rays are well developed.

Obligate vertebrate parasites with the exception of fish. Early larval stages are free-living and microbivorous.

Superfamily 5. Metastrongyloidea (Fig. 8.8)

Stoma capsule reduced or absent. However, oral opening may be surrounded by six well developed or rudimentary lips. Submedian papillae of external labial sensilla not fused. Body cuticle not adorned with longitudinal ridges. Female tail generally asymmetric and always without points or mucrons. Rays of bursa copulatrix somewhat fused and bursa itself reduced. First stage larva esophagus rhabditoid, but lacks rhabditoid valve in posterior bulb.

Order 3. Ascaridida
(Figs. 3.6F, 3.8B, 3.13, 4.7B, 4.19B, 8.11, 8.12)

Oral opening generally surrounded by three or six lips; in some lips absent. Outer circle of labial papillae usually composed of eight sensilla; sometimes submedians are fused so only four sensilla are seen. In addition there are paired porelike amphids. Esophagus variable, sometimes there is a short swollen region in stomatal area, followed by cylindrical to club-shaped region; often ending in a terminal bulb with a typical rhabditoid three lobed valve. Exceptionally there may be appendices (ceca) extending from posterior region of esophagus. Excretory system has lateral collecting tubules, sometimes extending both posteriorly and anteriorly, as such, system is H-shaped. Subventral accessory "excretory" glands absent. Males, generally, have two spicules but there may be only one or none. Gubernaculum may or may not be present. Females generally have two ovaries, but may have multiple ovaries and uteri numbering three, four, or six.

Parasites of vertebrates.

There is only one suborder currently recognized: Ascaridina and its characters are those of the order. In addition, the suborder Dioctophymatina is being tentatively placed in the order and its characters are discussed separately.

Superfamily 1. Ascaridoidea (Fig. 8.11)

Medium to very large nemas varying from one to forty centimeters. Cuticle proportionally thick and superficially annulated. Anterior, terminal oral opening generally surrounded by three distinct, well developed lips: one dorsal and two subventral. Porelike amphids located on subventral lips. Stoma poorly developed (collapsed) and surrounded by esophageal tissue. Esophagus cylindrical to clavate. From posterior esophagus a cecum may extend over anterior intestine. A cecum may also project forward past base of esophagus. One or both ceca may be present. Females most often have paired ovaries in multiples of three, four, or six. Males have two spicules only seldom accompanied by a small gubernaculum.

Parasites of vertebrates. Life cycle direct or indirect.

Superfamily 2. Cosmocercoidea

Either three or six lips; however, ventrolateral papillae always present. Stoma weakly developed and surrounded by esophageal tissue. Esophagus divisible into corpus, isthmus, and posterior bulb. Posterior bulb always valved. Esophageal glands uninucleate. No intestinal or esophageal ceca. Males may have a precloacal sucker; spicules are of equal length. Eggs oviparous or ovoviviparous.

Parasitic in a variety of animals including molluscs, amphibians, reptiles, and marsupial mammals.

Superfamily 3. Oxyuroidea (Fig. 8.12)

Lips either greatly reduced or absent and cephalic sensilla occur in a whorl of eight or four. Ventrolateral sensilla always absent. Stoma vestibular and of cheilostomal origin; esophastome not distinguished. Esophagus variable in form but posterior bulb always valved. Intestinal ceca unknown. Precloacal suckers may occur on male. Spicules may be greatly reduced.

Generally parasites of amphibians, reptiles, and mammals.

Superfamily 4. Heterakoidea

Labial region with small, well developed lips with eight cephalic sensilla paired into a circlet of four. Small vestibular or infundibular stoma formed primarily of esophostome. Esophagus consists of a clavate corpus, a short but not spheroid isthmus and a valved posterior bulb. Subventral esophageal glands binucleate. Rarely, esophagus cylindrical. Precloacal sucker present on male, surrounded by a cuticular ring. Males always have paired spicules.

Parasites of warm blooded vertebrates as well as reptiles and amphibians.

Superfamily 5. Subuluroidea

Lips small or completely absent and external circle of cephalic sensilla consists of four simple or weakly doubled papillae. Thick walled, well developed cheilostome occupies about one-third width of lip region and about as long as wide. Generally at base of cheilostome are three large teeth. Corpus clavate and narrower isthmus short; posterior bulb generally valved and subventral esophageal glands binucleate. Males have a precloacal sucker without a cuticular ring. Male spicules paired.

Parasitic in birds and mammals.

Superfamily 6. Seuratoidea (Figs. 3.8B, 3.13, 4.7B, 4.19B)

Lip region greatly reduced or absent. Ventrolateral cephalic sensilla may be present and submedian sensilla doubled, rarely single. Stoma variable, may be small and weakly developed and sometimes provided with teeth. When stoma large, then formed primarily of esophastome. Esophagus generally simple, being either cylindrical or somewhat clavate, always without posterior bulb. Only rarely, intestinal ceca present. Precloacal sucker may be present on posterior body of males. Spicules of equal length and generally accompanied by a well developed gubernaculum. If bursa present, generally very narrow.

Parasitic in a wide variety of vertebrates from fish to mammals.

Suborder b. Dioctophymatina (Fig. 3.6F)

The characteristics that place the group in Secernentea are based both on internal and external morphology. Cuticle lacks endocuticular layer. Basally only oblique fiber layer as in *Ascaris*. External cuticle demonstrates annulation but may also be ornamented with distinct hooklike spines. This character known only among Secernentea and Chromadoria. Stoma when well developed principally esophastomal and reminiscent of Seuratoidea. Esophagi cylindrical but internally the corpus and postcorpus can be distinguished. Postcorpus contains three ramifying glands whose orifices are anterior to nerve ring. Hypodermis multinucleate, a character unknown among Adenophorea. Females have one ovary and males a single testis. Male tail ends in an expanded and thickened bursa copulatrix with bordering papillae. Males have a single elongated spicule.

Parasites of mammals, birds, reptiles, and amphibians.

Subclass B. Spiruria

Oral opening surrounded by six apical lobes, referred to as pseudolabia; or lips modified into two lateral labia. Complement of cephalic sensilla variable but ventrolateral papillae always absent. When stoma well developed it is formed by cheilostome. Stoma may be spacious and globose or long and cylindrical. Some taxa stoma is not distinguished; in such instances esophagus extends to anterior extremity. When stoma armed, armature cheilostomal. Esophagus either divided into two cylindrical parts, anterior portion narrow and posterior swollen in a dorylaim fashion or esophagus may be cylindrical to clavate. Esophagus never divisible into corpus, isthmus, and postcorpus. Postcorpus never with a valve, even in first stage larva. Excretory system an inverted tuning fork, distinct subventral accessory glands not seen. Collecting tubules in lateral hypodermal chords. Vagina often greatly elongated and tortuous. Males have two spicules but often extremely unequal in length. Sometimes a bursa present but not supported by muscle associated papillae.

Order 1. Spirurida (Figs. 7.11, 8.13, 8.15–8.17, 8.21–8.24)
Nemas frequently provided with two lateral lips or pseudolabia. Some taxa have four or more lips and only rarely are lips absent. Variably shaped oral aperture may be encircled by teeth. Amphids laterally located on anterior extremity, sometimes disposed off labia or pseudolabia. Stoma varies from cylindrical and elongate to rudimentary. Esophagus generally divisible into narrow anterior portion and an expanded postcorpus enclosing multinucleate glands. Hatched larvae generally provided with cephalic hook and porelike phasmids on tail.
   Parasites of annelids or terrestrial and aquatic vertebrates.

Superfamily 1. Spiruroidea
(Figs. 8.13, 8.15–8.17, 8.21–8.24)
Lateral lips developed, in some four lips evident, rarely lips are absent. Lips often called pseudolabia. External circlet of cephalic papillae either four or eight sensilla. In some double papillae evident. Cephalic and cervical region may be ornamented with cordons, collarettes or cuticular rings. Oral opening round, hexagonal, or dorsolaterally extended. Stoma always well developed from cheilostome. Anteriorly, just inside oral aperture, stoma may be provided with teeth. Vulva generally near equatorial, rarely in posterior portion of body or near esophagus. Secondary male characters variable according to groups. However, just anterior to cloacal opening ventral surface of male often ornamented with incomplete longitudinal ridges.
   Parasites of arthropods and vertebrates.

Superfamily 2. Physalopteroidea
Large paired lateral lips (pseudolabia) generally provided with teeth on inner surfaces. Lateral lips not interspersed with interlabia. Inner whorl of circumoral sensilla reduced or absent and external circle four fused sensilla. Cuticle at anterior end of body often reflexed inwardly partially covering pseudolabia. Occasionally cephalic cuticle forms a bulbous structure which may be spined. Anterior never provided with cordons, collarettes, or rings.

Stoma reduced. Caudal papillae on males sometimes pedunculated, and caudal alae or bursa well developed, often merging ventrally. When caudal papillae not pedunculate caudal alae absent. Female genitalia often four or more; vulva either preequatorial or postequatorial.

Parasites of fish, amphibians, reptiles, birds, and mammals.

Superfamily 3. Filarioidea (Figs. 8.21–8.24)

Oral opening circular or oval, generally surrounded by eight sensilla of external circle. Internal circle absent or consists of two or four papillae. Stoma small and rudimentary. Esophagus with multinucleate glands, little difference noted between corpus and postcorpus. Generally vulva in anterior portion of body. Male spicules equal or unequal; caudal alae present or absent; spicules not accompanied by gubernaculum.

Parasites of vertebrates except fish.

Superfamily 4. Drilonematoidea (Fig. 7.11)

Stoma very reduced; may be surrounded by rudimentary lips. Generally lips not seen. Adult esophagus: an elongate corpus, a slight or no isthmus followed by pyriform glandular region, never valved. Sometimes esophagus short and clavate, reminiscent of Camallanoidea. Females always possess a single elongated ovary. Males have either paired spicules or none. Gubernaculum present or absent. Phasmids greatly enlarged, often occupying most of tail width.

Parasites of annelid worms and molluscs.

Order 2. Camallanida (Fig. 8.19)

Distinguished by simplicity of uninucleate esophageal glands. Lips absent but cephalic sensilla generally raised above head contour. Stoma varies from a massive globe to being completely reduced with only a small vestibular cheilostome. Larvae lack stomatal hook of Spirurids; phasmids large and pocketlike.

Utilizes copepods as obligate intermediate hosts; definitive host is terrestrial and aquatic vertebrates.

Superfamily 1. Camallanoidea

Distinguished by degree of stoma development. Though variable in size, always well developed; may be globose or transversally rectangular. Internally stoma may be supported by numerous longitudinal or oblique ridges. Well developed stoma always cheilostomal; esophastome plays no part in stomal formation beyond basal plates. Surrounding stoma are cephalic sensilla; internal circle minute and external circle eight partially fused papillae. The esophagus short and generally clavate.

Parasites of fish, reptiles and amphibians; copepods serve as intermediate hosts.

Superfamily 2. Dracunculoidea (Fig. 8.19)

Stoma generally very reduced; often only vestibular. Surrounding oral opening a full complement of sensilla. Internal circle six well developed sensilla and external circle eight separate and well developed sensilla. Vulva located in midbody region, atrophied in mature females. In females posterior intestine also atrophied. Males never have a well developed caudal alae; when present small and postcloacal.

Obligate tissue parasites of fish, reptiles, and mammals.

Subclass C. Diplogasteria

Most nematodes in this group small to medium sized, seldom exceeding 3 or 4 mm. External cuticle ornamented by annulations which may be transversed by longitudinal striae. Cuticle may be punctated. Labial region may not have well developed lips but hexaradiate symmetry almost always evident. Full complement of 16 cephalic sensilla often evident, especially on males. In derived forms inner circle of six may be lacking. In the external circlet externolateral sensilla take a ventrolateral position on lateral lips; the amphids are in dorsolateral position on lateral lips. Amphids generally porelike or at most small ovals, clefts, or slits. Rarely amphids not on labial region, a situation usually associated with dauerlarvae. Stoma variable in shape but always principally product of cheilostome, and often armed, but some forms without movable armature. Moveable stomatal armature, with extremely rare exception, limited to this subclass of Secernentea. Armature form of large teeth, opposable fossores, or an axial spear. Armature operable through modification of three anterior most muscles of esophagus. Function of esophastome, when developed as part of stylet, is to form apodemes for muscle attachment. Noncontractile portion muscles remain in anterior esophagus. Muscular corpus divided into an almost cylindrical procorpus and muscular, almost always, valved metacorpus followed by an isthmus and glandular postcorpus, which in derived groups is essentially deprived of musculature. Never is there a valve in postcorporal bulb. Esophagus called diplogasteroid in absence of an axial spear in stoma. When axial spear present esophagus is termed tylenchoid. As with other Secernentea, females have one or two ovaries and males characteristically have one testis with ductus ejaculatoris almost devoid of musculature, but usually with associated glands. Paired spicules most common but an associated gubernaculum may or may not be present. Males may have caudal alae.

Microbivorous omnivores, insect parasites, and plant parasites.

Order 1. Diplogasterida (Fig. 4.8C)

Lips seldom well developed though hexaradiate symmetry distinct. Cephalic (labial) sensilla often setose but always short, never long and hairlike. Stoma variable either slender, elongate or spacious with every transition between. Stoma often armed with movable teeth or fossores. Corpus always muscled and metacorpus, with few exceptions valved. Postcorpus divisible into a narrow elongate isthmus and swollen glandular bulb never valved. Female reproductive system paired or single and males may or may not have caudal alae.

Predaceous, omnivorous, bacterial feeders, fungal feeders, and often are consorts of insects.

Superfamily 1. Diplogasteroidea (Fig. 4.8C)

Lip region composed of six distinct lips or six fused lips, lip region never set off by a constriction. Cephalic sensilla of external circle may be setose or papilliform. Amphids generally porelike and dorsolateral; in some males and dauerlarvae amphids oval or circular and located postlabially. Stoma variable in shape and structure but never one-fourth the length of the length of the esophagus. Cheilostome major contributor to stoma; often complex and three regions discernible. Anteriorly may be supported by rugae-like structures; second section may be unarmed, often with straight walls, may be narrow or wide; base of stoma, at junction with esophastome, may be armed with a large tooth, or a combination of large teeth and small denticles. In a few cheilostome and esophastome modified into a conical, cylindrical spearlike structure. Esophastome in these shaft-like and posteriorly may be modified by one or more knobs for muscle attachment. Esophagus diplogasteroid and corpus with metacorporal bulb and valve clearly distinguishable from glandular postcorpus. Males have paired spicules either arcuate with a well developed capitulum (head) or straight with little or no capitulum. An accompanying gubernaculum always present, often well developed. Male tail commonly has nine pairs of genital papillae and one pair of phasmids. Three pairs of genital papillae preanal. Caudal alae, when present, reduced and narrow.

These nemas are predators, bacterial feeders, or omnivores.

Superfamily 2. Cylindrocorporoidea

Six lips well developed and bear cephalic sensilla. Small amphids located on lateral lips. Stoma very elongate, one-fourth or more esophageal length. Esophagus has corpus greatly enlarged into a distinct cylindroid muscular complex. Glandular postcorpus swells slightly at esophago-intestinal junction. Caudal alae, when present, always rudimentary and spicules elongate and thin.

Microbivorous or intestinal parasites of amphibians, reptiles, and a few mammals.

Order 2. Aphelenchida (Figs. 6.6, 6.23–6.25, 7.10, 7.12)

Labial cap distinguishable and often set off by constriction. Hollow axial spear seldom strongly developed and basally it may or may not have knobs and only rarely are these well developed. Esophagus almost always has a large valved metacorporal bulb, often squarish in outline. All esophageal glands open into metacorpus, never into procorpus. Dorsal gland opens anterior to valve and subventrals posterior to valve. Except for one genus glands always overlap anterior intestine. Females all have a single anteriorly directed ovary; a postuterine sac may or may not be present. When present, postuterine sac acts as a spermatheca. Vulva always posteriorly located. Males may or may not possess a bursa. When present, genital papillae form rays similar to those seen in Rhabditia. Males always have two or more pairs of caudal papillae. The spicules may be slender, slightly arcuate, or highly distinctive and thornlike. When gubernaculum present it apparently is forked.

Included are mycetophagous forms, higher plant parasites (generally above ground parts), predators, and obligate insect parasites.

Superfamily 1. Aphelenchoidea (Fig. 6.6)

Three regions of esophagus recognizable: corpus, isthmus, and posterior bulb, which may or may not overlay anterior intestine. Laterally body marked by six to 14 fine incisures. Males have slightly curved narrow spicules, somewhat similar to those of Tylenchida, accompanied by a narrow forked gubernaculum. Caudal alae present or absent.

Mycetophagous soil inhabitants.

Superfamily 2. Aphelenchoidoidea
(Figs. 6.23–6.25, 7.10, 7.12)

Labial cap always evident and often lip region expanded and set off from cervical region by an incisure. Individual lips amalgamated, but can be distinguished by faint cephalic framework. Stylet generally slight and knobs range from absent to well developed. Most distinctive character is lack of an isthmus in esophagus. Intestine joins esophagus immediately behind metacorpus and elongated esophageal glands overlap anterior intestine dorsally immediately from this juncture. Spicules very characteristic, they are robust and because of development of rostrum they are characteristically rosethorn-shaped. Caudal papillae only slightly raised above body contour, no bursa.

Range from phytophagus to mycetophagous to predaceous to invertebrate parasites.

Order 3. Tylenchida (Figs. 3.7B, 6.4, 6.5, 6.7–6.22, 7.1, 7.8)

Lip region may be smoothly rounded or well developed; almost always it is hexaradiate. With few exceptions amphids porelike and located on lips; in some taxa oval or extended clefts. Internal circle of sensilla may be lacking, 10 sensilla of outer or external circle usually evident. Stylet or spear formed from both cheilostome and esophastome. Cheilostome forms framework, guiding apparatus (original stoma), and forepart of spear. Spear shaft and three basal knobs (apodemes for muscle origin) formed from esophastome. Musculature for protrusion of spear modified from first three muscles of esophagus. In some adult insect parasites spear not evident, also true of males in some plant parasites. Esophagi divisible into three parts: corpus, isthmus, and posterior glandular bulb (postcorpus). Corpus further divisible into procorpus and metacorpus. Metacorpus commonly valved; however, valve may be absent in some taxa and is characteristic of some groups. In all instances, dorsal esophageal gland has its orifice in procorpus usually near base of spear. Excretory system asymmetrical, a single lateral collecting tube extends length of body. Females may have one or two ovaries. Males never have more than one pair of caudal papillae (= phasmids). Caudal alae may or may not be present. Spicules always paired, sometimes reportedly fused at tip; an accompanying gubernaculum may be present or absent.

Mycetophagous forms and higher plant and insect parasites.

Suborder a. Tylenchina

Universally, among females esophagus composed of a procorpus and metacorpus, together called corpus; and a glandular postcorpus with an intervening isthmus between corpus and posterior bulb. Males may have degenerate esophagi in some groups. Metacorpus, in Neotylenchidae, lacks a valve. Even when valve absent esophagus well developed and functional, not to be interpreted as aberrant. Lip region may be distinct or undifferentiated from general body contour. Generally, transverse body annulation evident and interrupted laterally by variable number of longitudinal incisures. In some taxa entire circumference of body has longitudinal incisures as well as transverse annulation. Except in aberrant males and larvae, stoma armed with a knobbed axial spear. Glandular postcorpus of esophagus shows varying degrees of development and enlargement; it may join intestine directly or overlap intestine for some distance. Females have one or two ovaries and males generally have, with few exceptions, caudal alae. Paired spicules may or may not be accompanied by a gubernaculum.

Mycetophagous forms and plant parasites.

Superfamily 1. Tylenchoidea
(Figs. 6.4, 6.5, 6.12, 6.13–6.22, 7.1)

Lip region generally hexaradiate and distinguishable from general body contour. Labial region may be supported internally by a cuticularized basketwork. Central cylinder of basket lines forepart of cheilostome. Procorpus set off from metacorpus; usually appears slender and cylindrical. In one group this distinction hampered by retraction of greatly elongated spear. Isthmus narrow and leads into expanded glandular region almost always wider than metacorpus. Glandular region may consist of a true bulb enclosing glands or glands may form a lobe and extend over anterior intestine. Females have one or two ovaries and males have either a leptoderan or peloderan bursa; in one family males lack a caudal alae. Phasmids common on tail; however, often obscure.

Vary from mycetophagous to plant parasitic.

Superfamily 2. Criconematoidea (Figs. 3.7B, 6.7–6.11)

Never is there a well developed labial cap internally supported by a heavily cuticularized skeleton. Skeletal development for anchorage of the spear muscles difficult to discern because of body annulation or body structure at anterior end. At anterior extremity "lips" represented by a disc generally with four submedian lobes; this feature, with minor variations, consistent. This character combined with form of esophagus becomes characteristic. Corpus can be divided into pro- and metacorpus; however, they merge together in such a fashion that no clear point of distinction exists between them. Glandular postcorpus varies from almost cylindrical with a slight terminal swelling to a postcorpus almost tylenchoid, that is, pyriform and clearly offset from intestine without overlap. Females always have one anteriorly directed ovary. In a few taxa a small postuterine sac may be evident. Females vary in shape from vermiform to obese saccate. Vermiform males never functional feeders and generally spicules are almost straight, with exception of one taxa where they are strongly recurved and hooklike. In rare instances caudal alae present.

Obligate ectoparasites which vary from migratory to highly specialized sessile forms.

Suborder b. Sphaerulariina (Figs. 6.26, 7.8)

Sphaerulariina lacks a valved median bulb, the absence is not a secondary derivation recurring within taxa. Generally these nematodes have three distinct forms: two free-living and one parasitic. Two free-living forms are found in habitat of young hosts, whereas parasitic form is found in host's hemocoel. Free-living phase of female generally possesses a stylet, esophagus lacks a valve in metacorpus. Dorsal gland orifice is always in corpus, but may be some distance behind base of stylet. Orifices of esophageal glands can be located by prominent ampullae near duct. Esophageal glands long and overlap anterior intestine. In free-living phase reproductive system of female is diagnostic. Single anteriorly directed ovary small and fingerlike with very few developing oocytes. Short oviduct followed by a prominent uterus often filled with sperm. This stage is infective. In parasitic phase of female's life, body either becomes grossly enlarged and degenerates to a reproductive sac or uterus prolapses and gonadal development takes place outside body. In this reduced state a stylet is generally evident; however, esophagus and intestine become degenerate. Commonly oocytes arranged around a rachis. Males remain free-living and not infective; generally slightly longer than free-living phase of adult female. Stylet often absent or obscure and esophagus degenerated. Male spicules are not always typical tylenchoid, they may be elaborate and accompanied by a gubernaculum. The caudal alae when present peloderan.

Parasites of invertebrates, primarily insects.

Selected References

Chapter 1

Chan, L., and Croll, N. A. 1979. Medical parasitology in China: An historical perspective. American Journal of Chinese Medicine, VII:39–51.

Chitwood, B. G., and Chitwood, M. B. 1974. Introduction to nematology. University Park Press, Baltimore, London, Tokyo. 334 pp.

Crofton, H. D. 1966. Nematodes. Hutchinson University Library, London. 160 pp.

Faust, E. C. 1949. Human helminthology, a manual for physicians, sanitarians and medical zoologists. Lea & Febiger, Philadelphia. 744 pp.

Hoeppli, R., and Ch'iang, I. H. 1940. Selections from old Chinese medical literature on various subjects of helminthological interest. Chinese Medical Journal, 57:373–387.

Mjuge, S. G. 1977. Historical development of nematology in Russia. Journal of Nematology, 9:1–8.

Thorne, G. 1961. Principles of nematology. McGraw-Hill Book Company, Inc., New York, Toronto, London. 533 pp.

Veith, I. 1972. Huang Ti Nei Ching Su Wen (The Yellow Emperor's classic of internal medicine), new edition, translated with an introductory study. University of California Press, Berkeley. 260 pp.

Chapter 2

Barnes, R. D. 1974. Invertebrate zoology. W. B. Saunders Company, Philadelphia, London, Toronto. 870 pp.

Borradaille, L. A., and Potts, F. A. 1956. The invertebrata—A manual for the use of students. Cambridge University Press. 725 pp.

Chitwood, B. G. 1958. The classification of the phylum Kinorhyncha (Reinhard, 1887) Pearse 1936. XVth International Congress of Zoology, Dec. 1, paper 22, pp. 1–3.

Chitwood, B. G., and Wehr, E. E. 1934. The value of cephalic structures as characters in nematode classification, with special reference to the superfamily Spiruroidea. Z. f. Parasitenkunde Bd. 7:273–335.

Chitwood, B. G., and Chitwood, M. B. 1974. Introduction to nematology. University Park Press, Baltimore, London, Tokyo. 334 pp.

Clark, C. B. 1963. The evolution of the coelom and metameric segmentation. Pages 91–107 in E. C. Dougherty, Ed. The lower metazoa comparative biology and phylogeny. University of California Press, Berkeley and Los Angeles.

Clark, R. B. 1964. Dynamics in metazoan evolution the origin of the coelom and segments. Oxford University Press, Ely House, London. 313 pp.

Gould, S. J. 1974. This view of life: Size and shape. Natural History. 83(1): 20–26.

Hyman, L. H. 1951. The invertebrates: Acanthocephala, Aschelminthes and Entoprocta. The pseudocoelomate bilateria. Vol. III. McGraw-Hill Book Company, Inc., New York, Toronto, London. 572 pp.

Maggenti, A. R. 1971. Nemic relationships and the origins of plant parasitic nematodes. Pages 65–81 in B. M. Zuckerman, Ed. Plant parasitic nematodes. Academic Press, New York, London.

Maggenti, A. R. 1976. Taxonomic position of nematoda among the pseudocoelomate bilateria. Pages 1–10 in N. A. Croll, Ed. The organization of nematodes. Academic Press, New York, London, San Francisco.

Remane, A. 1963. The systematic position and phylogeny of the pseudocoelomates. Pages 247–255 in E. C. Dougherty, Ed. The lower metazoa comparative biology and phylogeny. University of California Press, Berkeley and Los Angeles.

Steiner, G. 1960. The nematoda as a taxonomic category and their relationship to other animals. Pages 12–18 in J. N. Sasser, and W. R. Jenkins, Eds. Nematology fundamentals and recent advances with emphasis on plant parasitic and soil forms. The University of North Carolina Press, Chapel Hill.

Storer, T. I., and Usinger, R. L. 1957. General zoology. McGraw-Hill Book Company, Inc., New York, Toronto, London. 664 pp.

Chapter 3

Aboul-Eid, H. Z. 1969. Electron microscope studies on the body wall and feeding apparatus of *Longidorus macrosoma*. Nematologica, 15:451–463.

Anya, A. O. 1966. The structure and chemical composition of the nematode cuticle. Observations on some oxyurids and *Ascaris*. Parasitology, 56:179–198.

Baldwin, J. G., and Hirschmann, H. 1973. Fine structure of cephalic sense organs in *Meloidogyne incognita* males. Journal of Nematology, 5:285–302.

Baldwin, J. G., and Hirschmann, H. 1975. Fine structure of cephalic sense organs in *Heterodera glycines males*. Journal of Nematology, 7:40–49.

Baldwin, J. G., and Hirschmann, H. 1975. Body wall fine structure of the anterior region of *Meloidogyne incognita* and *Heterodera glycines* males. Journal of Nematology, 7:175–193.

Bird, A. F., and Rogers, G. E. 1965. Ultrastructure of the cuticle and its formation in *Meloidogyne javanica*. Nematologica 11:224–230.

Bird, A. F. 1971. The structure of nematodes. Academic Press, New York and London. 318 pp.

Byers, J. R., and Anderson, R. V. 1972. Ultrastructural morphology of the body wall, stoma, and stomatostyle of the nematode, *Tylenchorhynchus dubius* (Bütschli, 1873) Filipjev, 1936. Canadian Journal of Zoology, 50:457–465.

Chitwood, B. G., and Chitwood, M. B. 1974. Introduction to nematology. University Park Press, Baltimore, London, Tokyo. 334 pp.

Coomans, A. 1979. Article bibliographique. The anterior sensilla of nematodes. Revue de Nématologie, 2:259–283.

Croll, N. A. 1976. The organization of nematodes. Academic Press, London, New York, San Francisco. 439 pp.

Croll, N. A. 1977. Sensory mechanisms in nematodes. Annual Review of Phytopathology, 15:75–89.

Davey, K. G. 1965. Molting in a parasitic nematode, *Phocanema decipiens*. Canadian Journal of Zoology, 43:997–1003.

De Coninck, L. 1950. Les relations de symetrie, regissant la distribution des organes sensibles anterieurs chez les nematodes. Annales de la Société Royale Zoologique de Belgique, 81:25–32.

De Grisse, A. T. 1972. Body wall ultrastructure of *Macroposthonia xenoplax*. Nematologica, 18:25–30.

De Grisse, A. T. 1977. De ultrastruktuur van het zenuwstelsel in da kop van 22 soorten plantenparasitaire nematoden, behorende tot 19 genera (Nematoda: Tylenchida). Rijksuniversiteit Gent. 420 pp.

Epstein, J., Castillo, J., Himmelhoch, S. and Zuckerman, B. M. 1971. Ultrastructural studies on *Caenorhabditis briggsae*. Journal of Nematology, 3: 69–78.

Goldschmidt, R. 1903. Histologische untersuchungen an nematoden. I. Die sinnesorgane von *Ascaris lumbricoides* L. und *A. megalocephala* Cloque. Zoologische Jahrbuecher Abteilung für Morphologie, pp. 1–57.

Goldschmidt, R. 1904. Der chromidialapparat lebhaft funktionierender gewebszellen. (Histologische untersuchungen an nematoden II.) Verlag von Gustav Fischer in Jena. 100 pp.

Grassé, P.-P. 1965. Traité de Zoologie, Tome IV, Fascicule 2–3. Masson, Paris.

Grootaert, P., and Lippens, P. L. 1974. Some ultrastructural changes in cuticle and hypodermis of *Aporcelaimellus* during the first moult (Nematoda: Dorylaimoidea). Zeitschrift für Morphologie der Tiere, 79:269–282.

Harris, J. E., and Crofton, H. D. 1957. Structure and function in the nematodes: Internal pressure and cuticular structure in *Ascaris*. Journal of Experimental Biology, 34(1):116–130.

Hepburn, H. R., Ed. 1976. The insect integument. Elsevier Scientific Publication Company, Amsterdam and New York. 571 pp.

Hinz, E. 1963. Elektronenmikroskopische untersuchungen an *Parascaris equorium*. Sonderabdruck aus Band 56, Heft 2:202–241.

Hope, W. D. 1974. *Deontostoma timmerchioi* n. sp., a new marine nematode (Leptosomatidae) from Antarctica, with a note on the structure and possible function of the ventromedian supplement. Transactions of the American Microscopical Society, 93(3):314–324.

Inglis, W. G. 1964. The structure of the nematode cuticle. Proceedings of the Zoological Society of London, Vol. 1, Part 3, 465–502 pp.

Jamuar, M. P. 1966. Electron microscope studies on the body wall of the nematode *Nippostrongylus brasiliensis*. The Journal of Parasitology, 52:209–232.

Johnson, P. W., Van Gundry, S. D., and Thomson, W. W. 1970. Cuticle ultra-structure of *Hemicycliophora arenaria, Aphelenchus avenae, Hirschmanniella gracilis* and *Hirschmanniella belli.* Journal of Nematology 2:42–58.

Kisiel, M., Himmelhoch, S., and Zuckerman, B. M. 1972. Fine structure of body wall and vulva area of *Pratylenchus penetrans.* Nematologica, 18: 234–238.

Kondo, E., and Ishibashi, N. 1975. Ultrastructural changes associated with the tanning process in the cyst wall of the soybean cyst nematode, *Heterodera glycines* Ichinohe. Applied Entomology and Zoology, 10:247–253.

Lippens, P. L. 1974. Ultrastructure of a marine nematode, *Chromadorina germanica* (Buetschli, 1874). Zeitschrift für Morphologie der Tiere, 79: 283–294.

Lippens, P. L., Coomans, A., De Grisse, A. T., and Lagasse, A. 1974. Ultra-structure of the anterior body region in *Aporcelaimellus obtusicaudatus* and *A. obscurus.* Nematologica, 10:242–256.

Maggenti, A. R. 1964. Morphology of somatic setae: *Thoracostoma californicum* (Nemata: Enoplidae). Proceedings of the Helminthological Society of Washington, 31:159–166.

Maggenti, A. R. 1979. The role of cuticular strata nomenclature in the systematics of Nemata. Journal of Nematology, 11:94–98.

McLaren, D. J. 1976. Nematode sense organs. Advances in Parasitology, 14: 195–265.

Minier, L. N., and Bonner, T. P. 1975. Cuticle formation in parasitic nematodes: Ultrastructure of molting and the effects of Actinomycin-D. Journal of Ultra-structure Research, 53:77–86.

Raski, D. J., and Jones, N. O. 1973. Ultrastructure of the cuticle of *Bunonema* spp. (Nematoda: Bunonematidae). Proceedings of the Helminthological Society of Washington, 40:216–227.

Raski, D. J., Jones, N. O., and Roggen, D. R. 1969. On the morphology and ultrastructure of the esophageal region of *Trichodorus allius* Jensen. Proceedings of the Helminthological Society of Washington, 36:106–118.

Riemann, F. 1972. Corpus gelatum und ciliäre strukturen als lichtmikroskopisch sichtbare bauelemente des seitenorgans freilebender nematoden. Zeitschrift für Morphologie der Tiere, 72:46–76.

Roggen, D. R., Raski, D. J., and Jones, N. O. 1966. Cilia in nematode sensory organs. Science, 152:515–516.

Roggen, D. R., Raski, D. J., and Jones, N. O. 1967. Further electron micro-scopic observations of *Xiphinema index.* Nematologica, 13:1–16.

Sheffield, H. G. 1963. Electron microscopy of the bacillary band and stichosome of *Trichuris muris* and *T. vulpis.* The Journal of Parasitology, 49:998–1009.

Siddiqui, I. A., and Viglierchio, D. R. 1977. Ultrastructure of the anterior body region of marine nematode *Deontostoma californicum.* Journal of Nematology, 9:56–82.

Singh, R. N., and Sulston, J. E. 1978. Some observations on moulting in *Caenorhabditis elegans.* Nematologica 24:63–71.

Smith, L., and Stephenson, A. M. 1970. The hypodermis of *Enoplus brevis* and other related nematodes; No structural evidence for a peripheral nerve net. Nematologica 16:572–576.

Snodgrass, R. E. 1935. Principles of insect morphology. McGraw-Hill Book Company, Inc., New York and London. 667 pp.

Storch, V., and Riemann, F. 1973. Zur ultrastruktur der seitenorgane (amphiden) des limnischen nematoden *Tobrilus aberrans* (W. Schneider, 1925) (Nematoda, Enoplida). Zeitschrift für Morphologie der Tiere, 74:163–170.

Ward, S., Thomson, N., White, J. G., and Brenner, S. 1975. Electron microscopical reconstruction of the anterior sensory anatomy of the nematode *Caenorhabditis elegans.* Journal of Comparative Neurology, 160:313–338.

Watson, B. D. 1965. The fine structure of the body-wall in a free-living nematode, *Euchromadora vulgaris.* Quarterly Journal of Microscopical Science, 106:75–81.

Wergin, W. P., and Endo, B. Y. 1976. Ultrastructure of a neurosensory organ in a root-knot nematode. Journal of Ultrastructure Research, 56:258–276.

Wright, K. A. 1963. Cytology of the bacillary bands of the nematode *Capillaria hepatica* (Bancroft, 1893). Journal of Morphology, 112:233–259.

Wright, K. A. 1968. Structure of the bacillary band of *Trichuris myocastoris.* The Journal of Parasitology, 54:1106–1110.

Yuen, P. H. 1968. Electron microscopical studies on the anterior end of *Panagrellus silusiae* (Rhabditidae). Nematologica, 14:554–564.

Zuckerman, B. M., Himmelhoch, S., and Kisiel, M. 1973. Fine structure changes in the cuticle of adult *Caenorhabditis briggsae* with age. Nematologica, 19:109–112.

Chapter 4

Albertson, D. G., and Thomson, J. N. 1975. The pharynx of *Caenorhabditis elegans.* Philosophical Transactions of the Royal Society of London, Series B:299–325.

Baldwin, J. G., and Hirschmann, H. 1976. Comparative fine structure of the stomatal region of males of *Meloidogyne incognita* and *Heterodera glycines.* Journal of Nematology, 8:1–17.

Baldwin, J. G., Hirschmann, H., and Triantaphyllou, A. C. 1977. Comparative fine structure of the esophagus of males of *Heterodera glycines* and *Meloidogyne incognita.* Nematologica, 23:239–252.

Bird, A. F. 1971. The structure of nematodes. Academic Press, New York and London. 318 pp.

Bird, A. F., and Rogers, G. E. 1965. Ultrastructural and histochemical studies of the cells producing the gelatinous matrix in *Meloidogyne.* Nematologica, 11:231–238.

Chalfie, M., and Thomson, J. N. 1979. Organization of neuronal microtubules in the nematode *Caenorhabditis elegans.* Journal of Cell Biology, 82:278–289.

Chitwood, B. G., and Chitwood, M. B. 1974. Introduction to nematology. University Park Press, Baltimore, London, Tokyo. 334 pp.

Coomans, A. 1962. The spicular muscles in males of the subfamily Hoplolaiminae (Tylenchida, Nematoda). Biologisch Jaarboek, 30:313–315.

Coomans, A. 1963. Stoma structure in members of the Dorylaimina. Nematologica, 9:587–601.

Coomans, A., and De Coninck, L. 1963. Observations on spear-formation in Xiphinema. Nematologica, 9:85–96.

Croll, N. A. 1976. The organization of nematodes. Academic Press, London, New York, San Francisco. 439 pp.

Croll, N. A., and Maggenti, A. R. 1968. A peripheral nervous system in nematoda with a discussion of its functional and phylogenetic significance. Proceedings of the Helminthological Society of Washington, 35:108–115.

Davey, K. G. 1964. Neurosecretory cells in a nematode, *Ascaris Lumbricoides*. Canadian Journal of Zoology, 42:731–734.

Davey, K. G. 1966. Neurosecretion and molting in some parasitic nematodes. American Zoologist, 6:243–249.

Debell, J. T. 1965. A long look at neuromuscular junctions in nematodes. The Quarterly Review of Biology, 40:233–251.

De Grisse, A., Natasasmita, S., and B'Chir, M. 1979. L'ultrastructure des nerfs de la région céphalique chez *Aphelenchoides fragariae* (Nematoda: Tylenchida). Revue Nématology, 2:123–141.

Endo, B. Y., and Wergin, W. P. 1977. Ultrastructure of anterior sensory organs of the root-knot nematode, *Meloidogyne incognita*. Journal of Ultrastructure Research, 59:231–249.

Goodey, J. B. 1961. The nature of the spear guilding apparatus in Dorylaimidae. Journal of Helminthology, R. T. Leiper Supplement, pp. 101–106.

Hamada, G. S., and Wertheim, G. 1978. *Mastophorus muris* (Nematoda: Spirurina): Ultrastructure of somatic muscle development. International Journal for Parasitology, 8:405–414.

Hirumi, H., Raski, D. J., and Jones, N. O. 1971. Primitive muscle cells of nematodes: Morphological aspects of platymyarian and shallow coelomyarian muscles in two plant parasitic nematodes, *Trichodorus christiei* and *Longidorus elongatus*. Journal of Ultrastructure Research, 34:517–543.

Hope, W. D. 1969. Fine structure of the somatic muscles of the free-living marine nematode *Deontostoma californicum* Steiner and Albin, 1933 (Leptosomatidae). Proceedings of the Helminthological Society of Washington, 38: 10–29.

Hyman, L. H. 1951. The invertebrates: Acanthocephala, Aschelminthes, and Entoprocta. The pseudocoelomate bilateria. McGraw-Hill Book Company, Inc., New York, Toronto, London. Volume 3, 572 pp.

Inglis, W. G. 1964. The marine Enoplida (Nematoda): A comparative study of the head. Bulletin of the British Museum (Natural History) Zoology, 2: 263–376.

Inglis, W. G. 1967. The relationships of the nematode superfamily Seuratoidea. Journal of Helminthology, 41:115–136.

Kashio, T., Ishibashi, N., and Yokoo, T. 1975. Morphology of larval molting in the pin nematode, *Paratylenchus aciculus* Brown (Nematoda:Paratylenchidae) with the emphasis on stylet regeneration. Applied Entomology and Zoology, 10:96–102.

Kessel, R. G., Prestage, J. J., Sekhon, S. S., Smalley, R. L., and Beams, H. W. 1961. Cytological studies on the intestinal epithelial cells of *Ascaris lumbricoides suum*. Transactions of the American Microscopical Society, 80:103–118.

Lippens, P. L. 1974. Ultrastructure of a marine nematode, *Chromadorina germanica* (Buetschli, 1874). Zeitschrift für Morphologie der Tiere, 78: 181–192.

Lopez-Abella, D., Jimenez-Millan, F., and Garcia-Hidalgo, F. 1967. Electron microscope studies of some cephalic structures of *Xiphinema americanum*. Nematologica, 13:283–286.

Maggenti, A. R., and Allen, M. W. 1960. The origin of the gelatinous matrix in *Meloidogyne*. Proceedings of the Helminthological Society of Washington, 27:4–10.

Malakhov, V. V. 1975. On the cleavage of the eggs of Nematoda and Gastrotricha as a derivative of one-radial spiral cleavage. Vestnik Moskovskogo Universiteta Seriya Vi Biologiya, 2:14–17.

Malakhov, V. V., and Akimushkina, M. I. 1976. Embryogenesis of a free-living nematode *Enoplus brevis*. Zoologicheskii Zhurnal, 55:1788–1799.

Ohmori, Y. 1974. Arrangement of the somatic muscle cells of meromyarian nematodes (1) female worms of oxyurids and hookworms. Japanese Journal of Parasitology, 23:95–99.

Raski, D. J., Jones, N. O., and Roggen, D. R. 1969. On the morphology and ultrastructure of the esophageal region of *Trichodorus allius* Jensen. Proceedings of the Helminthological Society of Washington, 36:106–118.

Robertson, W. M. 1976. A possible gustatory organ associated with the odontophore in *Longidorus leptocephalus* and *Xiphinema diversicaudatum*. Nematologica, 21:443–448.

Rosenbluth, J. 1965. Ultrastructure of somatic muscle cells in *Ascaris lumbricoides*. Journal of Cell Biology, 26:579–591.

Rosenbluth, J. 1967. Ultrastructural organization of obliquely striated muscle fibers in *Ascaris lumbricoides*. Journal of Cell Biology, 25:495–510.

Rosenbluth, J. 1967. Obliquely striated muscle. III. Contraction mechanism of *Ascaris* body muscle. Journal of Cell Biology, 34:15–33.

Siddiqui, I. A., and Viglierchio, D. R. 1970. Ultrastructure of photoreceptors in the marine nematode *Deontostoma californicum*. Journal of Ultrastructure Research, 32:558–571.

Siddiqui, I. A., and Viglierchio, D. R. 1977. Ultrastructure of the anterior body region of marine nematode *Deontostoma californicum*. Journal of Nematology, 9:56–82.

Smith, J. M. 1974. Ultrastructure of the hemizonid. Journal of Nematology, 6:53–55.

Stretton, A. O. W., Fishpool, R. M., Southgate, E., Donmoyer, J. E., Walrond, J. P., Moses, J. E. R., and Kass, I. S. 1978. Structure and physiological activity of the motoneurons of the nematode *Ascaris*. Proc. Nat. Acad. Sci., USA, 75:3493–3497.

Van Der Heiden, A. 1974. The structure of the anterior feeding apparatus in members of the Ironidae (Nematoda: Enoplida). Nematologica, 20:419–436.

Waddell, A. H. 1968. The excretory system of the kidney worm *Stephanurus dentatus* (Nematoda). Parasitology 58:907–919.

Ward, S., Thomas, N., White, J. F., and Brenner, S. 1975. Electron microscopical reconstruction of the anterior sensory anatomy of the nematode *Caenorhabditis elegans*. Journal of Comparative Neurology, 160:313–338.

Ware, R. W., Clark, D., Crossland, K., and Russell, R. L. 1976. The nerve ring of the nematode *Caenorhabditis elegans*: sensory input and motor output. Journal of Comparative Neurology, 162:71–110.

White, J. G., Southgate, E., Thomson, J. N., and Brenner, S. 1976. The structure of the ventral nerve cord of *Caenorhabditis elegans*. Philosophical Transactions of the Royal Society of London Series B, 275:327–348.

Wright, K. A. 1964. The fine structure of the somatic muscle cells of the nematode *Capillaria hepatica* (Bancroft, 1893). Canadian Journal of Zoology, 42:483–490.

Wright, K. A. 1965. The histology of the oesophageal region of *Xiphinema index* Thorne and Allen, 1950, as seen with the electron microscope. Canadian Journal of Zoology, 43:689–700.

Wu, L.-Y. 1968. Morphological study of *Anguina calamagrostis* Wu, 1967 (Tylenchidae: Nematoda) with the electron microscope. Canadian Journal of Zoology, 46:467–468.

Yuen, P.-H. 1967. Electron microscopical studies on *Ditylenchus dipsaci* (Kuhn). Canadian Journal of Zoology, 45:1019–1032.

Chapter 5

Anya, A. O. 1976. Physiological aspects of reproduction in nematodes. Advances in Parasitology, 14:267–351.

Bird, A. F. 1971. The structure of nematodes. Academic Press. New York and London. 318 pp.

Chitwood, B. G., and Chitwood, M. B. 1974. Introduction to nematology. University Park Press, Baltimore, London, Tokyo. 334 pp.

Flegg, J. J. M. 1966. The Z-organ in *Xiphinema diversicaudatum*. Nematologica, 12:174.

Foor, W. E. 1970. Spermatozoan morphology and zygote formation in nematodes. Biology of Reproduction Supplement 2, 177–202.

Geraert, E. 1972. A comparative study on the structure of the female gonads in plant-parasitic Tylenchida (Nematoda). Annales de la Société Royale Zoologique de Belgique, Tome 102, fascicule 3, pp. 171–198.

Gysels, H., and Van Der Haegen, W. 1962. Post-embryonale ontwikkeling en vervellingen van de vrijlevende nematode *Panagrellus silusiae* (de Man 1913) Goodey 1945. Natuurwetenschappelijk Tijdschrift, 44:3–20.

Hirschmann, H. 1951. Über das vorkommen zweier mundhöhlentypen bei *Diplogaster lheritieri* maupas und *Diplogaster biformis* n. sp. und die entstehung dieser hermaphroditischen art aus *Diplogaster lheritieri*. Zoologische Jahrbucher, Abteilung für Systematik, Okologie und Geographie der Tiere, 80:1–188.

Hirschmann, H. 1962. The life cycle of *Ditylenchus triformis* (Nematoda: Tylenchida) with emphasis on post-embryonic development. Proceedings of the Helminthological Society of Washington, 29:30–43.

Hope, D. W. 1974. In Giese, A. C., and Pearse, J. S. Eds. Nematoda. Reproduction of Marine Invertebrates, Volume 1, Acoelomate and Pseudocoelomate Metazoans. Academic Press, London, New York, San Francisco. Chapter 8, pp. 391–469.

Hyman, L. H. 1951. The invertebrates: Acanthocephala, Aschelminthes, and Entoprocta. The pseudocoelomate bilateria. McGraw-Hill Book Company, Inc., New York, Toronto, London. Volume 3, 572 pp.

Ishibashi, N., and Kondo, E. 1976. Considerations on the role of male adults of root-knot nematodes in the reproduction. Japanese Journal of Nematology, 6:35–38.

Krieg, C., Cole, T., Deppe, U., Schierenberg, E., Schmitt, D., Yoder, B., and von Ehrenstein, G. 1978. The cellular anatomy of embryos of the nematode *Caenorhabditis elegans*. Developmental Biology, 65: 193–215.

Lamberti, F., and Taylor, C. E., Eds. 1979. Root-knot nematodes (*Meloidogyne* species), Systematics, biology and control. Academic Press, London, New York, San Francisco. 477 pp.

Lorenzen, S. 1978. New and known gonadal characters in free-living nematodes and the phylogenetic implications. Zeitschrift für Zoologische Systematik und Evolutionsforschung, 16:108–115.

Mayer, A. 1906. Zur konntnis der rhachis im ovarium und hoden der nematoden. Zoologischer Anzeiger, 30:289–297.

Paetzold, D. 1958. Bemerkungen zur "endotokia matricida" von Lordello 1951. Wissenschaftliche Zeitschrift der Martin Luther Universität Halle-Wittenberg, Mathematisch-Naturwissenschaftliche Reihe, 7:81–84.

Rachor, E. 1969. Das de mansche organ der Oncholaimidae, eine genito-intestinale verbindung bei nematoden. Zeitschrift für Morphologie der Tiere, 66:87–166.

Rachor, E. 1970. Systematische bemerkungen zur familie der Oncholaimidae (Nematoda: Enoplida). Veröff Inst. Meeresforsch Bremerh, 12:443–453.

Roman, J., and Triantaphyllou, A. C. 1969. Gametogenesis and reproduction of seven species of *Pratylenchus*. Journal of Nematology, 1:357–362.

Shepherd, A. M., Clark, S. A., and Kempton, A. 1974. Spermatogenesis sperm ultrastructure in some cyst nematodes, *Heterodera* spp. Nematologica, 19:551–560.

Steiner, G. 1923. Intersexes in nematodes. The Journal of Heredity, 14:147–158.

Sulston, J. E., and Horvitz, H. R. 1977. Post-embryonic cell lineages of the nematode, *Caenorhabditis elegans*. Developmental Biology, 56:110–156.

Triantaphyllou, A. C. 1966. Polyploidy and reproductive patterns in the root-knot nematode *Meloidogyne hapla*. Journal of Morphology, 118:403–414.

Triantaphyllou, A. C., and Hirschmann, H. 1964. Reproduction in plant and soil nematodes. Annual Review of Phytopathology, 2:57–80.

Triantaphyllou, A. C., and Hirschmann, H. 1980. Cytogenetics and morphology in relation to evolution and speciation of plant-parasitic nematodes. Annual Review of Phytopathology, 18:333–59.

Wen, G. Y., and Chen, T. A. 1974. Ultrastructure of the spicules of *Pratylenchus penetrans*. Journal of Nematology, 8:69–74.

Wright, K. A., Hope, W. D., and Jones, N. O. 1973. The ultrastructure of the sperm of *Deontostoma californicum*, a free-living marine nematode. Proceedings of the Helminthological Society of Washington, 40:30–36.

Chapter 6

Christie, J. R. 1959. Plant nematodes, their bionomics and control. H. and B. Drew Company, Jacksonville, Florida. 256 pp.

Cohn, Eli. 1974. Histology of the feeding site of *Rotylenchulus reniformis*. Nematologica, 19:455–458.

Croll, N. A. 1977. Biology of nematodes. John Wiley and Sons, New York, Toronto, 201 pp.

DuCharme, E. P. 1959. Morphogenesis and histopathology of lesions induced on citrus roots by *Radopholus similis*. Phytopathology, 49:388–395.

Filipjev, I. N., and Stekhoven, J. H. S., Jr. 1941. A manual of agricultural helminthology. E. J. Brill, Leiden. 878 pp.

Huang, C. S., and Maggenti, A. R. 1969. Mitotic aberrations and nuclear changes of developing giant cells in *Vicia faba* caused by root knot nematode, *Meloidogyne javanica*. Phytopathology, 59:447–455.

Huang, C. S., and Maggenti, A. R. 1969. Wall modifications in developing giant cells of *Vicia faba* and *Cucumis sativus* induced by root knot nematode, *Meloidogyne javanica*. Phytopathology, 59:931–937.

Ishibashi, N., and Kondo, E. 1977. Occurrence and survival of the dispersal forms of pine wood nematode, *Bursaphelenchus lignicolus*, Mamiya and Kiyohara. Applied Entomology and Zoology, 12:293–302.

Ishibashi, N., Aoyagi, M., and Kondo, E. 1978. Comparison of the gonad development between the propagative and dispersal forms of pine wood nematode, *Bursaphelenchus lignicolus* (Aphelenchoididae). Japanese Journal of Nematology, 8:28–31.

Jenkins, W. R., and Taylor, D. P. 1967. Plant nematology. Reinhold Publishing Corporation, New York, Amsterdam, London. 270 pp.

Lamberti, F., and Taylor, C. E. 1979. Root-knot nematodes (*Meloidogyne* species), systematics, biology and control. Academic Press, London, New York, San Francisco. 477 pp.

Lamberti, F., Taylor, C. E., and Seinhorst, J. W. 1975. Nematode vectors of plant viruses. Plenum Press, London and New York. 460 pp.

Maggenti, A. R. 1962. The production of the gelatinous matrix and its taxonomic significance in *Tylenchulus* (Nematoda: Tylenchulinae). Proceedings of the Helminthological Society of Washington, 29:139–144.

Maggenti, A. R. 1971. Nemic relationships and the origins of plant parasitic nematodes. Pages 65–81 in B. M. Zuckerman, Ed. Plant parasitic nematodes. Academic Press, New York, London.

Maggenti, A. R., and Allen, M. W. 1960. The origin of the gelatinous matrix in *Meloidogyne*. Proceedings of the Helminthological Society of Washington, 27:4–10.

Maggenti, A. R., Hart, W. H., and Paxman, G. A. 1974. A new genus and species of gall forming nematode from *Danthonia californica*, with a discussion of its life history. Nematologica, 19:491–497.

Mai, W. F., Bloom, J. R., and Chen, T. A. 1977. Biology and ecology of the plant-parasitic nematode *Pratylenchus penetrans*. Pennsylvania State University, Agricultural Experiment Station, University Park, Pennsylvania Bulletin 815. 64 pp.

Mamiya, Y., and Enda, N. 1979. *Bursaphelenchus mucronatus* n. sp. (Nematoda: Aphelenchoididae) from pine wood and its biology and pathogenicity to pine trees. Nematologica, 25:353–361.

Martin, G. C. 1969. Survival and infectivity of eggs and larvae of *Meloidogyne javanica* after ingestion by a rodent. Nematologica, 15:620.

Nickle, W. R. 1970. A taxonomic review of the genera of the Aphelenchoidea

(Fuchs, 1937) Thorne, 1949 (Nematoda: Tylenchida). Journal of Nematology, 2:375–392.

Norton, D. C. 1978. Ecology of plant-parasitic nematodes. John Wiley & Sons, New York, Chichester, Brisbane, Toronto. 268 pp.

Orion, D., and Minz, G. 1971. The influence of morphactin on the root-knot nematode, *Meloidogyne javanica*, and its galls. Nematologica, 17:107–112.

Orum, T. V., Bartels, P. G., and McClure, M. A. 1979. Effect of oryzalin and 1,1-dimethylpiperidinium chloride on cotton and tomato roots infected with the root-knot nematode, *Meloidogyne incognita*. Journal of Nematology, 11: 78–83.

Paramonov, A. A., edited by K. I. Skrjabin. 1962. Plant-parasitic nematodes, Volume I. Izdatel'stvo Akademii Nauk SSR Moskva. 390 pp. English translation.

Paramonov, A. A., edited by K. I. Skryabin [sic]. 1964. Plant parasite [sic] nematodes, Volume II. Izdatel'stvo "Nauka" Moskva. 570 pp. English translation.

Paramonov, A. A., edited by K. I. Skrjabin. 1970. Plant-parasitic nematodes, Volume III. Izdatel'stvo "Nauka" Moskva. 200 pp. English translation.

Robertson, W. M. 1975. A possible gustatory organ associated with the odontophore in *Longidorus leptocephalus* and *Xiphinema diversicaudatum*. Nematologica, 21:443–448.

Robertson, W. M. 1979. Observations on the oesophageal nerve system of *Longidorus leptocephalus*. Nematologica, 25:245–254.

Rohde, R. A., and McClure, M. A. 1975. Autoradiography of developing syncytia in cotton roots infected with *Meloidogyne incognita*. Journal of Nematology, 1:64–69.

Siddiqui, I. A., Sher, S. A., and French, A. M. 1973. Distribution of plant parasitic nematodes in California. State of California, Department of Food and Agriculture, Division of Plant Industry. 324 pp.

Southey, J. F. 1978. Plant nematology. Ministry of Agriculture, Fisheries and Food. GDI, London, H.M.S.O. 440 pp.

Sveshnikova, N. W., Ed. Nematode diseases of agricultural plants. Proceedings of the 6th All-Union Conference on Plant Nematology. Izdatel'stvo "Kolos" Moskva 1967. Israel Program for Scientific Translations, Jerusalem 1972. 198 pp.

Thorne, G. 1961. Principles of nematology. McGraw-Hill Book Company, Inc., New York, Toronto, London. 553 pp.

Viglierchio, D. R. 1971. Race genesis in *Ditylenchus dipsaci*. Nematologica, 17:386–392.

Viglierchio, D. R., and Yu, P. K. 1966. On the nature of hatching of *Heterodera schachtii*. III. Principles of hatching activity. Journal of the American Society of Sugar Beet Technologists, 13:698–715.

Webster, J. M., Ed. 1972. Economic nematology. Academic Press, London, New York. 563 pp.

Wyss, U., Jank-Ladwig, R., and Lehmann, H. 1979. On the formation and ultrastructure of feeding tubes produced by trichodorid nematodes. Nematologica, 25:385–390.

Zuckerman, B. M., Mai, W. F., and Rohde, R. A., Eds. 1971. Plant parasitic nematodes, Volume I. Academic Press, New York and London. 345 pp.

Zuckerman, B. M., Mai, W. F., and Rohde, R. A., Eds. 1971. Plant parasitic nematodes, Volume II. Academic Press, New York and London. 347 pp.

Chapter 7

Bedding, R. A. 1967. Parasitic and free-living cycles in entomogenous nematodes of the genus *Deladenus*. Nature, 214:174–175.

Bedding, R. A. 1968. *Deladenus wilsoni* n. sp. and *D. siricidicola* n. sp. (Neotylenchidae), entomophagous-mycetophagous nematodes parasitic in siricid woodwasps. Nematologica, 14:515–525.

Bedding, R. A. 1973. Biology of *Deladenus siricidicola* (Neotylenchidae) an entomophagous-mycetophagous nematode parasitic in siricid woodwasps. Nematologica, 18:482–493.

Blinova-Lasarevskaja, S. L. 1970. The development of parasitism in xylobiont nematodes and evolutionary relations in the Rhabditidae family. In Proceedings IX International Nematological Symposium Warsaw. Zesz. Probl. Postep. Nauk. Roln., 92:339–347.

Chitwood, B. G., and Chitwood, M. B. 1974. Introduction to nematology. University Park Press, Baltimore, London, Tokyo. 334 pp.

Cobb, N. A. 1928. Zoology.-*Ungella secta* n. gen., n. sp.; A nemic parasite of the Burmese Oligochaete (earthworm), *Eutyphoeus rarus*. The Washington Academy of Sciences, 18:197–200.

Croll, N. A. 1966. A contribution to the light sensitivity of the "chromatrope" of *Mermis subnigrescens*. Journal of Helminthology, 40:33–38.

Dale, P. S. 1970. Dispersal and phylogeny of some oxyuroid nematodes. In Proceedings IX International Nematological Symposium Warsaw. Zesz. Probl. Postep. Nauk. Roln., 333–337.

Ebsary, B. A., and Bennett, G. F. 1975. The occurrence of some endoparasites of blackflies (Diptera: Simulidae) in insular Newfoundland. Canadian Journal of Zoology, 53:1058–1062.

Ferris, J. M., and Ferris, V. R. 1966. Observations on *Tetradonema plicans*, an entomoparasitic nematode with a key to the genera of the family Tetradonematidae (Nematoda: Trichosyringida). Annals of the Entomological Society of America, 59:964–971.

Filipjev, I. N., and Stekhoven, J. H. S., Jr. 1941. A manual of agricultural helminthology. E. J. Brill, Leiden. 878 pp.

Fisher, J. M., and Nickle, W. R. 1968. On the classification and life history of *Fergusobia curriei* (Sphaerulariidae: Nematoda). Proceedings of The Helminthological Society of Washington, 35:40–46.

Hungerford, H. B. 1919. Biological notes on *Tetradonema plicans*, Cobb, a nematode parasite of *Sciara coprophila* Lintner. Journal of Parasitology, 5: 186–192.

Hyman, L. H. 1951. The invertebrates: Acanthocephala, Aschelminthes, and Entoprocta. The pseudocoelomate bilateria. Vol. III. McGraw-Hill Book Company, Inc., New York, Toronto, London. 572 pp.

Johnson, G. E. 1913. On the nematodes of the common earthworm. Quarterly Journal of Microscopical Science, 58:605–652.

Mulvey, R. H., and Nickle, W. R. 1978. Taxonomy of mermithids (Nematoda: Mermithidae) of Canada and in particular of the Mackenzie and Porcupine river systems, and Somerset Island, N.W.T., with descriptions of eight new species and emphasis on the use of the male characters in identification. Canadian Journal of Zoology, 56:1291–1329.

Nickle, W. R. 1967. On the classification of the insect parasitic nematodes of the Sphaerulariidae Lubbock, 1861 (Tylenchoidea: Nematoda). Proceedings of The Helminthological Society of Washington, 34:72–94.

Nickle, W. R. 1967. *Heterotylenchus autumnalis* sp. n. (Nematoda: Sphaerulariidae), a parasite of the face fly, *Musca autumnalis* De Geer. Journal of Parasitology, 53:398–401.

Nickle, W. R. 1968. Observations on *Hexatylus viviparus* and *Neotylenchus abulbosus* (Neotylenchidae: Nematoda). Proceedings of The Helminthological Society of Washington, 35:154–160.

Nickle, W. R. 1972. A contribution to our knowledge of the Mermithidae (Nematoda). Journal of Nematology, 4:112–146.

Nickle, W. R. 1974. Nematode infections. Insect diseases, Volume II, G. E. Cantwell, Ed. Marcel Deeker, Inc., New York, pp. 327–376.

Poinar, G. O., Jr. 1969. Aphelenchoidea parasiting *Rhynchophorus bilineatus* (Montrouzier) (Coleoptera: Curculionidae) in New Britain. Journal of Nematology, 1:227–231.

Poinar, G. O., Jr. 1975. Entomogenous nematodes. A manual and host list of insect-nematode associations. E. J. Brill, Leiden. 317 pp.

Poinar, G. O., Jr. 1978. *Mesidionema praecomasculatis* gen. et sp. n.; Mesidionematidae fam. n. (Drilonematoidea: Rhabditida), a nematode parasite of earthworms. Proceedings of the Helminthological Society of Washington, 45:97–102.

Poinar, G. O., Jr. 1979. Nematodes for biological control of insects. CRC Press, Inc., Boca Raton, Florida. 277 pp.

Poinar, G. O., Jr., and Hess, R. T. 1977. Immune responses in the earthworm, *Aporrectodea trapezoides* (Annelida), against *Rhabditis pellio* (Nematoda). Comparative Pathobiology, 3:69–84.

Poinar, G. O., Jr., and Thomas, G. M. 1975. *Rhabditis pellio* Schneider (Nematoda) from the earthworm, *Aporrectodea trapezoides* Duges (Annelida). Journal of Nematology, 7:374–379.

Poinar, G. O., Jr., and van der Laan, P. A. 1972. Morphology and life history of *Sphaerularia bombi*. Nematologica, 18:239–252.

Quentin, J. C., and Poinar, G. O., Jr. 1973. Comparative study of the larval development of some heteroxenous subulurid and spirurid nematodes. International Journal for Parasitology, 3:809–827.

Rubzov, I. A. 1977. Mermithidae, origin, distribution, and biology. Nauka, Leningrad. 191 pp.

Siddiqi, M. R., and Goodey, J. B. 1963. The status of the genera and subfamilies of the Criconematidae (Nematoda); with a comment on the position of *Fergusobia*. Nematologica, 9:363–377.

Steiner, G. 1929. Zoology-*Neoaplectana glaseri*, n.g., n. sp. (Oxyuridae) a new nemic parasite of the Japanese beetle (*Popillia japonica* Newm.). Journal of The Washington Academy of Sciences, 19:436–440.

Stekhoven, J. H. S., Jr. 1950. *Alloionema appendiculatum* Schneider 1859. Nematodo parasitario de *Arion ater* L. De Acta Zoologica Lilloana, 9:481–485.

360 Selected References

Thorne, G. 1941. Some nematodes of the family Tylenchidae which do not possess a valvular median esophageal bulb. The Great Basin Naturalist, Brigham Young University, Provo, Utah. 2:37–85.

Timm, R. W. 1966. Nematode parasites of the coelomic cavity of earthworms III. *Homungella* new genus (Drilonematoidea: Homungellidae new family). Biologia, 12:1–5.

Timm, R. W., and Maggenti, A. R. 1966. Nematode parasites of the coelomic cavity of earthworms V. *Plutellonema, Iponema,* and *Filiponema,* new genera (Drilonematidae). Proceedings of The Helminthological Society of Washington, 33:177–184.

Van Waerebeke, D., and Remillet, M. 1973. Morphologie et biologie de *Heterogonema ovomasculis* n. sp. (Nematoda: Tetradonematidae) parasite de nitidulidae (Coleoptera). Nematologica, 19:80–92.

Chapter 8

Alicata, J. E., and Jindrak, K. 1970. Angiostrongylosis in the Pacific and Southeast Asia. Charles C Thomas, Springfield, Illinois. 105 pp.

Chabaud, A. G. 1971. Evolution of host-parasite adaptation in nematodes of vertebrates. International Journal for Parasitology, 1:217–221.

Croll, N. A. 1966. Ecology of parasites. Harvard University Press, Cambridge, Massachusetts. 136 pp.

Croll, N. A. 1970. The behaviour of nematodes. Edward Arnold, Ltd., London. 117 pp.

Desmond, A. J. 1977. The hot-blooded dinosaurs. A revolution in palaeontology. Warner Books, Inc., New York. 352 pp.

Dogiel, V. A., Petrushevsak, G. K., and Polyanski, Y. I. 1958. Parasitology of fishes. Leningrad University Press. Translation available from T. F. H. Publications, Neptune, New Jersey, 1971. 384 pp.

Dougherty, E. C. 1951. Evolution of zoöparasitic groups in the phylum nematoda, with special reference to host-distribution. Journal of Parasitology, 37:353–378.

Faust, E. C., Russell, P. F., and Jung, R. C. 1974. Clinical parasitology. Lea & Febiger, Philadelphia. 890 pp.

Georgi, J. R. 1980. Parasitology for veterinarians. W. B. Saunders Company, Philadelphia, London, Toronto. 460 pp.

Hiscox, J. I., and Brocksen, R. W. 1973. Effects of a parasite gut nematode on consumption and growth in juvenile Rainbow Trout (*Salmo gairdneri*). Journal of the Fisheries Research Board of Canada, 30:443–450.

Hoffman, G. L. 1967. Parasites of North American freshwater fishes. University of California Press, Berkeley and Los Angeles. 486 pp.

Inglis, W. G. 1967. The relationships of the nematode superfamily Seuratoidea. Journal of Helminthology, 41:115–136.

Karmanova, E. M. 1960. The life cycle of the nematode *Dioctophyme renale* (Goeze, 1782), a parasite in the kidneys of carnivora and of man. Parasitology (Translated from Doklady Akademii Nauk SSR, 132:1219–1220, June, 1960), 132:456–457.

Karmanova, E. M. 1968. Dioctophymatida of animals and man and diseases

they cause. Osnovy Nematodologii 20. Gel'mint. Lab. Akad. Nauk. SSSR. 262 pp.

Levine, N. D. 1968. Nematode parasites of domestic animals and of man. Burgess Publishing Company, Minneapolis, Minnesota. 600 pp.

Maggenti, A. R. 1978. Influence of morphology, biology, and ecology on evolution of parasitism in Nematoda. In J. A. Romberger, Ed. Biosystematics in Agriculture. Allanheld, Osmun and Co., Montclair; a Halsted Press Book, John Wiley and Sons, New York, Chichester, Brisbane, Toronto, pp. 173–191.

Muller, R. 1971. *Dracunculus* and Dracunculiasis. Advances in Parasitology, 9:73–151.

Olsen, O. W. 1974. Animal parasites, their life cycles and ecology. University Park Press, Baltimore, London, Tokyo. 562 pp.

Osche, G. 1963. Morphological, biological and ecological considerations in the phylogeny of parasitic nematodes. The lower metazoa comparative biology and phylogeny, E. C. Dougherty, Ed. University of California Press, Berkeley and Los Angeles.

Ostrom, J. H. 1979. Bird flight: How did it begin? American Scientist, 67: 46–56.

Petter, A. J. 1976. Essai d'interpretation de la répartition des Ascaridoidea chez les Mammifères. Annales de Parasitologie (Paris), 52:151–158.

Platzer, E. G., and Adams, J. R. 1967. The life history of a dracunculoid *Philonema oncorhynchi*, in *Oncorhynchus nerka*. Canadian Journal of Zoology, 45:31–43.

Rogers, W. P. 1962. The nature of parasitism. The relationship of some metazoan parasites to their hosts. Academic Press, New York and London. 287 pp.

Skrajabin, K. I., Shikobalova, N. P., and Orlov, I. V. 1957. Trichocephalida and Capillariidae of animals and man and diseases they cause. Osnovy Nematodologii 4. Gel'mint. Lab. Akad. Nauk. SSSR. 587 pp.

Chapter 9

Andrassy, I. 1965. Evolution as a basis for the systematization of nematodes. Pitman Publishing, London. 288 pp.

Chitwood, B. G. 1958. The designation of official names for higher taxa of invertebrates. Bulletin of Zoological Nomenclature, 15:860–895.

Chitwood, B. G., and Chitwood, M. B. 1974. Introduction to nematology. University Park Press, Baltimore, London, Tokyo. 334 pp.

Chitwood, B. G. 1977. Nematoda. McGraw-Hill Encylopedia of Science and Technology, 9:44–48.

Filipjev, I. N. 1934. The classification of the freeliving nematodes and their relation to the parasitic nematodes. Smithsonian Miscellaneous Collections, 89:1–63.

Goodey, T. 1963. Soil and freshwater nematodes (2nd ed. revised by Goodey, J. B.). Wiley & Sons, Inc., New York. 544 pp.

Grassé, P.-P. 1965. Traité de Zoologie, Tome IV, Fascicule 2–3. Masson, Paris.

Maggenti, A. R. 1982. Nemata. In: Sybil P. Parker (ed), Synopsis and classification of living organisms. McGraw-Hill, New York. In press.

Thorne, G. 1949. On the classification of the Tylenchida, new order (Nematoda, Phasmidia). Proceedings of the Helminthological Society of Washington, 16:37–73.

Index

* Numbers in **boldface** type refer to pages on which there are illustrations; G= Gastrotricha; Nt=Nematomorpha; K=Kinorhyncha; and R=Rotifera.

Springer Series in Microbiology

Editor: **Mortimer P. Starr,** Department of Bacteriology, University of California, Davis, California, U.S.A.

Thermophilic Microorganisms and Life at High Temperatures

T.D. Brock, University of Wisconsin, Madison
1978/xi, 465pp./195 illus./cloth
ISBN 0-387-**90309**-7

Bacterial Metabolism

G. Gottschalk, Universität Göttingen, Federal Republic of Germany
1979/xii, 281pp./161 illus./cloth
ISBN 0-387-**90308**-9

Ascomycete Systematics: The Luttrellian Concept

R.D. Reynolds, Natural History Museum, Los Angeles
1981/ix, 242pp./122 illus./cloth
ISBN 0-387-**90488**-3

An Introduction to Bacterial and Bacteriophage Genetics

E.A. Birge, Arizona State University, Tempe
1981/xiv, 359pp./111 illus./cloth
ISBN 0-387-**90504**-9

General Nematology

A. Maggenti, University of California, Davis
1981/384pp./135 illus./cloth
ISBN 0-387-**90588**-X

Advances in Studies of Basidiomycetes

K. Wells and **E.K. Wells,** University of California, Davis
1982. In preparation.